Deep Space Propulsi

K.F. Long

Deep Space Propulsion

A Roadmap to Interstellar Flight

 Springer

K.F. Long Bsc, Msc, CPhys
Vice President (Europe), Icarus Interstellar
Fellow British Interplanetary Society
Berkshire, UK

ISBN 978-1-4614-0606-8   e-ISBN 978-1-4614-0607-5
DOI 10.1007/978-1-4614-0607-5
Springer New York Dordrecht Heidelberg London

Library of Congress Control Number: 2011937235

Printed on acid-free paper

Springer is part of Springer Science+Business Media (www.springer.com)

This book is dedicated to three people who have had the biggest influence on my life. My wife Gemma Long for your continued love and companionship; my mentor Jonathan Brooks for your guidance and wisdom; my hero Sir Arthur C. Clarke for your inspirational vision – for Rama, 2001, and the books you leave behind.

Foreword

We live in a time of troubles. It is easy to name a few of them: endless wars, clashes of cultures, nuclear proliferation, overpopulation, global climate warming, economic recession, political disarray. It is also easy to give up and conclude that all is hopeless for our civilization and possibly for all humanity.

However, is such pessimism truly warranted? The year 2010 witnessed the first reasonably solid detection of potentially habitable worlds circling nearby stars. Our understanding of the variety of life forms that could inhabit these worlds has also been broadened.

Perhaps more significantly, another milestone of 2010 was the first successful flight test of a non-rocket propulsion system that could someday evolve into a star drive. True – the solar sails that may evolve from the successful interplanetary test flight of the Japanese Ikaros probe will never achieve the performance of *Star Trek's* mythical Enterprise – but they represent a start.

Progress towards controlled thermonuclear fusion continues. And at the CERN laboratory on the Italian-Swiss border, the Large Hadron Collider is now operational. It is not impossible that advances in our understanding of particle physics prompted by experiments at this new facility will ultimately lead to our taming of the antimatter/matter annihilation. Maybe some theoretical breakthrough will actually lead to a warp drive or a means of tapping the enormous energies of the universal vacuum.

It is impossible to know which of these new technologies – sails, fusion, or antimatter – will lead to the first human probes and ships capable of crossing the interstellar gulfs. But if you are interested in participating in the adventure of expanding the terrestrial biosphere into new realms, this book is an excellent place to hone your skills.

The author is a major contributor to the British Interplanetary Society Project Icarus, which is investigating near-term techniques that might lead to probes capable of reaching our Sun's nearest stellar neighbors. As well as reviewing the current state of interstellar propulsion technologies, Kelvin Long provides an extensive bibliography that will be an invaluable aid to the novice researcher.

Chapter exercises are also included so that young engineers and physicists can practice their skills.

Historical discussions regarding the evolution of flight within and outside Earth's atmosphere will be of great interest to the casual reader. References to science fiction demonstrate the significance of literature in inspiring much of the scientific inquiry that has led us to the threshold of galactic citizenship. And the color plates present the beautiful work of space artists and illustrators.

Only a few humans have felt the crunch of lunar soils beneath their booted feet. A somewhat larger number have remotely controlled their roving vehicles as they cross the deserts of Mars or communicated with Voyager probes now on the fringe of galactic space. In all likelihood, only a few will control our interstellar robots and fewer still will ride the galactic high road to our next stellar homes. But this book allows many to contribute to and share in the adventure.

Just maybe today's interstellar pioneers will help stem the tide of pessimism. Perhaps, just perhaps, books like this may help develop a welcome sense of optimism and feeling of hope among the readers.

New York City College of Technology, CUNY   Dr. Greg Matloff

Preface

Two lights for guidance. The first, our little glowing atom of community, with all that it signifies. The second, the cold light of the stars, symbol of hypercosmical reality, with its crystal ecstasy. Strange that in this light, in which even the dearest love is frostily assessed, and even possible defeat of our half-waking world is contemplated without remission of praise, the human crisis does not lose but gains significance. Strange, that it seems more, not less, urgent to play some part in this struggle, this brief effort of animalcules striving to win for their race some increase of lucidity before the ultimate darkness.

Olaf Stapledon

This book is about robotic exploration of the nearby stars and the prospects for achieving this within the next century. In particular, we will focus on the propulsion technology that will be used to accomplish such an ambitious objective. This is so called 'game changing' technology that goes beyond conventional chemical rockets, using exotic fuels and more efficient engines for the purpose of an interstellar mission. This includes ideas for engines based upon harnessing the emitted energy of the Sun, using fusion reactions or even tapping the energy release from matter-antimatter annihilation reactions. This book serves as an essential primer for anyone new to this field who wishes to become familiar with the ideas that have already been developed and how to attain the necessary theoretical tools to create similar ideas for themselves. If, by the end of this book, you are thinking of your own ideas for how machines can be propelled across the vastness of space, then this book will have been successful in its ultimate aim.

At the dawn of this new millennium we can look back on the previous century with pride, knowing that humanity took its first steps into space and even walked upon the surface of the Moon. We have collected a wealth of data on the many planets and moons of our Solar System, and our robotic ambassadors are still pioneering the way so that one day humans may hope to follow where they lead. The Moon and Mars are the current focus of human ambitions in space, and this is right for the short-term goals of our species. But in the long term, missions to places much further away will become not just aspirations but vital to our survival.

From the outset of this text, we must be truthful and admit that the technology to enable human transport to other stars is currently immature. The physics, engineering and physiological requirements are unlike anything we have ever encountered, and this unique problem demands our full attention if we are to ever cross the enormous gulfs of space that separate the stars in our galaxy and become a truly spacefaring civilization. But if we are bold and eventually attempt this, the scientific, economic and spiritual rewards will be many, and our civilization will become enriched for the attempt. Until then, we must be content with robotic exploration and to push that technology to its limit. We must continue to launch missions to the outer planets of our Solar System to explore those cold but interesting worlds. Eventually, our robotic probes will break through the termination shock of the solar heliosphere and pass out into the Kuiper Belt to explore many strange new dwarf planets, some perhaps yet undiscovered. Then they will be sent to much further distances out into the Oort Cloud to investigate the myriad of comets that orbit our Solar System in large period trajectories. Finally those same robots will enter the outer reaches of the interstellar medium, the diffuse nebula of space that is dispersed between the stars, and for the first time in history a human made machine would have fully left the gravitational cradle of our Sun.

By this time, the technology performance of our machines should have improved by many orders of magnitude so that missions to the nearest stars will become possible and scientific data return will become common. What will those probes discover? Perhaps unusual planets with oceans made of materials thought impossible where life might be swimming among its depths. The astronomical knowledge gained will be highly valuable; the chance to be so close to another star and its orbiting worlds will enrich our knowledge of the universe and give us a better understanding of its structure, evolution and origin.

A few moments spent thinking about the interstellar transport problem quickly leads to the realization that there are two main extremes to reaching the stars. We can build very lightweight vehicles with a limited payload size in the hope that their small mass allows for large speeds, such as a solar sail. Alternatively, we can build massive vehicles the size of small moons, which will move slowly and take perhaps thousands of years to reach their destination; these are so called world ships. For any form of human exodus to another world, clearly the latter option is the only credible approach. However, as one digs into the interstellar literature we find that there are approaches to this problem that lay between these two extremes. We find that physicists have invented clever theoretical (and some practical) means of allowing a large mass scientific payload to be delivered to a destination at a speed of a few to tens of percent of light speed, thus getting to the target within decades. It is then just a matter of the engineering technology progressing to an acceptable readiness level. Many of these schemes are discussed in this book.

In reading this book it will be clear that the author favors the nuclear pulse propulsion approach for interstellar missions. This is along the lines of the historical Orion and Daedalus concepts. It should be noted that this is not because of a belief that other concepts do not meet the requirements for interstellar missions. In fact the opposite is true, and technology such as solar sails, laser sails and microwave sails in particular do offer great potential for near term demonstration. However, it is a personal belief that nuclear pulse technology is nearly ready for use now, if not already available in some form, and is the most appropriate route for an interstellar flight. Power is what will take us to the stars, and sending something there fast requires powerful engines as provided by the nuclear pulse options. In the end, it is likely that the first interstellar probe will be a combination of propulsion technologies – a hybrid design utilizing nuclear electric, sails and nuclear pulse to augment different parts of the mission. When this happens, all of the individual efforts over the decades of research that have gone into making each of these technologies ready will have been worth the effort.

Another personal view that has been taken by this author is that Mars should be the next destination for human exploration. This will push our technology forward while also beginning the establishment of a human colony on another world. Contrary to some claims, the exploration of Mars is not prohibitively expensive if done in a manner similar to the proposed Mars Direct plan. National and international space agencies need a common focus and a common goal – Mars is the logical candidate and a clear springboard upon which a full program of interstellar exploration may begin. Indeed, there are no technological reasons preventing us from starting direct colonization of Mars today.

Many people believe that interstellar travel, even robotic exploration, is so difficult and the obstacles so unsurpassable that it will be many millennia before we can attempt it. However, it is the hope of this book to demonstrate to the reader that not only is interstellar travel perfectly possible, it is within our grasp, and the first unmanned mission will likely be launched by the end of the twenty-first century (a bold prediction) and certainly by the mid twenty-second century. As will be shown, many feasibility design studies have already been undertaken historically, often involving teams of physicists and engineers, producing study reports that demonstrate the engineering problems and potential solutions. These people are among a growing network of optimists that share in a single vision that the human destiny in space lies not just with the Moon and nearby planets, but much further to other worlds around other stars. History will show which one of these concepts becomes the true template for how our robotic ambassadors first enter the orbit of another star and achieve this seemingly impossible and long desired ambition.

Berkshire, UK   K.F. Long

Acknowledgments

There are many people I would like to thank in preparing the manuscript for this book. Firstly, a big thank you to my wife Gemma Long for the patience you have shown during the hours I was working on this book. Many other members of my family also had to put up with long periods of absence during this busy writing period. I would especially like to thank my parents Susan Kelly, Kevin Long, Maureen Perrin and Michael Perrin. For reviewing parts of the manuscript and giving constructive comments on earlier drafts I would like to thank Paul Gilster, Greg Matloff, Marc Millis, Richard Obousy, Adam Crowl, Pat Galea, Paul Thompson, Gemma Long and Jonathan Brooks. Any errors are solely this author's responsibility. A special thanks to Greg Matloff for writing the foreword and supporting me in this project. Discussions with many other people over the last 2 years have helped me to understand what a difficult and wide scoping academic subject interstellar research is. Other than those already mentioned above, I would also like to thank the following for constructive discussions or for helping me in some capacity during this work, including Claudio Maconne, Bob Parkinson, Martyn Fogg, Jim French, Alan Bond, Gerry Webb, Penny Wright, Friedwardt Winterberg, Richard Osborne, Andreas Tziolas, Jim Benford, Rob Swinney, Robert Freeland, Andy Presby, Andreas Hein, Philip Reiss, Ian Crawford, Stephen Baxter, Stephen Ashworth, Gerald Nordley, Giovanni Vulpetti, Claudio Bruno, Tibor Pacher, Rich Sternbach, David Johnson, Eric Davies, Ralph McNutt, George Schmidt, Claudio Bruno, Giovanni Vulpetti, Charles Orth, Dana Andrews, John Hunt, Remo Garattini, Mike McCulloch, Stephen Ashworth. Many discussions with the ever insightful Adam Crowl have especially been educational in helping me to learn this subject, with humility. Several people mentioned above are associated with the *Project Icarus Study Group* – we are pioneers together. In the course of my research, I would like to pay tribute to the evolving source that is the internet and the superb 'Atomic Rockets' web site run by Winchell Chung Jr. I must also mention the staff at the British Interplanetary Society, particularly the Executive Secretary Suszann Parry as well as Mary Todd, Ben Jones, Mark Stewart (BIS Honorary Librarian), Chris Toomer (JBIS Editor), Clive Simpson (Spaceflight

Editor), for their continued efforts in promoting space exploration and providing excellent publications which I relied upon for information. For supplying graphics I would like to thank David Hardy, Adrian Mann, Alexandre Szames and Christian Darkin. Jack Weedon and Peter Johnson are thanked for their long time encouragement and support. Acknowledgement is also given to the NASA images library for kindly allowing free use of their wonderful pictures for public use. I would like to thank my editor at Springer Maury Solomon for being patient with the manuscript delivery and publishing agent John Watson who introduced me to the team. In the review of propulsion concepts I have been selective in what projects to discuss. If I have left out some important schemes or research it is not because I don't think they are relevant but either due to page/time limitations or simply because I did not know of the existence of such methods. Finally, my thanks go to Sir Arthur C Clarke with whom I had the honor of meeting just the once. His literature in both science fiction and science writing has changed my view of the world and the possibilities for the future of our species in the exploration of space. My own vision is his inheritance and an inspiration for the generations ahead.

Contents

1 **Reaching for the Stars** 1
   1.1 Introduction 1
   1.2 An Overview of Propulsion Schemes for Space 1
   1.3 Practice Exercises 9
   References 10

2 **The Dream of Flight and the Vision of Tomorrow** 11
   2.1 Introduction 11
   2.2 Becoming a Spacefaring Civilization 11
   2.3 The Promise of the Future 14
   2.4 Why We Shouldn't Go Out into Space 17
   2.5 Why We Should Go Out into Space 20
   2.6 Practice Exercises 26
   References 26

3 **Fundamental Limitations to Achieving Interstellar Flight** 27
   3.1 Introduction 27
   3.2 How Much Speed Is Required to Reach the Stars? 27
   3.3 How Much Energy Is Required to Reach the Stars? 35
   3.4 Practice Exercises 37
   References 38

4 **Aviation: The Pursuit of Speed, Distance, Altitude, and Height** 39
   4.1 Introduction 39
   4.2 The History of Propulsion in Aviation 39
   4.3 The Physics of Aviation Propulsion 43
   4.4 Practice Exercises 47
   References 48

5 **Astronautics: The Development and Science of Rockets** 49
   5.1 Introduction 49
   5.2 The History of Propulsion in Rocketry 49

5.3   The Physics of Rocket Propulsion............................   54
5.4   Rockets for Space...   61
5.5   Practice Exercises..   74
References..   75

6   **Exploring the Solar System and Beyond**   77
6.1   Introduction...   77
6.2   Near Earth   78
6.3   The Colonization of Mars   81
6.4   Other Planetary Objects   88
6.5   Mining He-3 from the Gas Giants...............................   93
6.6   The Outer Solar System   94
6.7   Practice Exercises..   97
References..   97

7   **Exploring Other Star Systems**...................................   99
7.1   Introduction...   99
7.2   The Stars and the Worlds Beyond...............................   99
7.3   The Discovery and Evolution of Life   109
7.4   Practice Exercises..   114
References..   115

8   **Solar System Explorers: Historical Spacecraft**   117
8.1   Introduction...   117
8.2   Precursor Mission Probes......................................   117
8.3   Pioneer Probes   122
8.4   Voyager Probes   123
8.5   Galileo   125
8.6   Ulysses..   126
8.7   Cassini-Huygens...   126
8.8   Deep Space 1   127
8.9   Cosmos I..   128
8.10   New Horizons...   129
8.11   Dawn..   130
8.12   Interstellar Boundary Explorer   131
8.13   Summary Discussion on Probe Design...........................   132
8.14   Practice Exercises..   136
References..   137

9   **Electric and Nuclear-Based Propulsion**...........................   139
9.1   Introduction...   139
9.2   Electric Propulsion..   139
9.3   Nuclear Thermal and Nuclear Electric Propulsion...............   144
9.4   The Interstellar Precursor Probe................................   146
9.5   The Thousand Astronomical Unit Mission.......................   147

9.6 The Innovative Interstellar Explorer.............................. 149
9.7 Project Prometheus................................... 151
9.8 Practice Exercises................................ 153
References... 153

10 Sails & Beams... 155
10.1 Introduction.................................. 155
10.2 Solar Sailing.................................. 156
10.3 The Interstellar Heliopause Probe.............................. 164
10.4 Interstellar Probe................................ 166
10.5 Beamed Propulsion................................. 167
10.6 Beamed Microwaves and Starwisp............................ 172
10.7 Problems.................................... 175
References... 175

11 Nuclear Fusion Propulsion................................. 177
11.1 Introduction.................................. 177
11.2 Fusion: The Holy Grail of Physics............................ 178
11.3 Fusion Power for Space Exploration........................... 184
11.4 The Enzmann Starship................................. 189
11.5 Project Daedalus................................ 190
11.6 Project Longshot................................ 198
11.7 Project VISTA.................................. 201
11.8 Discovery II.................................... 203
11.9 Problems.................................... 205
References... 205

12 External Nuclear Pulse Propulsion.......................... 207
12.1 Introduction.................................. 207
12.2 Nuclear Pulse Detonation and Project Orion.................... 207
12.3 The Medusa Concept............................. 215
12.4 Problems.................................... 217
References... 218

13 Towards Relativistic Propulsion: Antimatter
and the Interstellar Ramjet................................. 219
13.1 Introduction.................................. 219
13.2 Relativity in Space Flight....................... 220
13.3 The Interstellar Ramjet.......................... 223
13.4 The Bussard Ramjet Spacecraft Concept....................... 226
13.5 Matter-Antimatter Annihilation................................ 227
13.6 The AIMStar Spacecraft Concept.............................. 229
13.7 The ICAN-II Spacecraft Concept.............................. 231
13.8 Practice Exercises.............................. 232
References... 233

14 Aerospace Design Principles in Interstellar Flight 235
14.1 Introduction 235
14.2 Principles of Aerospace Design 235
14.3 Systems Engineering Approach to Concept Design Studies 238
14.4 Technology Readiness Levels 243
14.5 A Concept Design Problem for a Precursor Mission Proposal .. 244
    14.5.1 Scoping Concept Design Space 245
    14.5.2 Concept Solution 248
14.6 Problems 261
References 261

**15 The Scientific, Cultural and Economic Costs
of Interstellar Flight** 263
15.1 Introduction 263
15.2 The Advance of Science 263
    15.2.1 Planetary Physics, Including Terrestrial
        and Giant Planets 265
    15.2.2 Stellar Physics, Including Different Spectral Types 265
    15.2.3 The Presence and Evolution of Life 265
    15.2.4 The Physics of the Interstellar Medium 266
    15.2.5 The Physics of the Kuiper Belt and Oort Cloud
        Layers of Different Stars 266
    15.2.6 Galactic Structure, Cosmology
        and the Global Picture 266
    15.2.7 Exotic Physics, Including Gravitational Issues 267
    15.2.8 Engineering Design Issues Such as Spacecraft
        Reliability 267
15.3 Cultural Growth and the Rise and Fall of Civilization 271
15.4 The Economics of Space Exploration 277
15.5 Practice Exercises 284
References 285

16 The Role of Speculative Science in Driving Technology 287
16.1 Introduction 287
16.2 The Connection between Science and Fiction 288
16.3 Breakthrough Propulsion Physics and the Frontiers
    of Knowledge 291
16.4 NASA Horizon Mission Methodology 300
16.5 Problems 302
References 303

**17 Realising the Technological Future and the Roadmap
to the Stars** 305
17.1 Introduction 305
17.2 International Co-operation in the Pursuit of Space 306

17.3   Establishing Precursor Missions and the Technological
Roadmap for Interstellar Exploration 309

17.4   Project Icarus, Son of Daedalus, Flying Closer
to another Star 312

17.5   The Alpha Centauri Prize... 318

17.6   Problems 322

References 323

18   From Imagination to Reality 325

Epilogue 345

Appendix A... 349

Appendix B... 351

Appendix C... 353

Appendix D... 357

Index.. 361

About the Author

K.F. Long was born in Cambridge, England. He holds a Bachelors degree in Aerospace Engineering and a Masters degree in Astrophysics, both from Queen Mary College University of London. He has been a college and university teacher and served for several years in the Parachute Regiment Volunteer Reserve of the British Army. He currently works as a physicist in industry. He has published numerous papers and articles on the subject of astrophysics and space exploration and is currently the Assistant Editor of the *Journal of the British Interplanetary Society*. He has also contributed to many other publications on interstellar travel. He has appeared on the Austrian Broadcasting Radio Station *FM4*, as well as in the documentary *How To Colonize the Stars* produced by Christian Darkin. He is a Fellow of the Royal Astronomical Society, a chartered member of the Institute of Physics, a member of The Planetary Society, a member of the American Institute of Aeronautics & Astronautics, Fellow of the British Interplanetary Society, and an associate member of the International Association for Astronomical Artists. He is a Practitioner of the Tau Zero Foundation, which actively promotes and coordinates international research into interstellar flight and within this capacity was the co-founder of an interstellar research initiative called Project Icarus. He is also the Vice President (Europe) and co-founder of the non-profit Icarus Interstellar. He is happily married to Gemma and likes to spend his spare time reading and writing science fiction, spacecraft model building and using his Schmidt-Cassegrain telescope.

Chapter 1
Reaching for the Stars

So far, Astronautics has been considered a complex set of technological disciplines, each driven by basic or fundamental physics, to allow humankind to accomplish space exploration. Even Astrodynamics is related to technology. Nevertheless, may Basic Physics be driven also by Astronautical Principles, Ideals, and Research? I do think so.

Giovanni Vulpetti

1.1   Introduction

In this first chapter we learn that humans have developed many concepts for how we may someday reach the distant stars. These range from large slow World Ships to spacecraft that can travel faster than the speed of light. We learn that far from interstellar travel being a vast unsolvable problem, instead multiple solutions have already been proposed. Some are practical today and some are speculative and belong in the far future. One day, perhaps some of the proposals may represent the mechanics of human expansion into the cosmos. But today, we must be content to wonder at the marvel of these ideas. In this chapter, for reader familiarization we are introduced to many propulsion schemes that have the purpose of propelling a vehicle across space. These are then discussed in greater detail in later chapters.

1.2   An Overview of Propulsion Schemes for Space

When the American rocketeer Robert Goddard first considered the problem of lifting a body up above the atmosphere and into space, he went through a series of considerations. He considered gyroscopes, magnetic fields and even made wooden models with lead weights to provide lift as the weights moved back and forth in vertical arcs. He designed a machine gun device that fired multiple bullets

K.F. Long, *Deep Space Propulsion: A Roadmap to Interstellar Flight,*
DOI 10.1007/978-1-4614-0607-5_1, © Springer Science+Business Media, LLC 2012

downward. All of these ideas were to provide the lifting force for a body up into space. Eventually, he settled on the simple idea of using Newton's action-reaction principle by combustion of propellants in a chamber, the exhaust products of which would be directed rearwards to propel the vehicle in the opposite direction. This is the basic principle of the rocket. The history of invention is littered with the graves of ideas that never succeeded. This is how machines are first created – from imagination to reality.

Although the problem of interstellar travel is one that pales all other historical technical challenges in comparison, mainly due to the vast distances involved, the same techniques of invention are applied. Clever physicists come up with theoretical ideas, develop them and if they are lucky get to conduct some experiments. In the end the most practical solution will win out and a method of reaching the stars will be derived.

Les Shepherd in the *Journal of the British Interplanetary Society* published one of the first technical papers addressing interstellar flight [1]. He considered that interstellar travel would be deemed possible when travel times were reduced to between 100 to 1,000 years and not before then. Shepherd said:

> There does not appear to be any fundamental reason why human communities should not be transported to planets around neighboring stars, always assuming that such planets can be discovered. However, it may transpire that the time of transit from one system to another is so great that many generations must live and die in space, in order that a group may eventually reach the given destination. There is no reason why interstellar exploration should not proceed along such lines, though it is quite natural that we should hope for something better.

Over the years people have devised 'better' methods for sending space vehicles to the outer reaches of space. In order for a vehicle to move across space from one place to another, it must gather velocity by accelerating. Some form of propulsive engine enables this acceleration. In general, one can conceive of three types of propulsive engines for this purpose. The first is an *internally driven engine*, that is, one in which the fuel products are somehow combusted internally within the vehicle and then ejected rearwards to produce a reaction force. This is how conventional rockets work. The second is an *externally driven engine*, that is, where some energetic reaction occurs external to a vehicle, but this reaction produces a force that can push the vehicle when it finally reaches it by a transfer of momentum.

There is a third type, which is more speculative and has never been used in any actual spacecraft designs to date, or demonstrated in a laboratory; this is what one may call a *locally driven engine*, that is, where the reaction force is derived from the properties of the local medium upon which the vehicle 'rests.' This may be some property of space itself or some other source of subatomic origin such as the quantum vacuum energy. All historically proposed methods of propulsion for spaceflight fit into one of these categories. We briefly discuss some of these so as to prepare the reader for the wealth of ideas that are to be discussed in later chapters. It is left to the reader to decide if each specific method is based upon an internal, an external or a locally driven engine.

The most basic rocket engine is one that uses chemical fuels such as liquid oxygen and hydrogen, and so we refer to it as a *chemical rocket*. However, this fuel has an inadequate performance for interstellar missions because it doesn't produce enough energy per mass of propellant. More efficient chemical fuels can be derived with different fuel mixtures, but ultimately it will not achieve the required mission in an appropriate timescale. Ideas do exist for using highly compressed gaseous hydrogen to produce metallic solid hydrogen, which if kept stable at normal atmospheric conditions can then be used to power an engine with an improved performance over a conventional hydrogen-based one. However, this technology has not yet been sufficiently demonstrated.

One could build an enormous spacecraft housing a crew of many hundreds or thousands of people and with a moderate amount of thrust allow that vehicle to head off on a very long journey towards the nearest stars. This would necessitate that the crew will produce children along the way and the crew that eventually arrives at the star system will be many generations from the original crew. For example, imagine a crew that embarks on a 1,000 year journey, producing children around the age of 25. The astronauts that actually witness the arrival at the star will be the 40th generation from those original astronauts that left Earth orbit. For this reason, we refer to this as the *Generation Ship* or *World Ship*. These have been studied extensively by the British engineers Alan Bond and Tony Martin, where they designed a wet (water containing) world concept with a radius of up to 10 km and a length of up to 220 km [2, 3].

The American Gerald O'Neill has also pioneered the idea of large-scale interstellar colonization using artificial habitats [4]. His visions typically described large cylindrical constructions 8 km in diameter and 32 km in length with entire land areas devoted to living space, parkland, forests, lakes and rivers. We may find that there are many people who are willing to embark on such an exodus. Many would desire a fresh opportunity to start the human civilization again, with a different set of values. In reality, this is the only way that interstellar colonization can really be achieved, launched on the basis of high quality information supplied from previous unmanned interstellar probe missions. True interstellar colonization depends not on the ability to transport only a few individuals in our engineering machines but on our ability to move many hundreds to thousands of people in only a few flights. Thinking through this argument and the requirements for infrastructure in order to build such large colony ships leads to the logical conclusion that human colonization missions to the stars are at the very least centuries away (Fig. 1.1).

An alternative is to place a small crew in a state of hibernation and awaken the crew once they reach their destination after many hundreds or even thousands of years. We know that animals can lower their body temperature and heart rate for an annual state of hibernation, so with sufficient understanding and the application of science we could learn how to do this, too. The *Hibernation Ship* is an interesting idea, but who would volunteer? Perhaps it is a more desirable option for some than facing a physical death.

One of the common forms of propulsion used in orbital spacecraft is electric propulsion. However, these engines simply don't have the performance for the

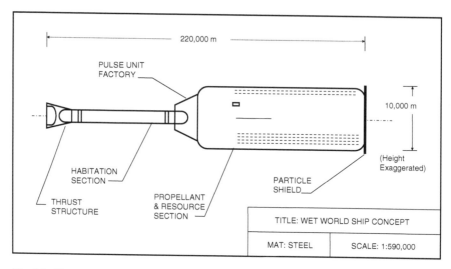

Fig. 1.1 Illustration of a World Ship

sorts of missions that this book is considering. An alternative, however, is the *Nuclear-Electric* propulsion scheme, which uses a nuclear reactor to generate the essential power supply to produce heat for the generation of electricity. This has the advantage that high power levels can still be obtained at large solar distances where solar energy is no longer available. These sorts of engines would certainly allow a spacecraft to go to a significant distance from the Sun, right out deep into the Oort Cloud at the outer edge of our Sun's gravitational influence.

Are there natural sources of power in space that can be employed to propel a vehicle, without the need to carry a propellant? The Sun ejects millions of tons of solar wind particles every year out into space, charged particles of ions and electrons. Directly employing these particles for thrust generation is possible in a *Solar Wind Spaceship*. Instead, it may be better to use the intensity of light photons in a *Solar Sail Spaceship*. One only has to go outside on a sunny day to feel the heat of solar photons, which carry momentum and can generate an intensity of around 1,400 W/m^2 at the orbit of Earth, where a watt is a unit of power measured as energy (Joules) per time (seconds).

Of course, solar intensity falls off inversely with distance squared, so instead, we could build enormous lasers that tap the energy of the Sun, and using Fresnel lenses send a monochromatic narrow collimated beam continuously out into space. Such a *Laser Sail Spaceship* would be sustained with power for much longer. Another suggestion is to send out beams of particles in the form of protons, for example, which impinge on a surface and propel it forward. Alternatively, the particles can be captured upon arrival and replenish any dwindling fuel supply that the vehicle needs. Both these ideas are what are called *Beamed Power Spaceship* designs.

Given that chemical fuels are performance limited, we can turn to more exotic fuels such as atomic-based ones. We know that atomic energy release is many

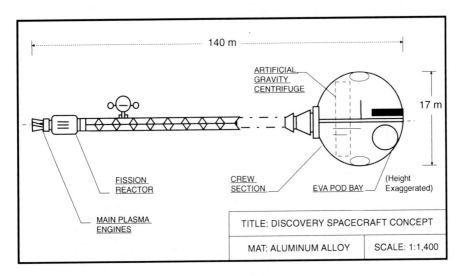

Fig. 1.2 *Discovery*-type spacecraft using nuclear engines

orders of magnitude greater than a typical chemical reaction, so this is a credible suggestion. The first proposal along these lines is the use of fission energy in a *Fission Reactor Rocket*, where the energy is released slowly in a highly controlled manner from radioactive particles. The Sun works mainly on the basis of hydrogen fusion reactions so the suggestion of a *Fusion Rocket* is also highly credible. After all, the world has plenty of hydrogen and deuterium locked up in the oceans, so there is a potentially unlimited fuel supply. Fusion is the holy grail of physics, utilizing the power of the Sun. There is a massive program of development underway on Earth to use fusion power generation for the electrical supply in cities. Great progress has been made since President Eisenhower's 'Atoms for Peace' conference in 1953.

The *Discovery* spacecraft illustrated above as used in the Stanley Kubrick film *2001: A Space Odyssey* and in the science fiction novel of the same name [5], may have been powered by nuclear fusion engines, although there are some indications [6] that it may have been powered by a gas core fission reactor. The basic idea for this design originated from a dumbbell-shaped configuration originally proposed by Arthur C. Clarke as a way of separating the human habitat module from the nuclear reactor located near the engine (Fig. 1.2).

Instead of carrying vast quantities of fusion fuel, the vehicle can be equipped with a large magnetic scoop and use this to 'scoop up' diffuse quantities of interstellar hydrogen that is distributed throughout space. Other materials could also be collected by such an *Interstellar Ramjet* design. If this could be done it would increase the performance of the engine in such a way that the spacecraft could travel close to the speed of light.

An alternative is to use atomic energy but in an uncontrolled manner like in the explosion of an atom bomb. This leads us to the proposal for an *Atomic Bomb Rocket*.

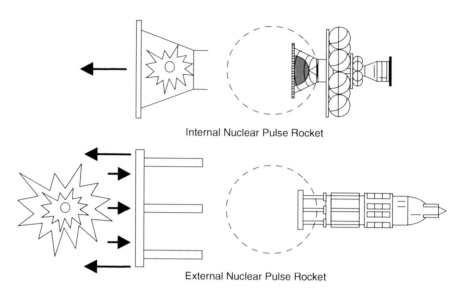

Fig. 1.3 Illustration of nuclear pulse rockets

Because any engine will be performance limited by the material melting temperatures, it would be better to have the fundamental 'combustion' cycle of the engine take place outside of its structure, hence the origin of this idea. The bombs are detonated behind the vehicle in succession, the blast wave of which imparts momentum to the ship. If ever there was a sudden asteroid threat to Earth, this idea may be the only one that governments could put together in a short timescale, not requiring any significant technology leaps or new physics breakthroughs. In a foreboding irony this type of propulsion may turn out to be the only salvation of humankind from the creation of its own demons. Both the fusion rocket and atomic bomb rocket are types of nuclear pulse engines and so in the literature they are referred to as an *Internal Nuclear Pulse Rocket* and an *External Nuclear Pulse Rocket,* respectively (Fig. 1.3).

Are there other types of energy that can be used for space propulsion? In 1928 the British physicist Paul Dirac predicted that the electron must have a twin particle that is identical except for its direction of charge and rotation. Since then it has been confirmed in experiments that indeed many particles do have a so-called antimatter equivalent known as antiprotons, antineutrons and positrons (antielectrons). An annihilation reaction between a matter and an antimatter particle pair produces around 1,000 times the energy release of a nuclear fission reaction and around 100 times that of nuclear fusion reactions. This leads to the concept of an *Antimatter Rocket,* provided we could someday produce sufficient quantities of it. One of the possible uses of antimatter is as a catalyst to initiate fusion in a fuel. This is known as *Antimatter Catalyzed Fusion Rocket.* This is achieved by injecting a beam of antiprotons that react with a fusion capsule wall and the annihilating protons produce a hot plasma that then ignites. A combination of fusion and antimatter may be both credible and give large performance gains, which are well ahead of many other propulsion schemes, although this research is still at an early theoretical stage.

If ordinary matter can be considered a form of positive matter, then if matter exists in an opposite state it must be negative. Negative matter is the same as ordinary matter (it is not antimatter) but may generate a negative pressure so that instead of obeying an attractive force law, it would obey a repulsive force law. This has led to the proposal of a *Negative Matter Spaceship*, which involves a positive matter-negative matter pairing, and the forces are such to propel the vehicle in the desired direction of motion [7]. Alternatively, if we could develop negative energy or its effects, the consequence would be that we could manipulate the very nature of space itself, by bending (inversely) gravitational fields. This would be required if we were to ever manipulate space to such an extent so as to create shortcuts through it in what are termed *wormholes*. There is currently no scientific evidence that wormholes exist naturally in space, except perhaps at the quantum level where the spatial dimensions become chaotic.

We must also mention the prospect of *white holes*, which are related to the concept of wormholes. Scientists know that a black hole is a collapsed star, which attracts all matter into it, which can never escape. It has been postulated that white holes would have the opposite property where all matter is repelled from it, so it could act as a kind of tunnel exit for any would be space travelers, provided they could survive the infinite compression and 'spagettification' during the entrance to the black hole in the first place.

Finally, if we could manipulate space and gravitational fields to such an extent, then perhaps we could cause space to massively collapse in one direction (like in a black hole) and expand in the opposite direction (like in a Big Bang explosion). If we could do this, then we would be able to create a *Warp Drive Spaceship* so beloved of science fiction fans, where the fabric of space and time are altered for the purposes of enabling transport. It will be a surprise to many to discover that the concept of a warp drive has been discussed openly in the academic literature for some time since the first seminal paper in 1994 [8]. This, too, would require negative energy. The warp drive concept is similar to the concept of *The Space Drive*, which is any idealized form of space propulsion where the fundamental properties of matter and space are used to create propulsive forces in space without having to carry or expel a reaction mass, but limited to sub-light speeds. Other than space itself, this may be some property such as dark energy and the quantum vacuum energy, both of which may be related to negative energy and the expansion of the universe.

The overall types of concepts explored for interstellar travel are shown in Table 1.1. Most of the current research is focused on medium and fast propulsion modes, although substantial research is increasingly being conducted into the potential for ultrafast systems. Current manned mission research efforts are focused on the slow (such as generation ship) and medium (such as fusion propulsion) modes, mainly due to the acceleration constraints. To launch an interstellar mission is also a very complex exploration program. There are many phases to the mission

Table 1.1 Propulsion modes
for interstellar travel

Propulsion mode	Mission duration (years)	Speed of light (%)
Slow, low acceleration	>400	<1
Medium, moderate acceleration	50–400	1–10
Fast, high acceleration	5–50	10–99
Ultrafast, superluminal	<5	>c

and considerable project management, which has to be considered; just look Project Apollo. This includes the design, manufacture, launch, assembly, fuel acquisition, boost stages, en route science, deceleration, sub-probe deployment, target system science and science data return. It may also be necessary to think about how to decontaminate the probe safely, in line with the Planetary Protection Protocol to ensure that no Earth bound bacteria are transmitted to the surface of another biologically defined world. Mission planners will also need to think about the overall science goals, the mission concepts and technology requirements prior to embarking on such an ambitious program of interstellar exploration.

Many of the concepts discussed above are the subject of an enormous amount of academic literature and it would take a single person a lifetime to fully explore the range of options for interstellar travel [9]. Instead, most researchers working in the field are dedicated to a specific concept or spend some years on one concept before moving onto another. International conferences with academic papers on interstellar propulsion are an exciting experience for any young student. The optimism and boldness of the appraisals is truly breathtaking. Unfortunately, there is also a negative side to conducting research in this field – one of credibility. Because many academics do not see interstellar travel as either near possible or relevant to today's problems, it is considered a 'hobby' and thus much research is done privately and on a volunteer basis. It is a sad fact that the majority of researchers working on interstellar propulsion are doing so as a spin-off from their main research interests, quietly tolerated by their colleagues.

One exception to this has been the inspirational leadership provided by the American space agency NASA. It has set up two programs in particular that have allowed free consideration of interstellar ideas. This includes the Institute for Advanced Concepts and the Breakthrough Propulsion Physics Project. Unfortunately, both of these programs were canceled. These exciting research programs will be discussed further in Chap. 16 as well as the work of other private institutes involved with interstellar research such as The British Interplanetary Society and the Tau Zero Foundation.

It is the aim of this book to demonstrate the high quality of research that has taken place in previous decades and to clearly show that a real engineering solution to reaching the nearest stars is just around the corner if not already available at a low technology readiness stage. In particular, several actual concept designs will be discussed in detail. Some of this work is the product of individuals and some the product of years of effort and teams of people. One of these designs may represent a blueprint for how both robotic and eventually human missions to the nearest stars are first accomplished. History will be the final judge of the true visionaries of these ideas, and this knowledge is a comfort for those that continue to pursue the most unworldly of intellectual passions which has only really be turned into a credible possibility in the last century. For it is an infectious addiction that grips people like the visionaries Arthur C. Clarke and Robert Forward as well as all those who follow the inspiration that they and others leave behind when they dare to dream of sending spacecraft to distant worlds.

In this chapter, we have briefly discussed some of the different propulsion systems for getting to the stars. Some of these are discussed in more detail in later chapters and some will not be mentioned again. The purpose here was to present the reader with a clear understanding that contrary to popular belief, which considers interstellar travel an insurmountable problem, in fact there are many ideas for how we can go to the stars in future years, and these have been discussed in books such as by the authors Paul Gilster [10] and Iain Nicholson [11]. Because this book is mainly concerned with the near-term prospects for sending a robotic probe to the nearest stars, manned space travel will not feature much further in this book. There is little point attempting a human exploration mission until we have at least demonstrated a robotic mission first and properly evaluated the benefits, risks and mission requirements. Instead, we shall concentrate on propulsion schemes that can be used in particular to propel a probe to one of the nearest stars in a time frame of a century or less. But first, we must consider some of the basic science needed to understand the design of such engines. For this we will need to learn something about the physics and engineering associated with spacecraft design.

1.3   Practice Exercises

1.1. List all of the propulsion concepts mentioned in this chapter. Using the Internet, do a search for each concept and list the main features of each one, including the potential exhaust velocity or specific impulse. Classify each propulsion scheme in your own assessment as being speculative ($>$1,000 years), far future ($>$200 years), near future ($>$100 years) or practical ($<$100 years). Once this exercise is completed file this away somewhere so you can consult it while reading the rest of the book. This list will act as your reference points to becoming familiar with the topic.

1.2. Read a copy of one of the following as a good background to this chapter (1) *A Program for Interstellar Exploration* by Robert Forward, published in the Journal of the British Interplanetary Society, Volume 29, pp. 611–632, 1976 (2), "Ad Astra" by Robert Forward, published Journal of the British Interplanetary Society, Volume 49, pp.23–32, 1996 (3) "Interstellar Travel: A Review for Astronomers" by Ian Crawford, published by O. J. R. astr. Soc, 31, pp. 377–400, 1990.

1.3. Read one of the following science fiction books (1) *Tau Zero* by Poul Anderson (2) *2001: A Space Odyssey* by Arthur C Clarke (3) *Rendezvous with Rama* by Arthur C. Clarke. When reading these books you should concentrate mostly on the science, which is generally representative of current knowledge. These three books illustrate the extremes of interstellar travel – from the relativistic Interstellar Ramjet to a slow form of World Ship.

References

1. Shepherd, L. R. (1952) Interstellar Flight, JBIS, 11.
2. Bond, A., and A. R. Martin (1984) World ships – An Assessment of the Engineering Feasibility, JBIS, 37, 6.
3. Martin, A. R. (1984) World Ships – Concept, Cause, Cost, Construction and Colonization, JBIS, 37, 6.
4. O'Neill, GK (1978) The High Frontier, Corgi Books.
5. Clarke, A. C., (1968) 2001: A Space Odyssey, New American Library.
6. Williams, CH et al., (2001) Realizing '2001; A Space Odyssey': Piloted Spherical Torus Nuclear Fusion Propulsion, NASA/TM-2005-213559.
7. Zubrin, R et al., (1996) Islands in the Sky, Bold New Ideas for Colonizing Space, Wiley.
8. Alcubierre, A (1994) The Warp Drive: Hyper-fast Travel within General Relativity, Class. Quantum Grav., 11, L73–L77.
9. Long, FF (2009) Fusion, Antimatter & The Space Drive: Charting a Path to the Stars. Presented at 59th IAC Glasgow October 2008, JBIS, 62, 3.
10. Gilster, P (2004) Centauri Dreams, Imagining and Planning Interstellar Exploration, Springer.
11. Nicholson, I (1978) The Road to the Stars, Book Club Associates.

Chapter 2
The Dream of Flight and the Vision of Tomorrow

Sometimes, flying feels too godlike to be attained by man.
Sometimes, the world from above seems too beautiful, too won-
derful, too distant for human eyes to see...

Charles A. Lindbergh

2.1 Introduction

We consider the possibility that our species may develop into a spacefaring
civilization. This represents a future where we rise to the challenges that the universe
presents us while enjoying the rewards of discovery and new sources of energy
production. In this possible future access to space is not limited to a few but enabled
for many, beginning an era of interplanetary and eventually interstellar migration of
our species into the cosmos. Our species would be free from the confines of a single
planetary body and instead have the opportunities of countless worlds and resources.
It is the observations of nature, the flight of the birds, and beyond to the depths of
space that drove us towards this vision. Plato said it was astronomy that compelled
the soul to look upwards and lead us from this world to another.

2.2 Becoming a Spacefaring Civilization

Let us begin by asking the question of what is a spacefaring civilization? Here is
one possible answer: a spacefaring civilization is one with many orbiting space
stations, active colonies on all local moons and nearby planetary bodies as well as
remote outposts in the outer parts of the Solar System with a Solar System wide
trade economy. To date the state of human civilization cannot be described by this

K.F. Long, *Deep Space Propulsion: A Roadmap to Interstellar Flight,*
DOI 10.1007/978-1-4614-0607-5_2, © Springer Science+Business Media, LLC 2012

simple definition. However, our science-based technological society is heading towards this state and there is reason to be optimistic about the future.

In the 1960s the Russian scientist Nikolai Kardashev considered the possibility of advanced civilizations existing in the universe [1]. This seemed a credible idea, considering that there were many stars much older than our own star, the Sun. Kardashev came up with a description for galactic civilizations based around energy consumption. He defined three possible types. A Type 1 civilization is one that has achieved control over its planet's entire resources. A Type 2 civilization is one that has achieved control over the resources of its whole Solar System, including the Sun. A Type 3 civilization is one that has achieved control over its entire galaxy, including the core. Clearly, passing into the twenty-first century humankind has not even achieved a Type 1 status by these definitions. But to take an optimistic perspective, it should also be clear that we are on the brink of moving towards Type 1 if we embrace our destiny (it is believed that we are currently at a level of 0.7), build on the achievements of the Apollo missions and pioneer the outer boundaries of space as the final frontier. It is quite possible that the current period in our history is a very critical one in which we either succeed in winning the Solar System and continue our expansion into the cosmos, or we fail to reach our full potential and the possible collapse of our current civilization is forced upon us.

If this assessment is true then the cold facts of reality should force us to embrace the greater challenges ahead, and in the words of the former U.S. President John F. Kennedy: "We choose... to do these things not because they are easy, but because they are hard."

For this is the challenge of our times; the route we take into the future will determine the ultimate fate of our species. Traveling into interstellar space may be the best way we can ensure our future survival, dispersing the species over a wide area, maximizing the resources available to us and progress our scientific knowledge. This is a highly productive way for our species to direct its energies. And if life from this world is indeed unique, as some may claim – the only instance of intelligence in this vast universe – than ever the more important that we spread that life outwards to ensure the survival and growth of that intelligence.

The first artificial satellite reached Earth orbit in October 1957 and was called Sputnik 1. This achievement from the former Soviet Union had such dramatic consequences on the world that it started what history now records as 'the space race.' For a while, the Soviets dominated the early achievements in space exploration with the first mammal in space, Laika the dog, in November 1957, the first man, Yuri Gagarin, in April 1961, the first woman, Valentina Tereshkova, in June 1963 and then the first space walk by Alexey Leonov in March 1965. These were tremendous accomplishments that would have been welcomed by the world warmly if it weren't for the suspicious motivations behind them. America eventually caught up, and the first American in space was Alan Sheppard in May 1961 followed by Virgil Grissom in July of the same year, and then that historic first orbit by John Glenn in February 1962. Events were moving at a fast pace, and in June 1983 Sally Ride became the first American woman in space.

After the initial Soviet achievements, the huge industry of America woke up and inspired by President Kennedy's vision on May 25, 1961, aimed for more ambitious missions than floating around in Earth orbit:

> I believe that this nation should commit itself to achieving the goal, before this decade is out, of landing a man on the Moon and returning him safely to the Earth. No single space project in this period will be more impressive to mankind, or more important in the long-range exploration of space; and none will be so difficult or expensive to accomplish.

The decision of the American political leaders to attempt the seemingly impossible and place a man on the Moon before the end of the decade was a courageous one. This was an open challenge to the Soviets to 'race' in the biggest peacetime competition in human history. Nine years later, in July 1969, the first man, Neil Armstrong, set foot on the lunar surface. A Soviet lander was en route to the lunar surface at the same time as Apollo 11, hoping to be the first to return a soil sample back to Earth, thereby claiming some form of technological and cultural victory over the United States. Although this caused some anxiety for the mission it ultimately had no real effect on the outcome, as Luna 15 crashed and Apollo 11 made a successful landing. America had won and was to be considered the more technologically advanced and thereby ideologically superior nation from a public relations perspective.

During the Project Apollo missions to the Moon twelve American astronauts walked on the lunar surface over six landing missions, the last of which was in December 1972. Around 380 kg of Moon rocks were returned back to Earth. If we had gradually built up a lunar colony over the last four decades, today there would be a permanently manned arctic-like station with spacecraft cycling back and forth between Earth and lunar orbit. Sadly, that was not how history turned out, and we are still waiting for this dream to be fulfilled. Since then, mainly due to a lack of political will, we have withdrawn from the Moon and concentrated on less ambitious missions, low Earth orbit (LEO) operations. This is not to say that the achievements in space over the last few decades have not been stupendous. The establishment of the International Space Station and over 100 launches of the space shuttle have been inspiring and kept the dream of spaceflight alive for many millions of people. But let us be frank in our evaluation – Earth orbit is just that, it is not outer space, with its very thin atmosphere and microgravity environment.

Astronauts are tremendously capable people who have demonstrated 'the right stuff' and earned their place to fly into space. There are many millions of people on this planet who are sufficiently fit to travel in a spacecraft. They do it every day in an airliner over the oceans of the world but at much lower altitudes. OK, so they have gravity to contend with and perhaps would struggle in zero gravity. But is that even true? Wouldn't it be easier to operate in a near-zero gravity environment, provided you allow for corrections in physical movement? The majority of Earth's population is physically capable of undergoing short space missions. So why choose only a select few? The main reason is because currently access to space is expensive, so if you are going to send up an expensive satellite mission you want to guarantee near 100% success, which means only selecting the most suitably qualified individuals.

Consider the opportunities if access to space wasn't expensive – this would change everything. Indeed, today there is an international effort, largely privately financed, to open up space to the world. This is now being called the commercial space market, and tourism is the main reason people want to go into space – for potential leisure activities. Although it is true that leisure in space is a driving factor, it should not be the dominant one. Instead, it should be driven by exploration and business incentives, and this is where the rest of the Solar System comes into its true value for our species. The return to Earth of rare minerals or gases presents a clear business motivation for opening up space to all of industry, and this motivation should be the driving force for future space exploration, opening up new avenues of knowledge along the way and pioneering the boundaries of technological innovation.

2.3 The Promise of the Future

When people are asked for their views on space travel, there are generally two viewpoints that emerge. The first are those people that see the 'heavenly bodies' (the planets and Moons) as almost mystical and take the view that humanity should not spoil them or alter them for self gain in any way. This may be motivated by spiritual reasons or a genuine acknowledgement that we haven't looked after our own planet that well, which sets a precedent for how we will behave on other worlds. The second viewpoint states that humanity should move out into space, exploiting the material resources along the way, purely for self gain and probably profit. But surely there is a third way, somewhere in between these two viewpoints, which states that indeed we should move out into space and exploit the material resources as needed to sustain our expansion and continued survival, but we should do so in a responsible manner. In particular, if we find life on other worlds, we should do our best to preserve it and minimize our potential contamination.

If we think about how humanity will behave when we first move out into space, we are led to inquire which of these three viewpoints are we likely to take in our journey? To address this, let us first examine our behavior on our own home world – Earth.

Our need for energy has resulted in a depletion of many natural resources. This includes the continuous destruction of the rainforests, the poisoning of the world's lakes and oceans, and the extensive burning of fossil fuels. So we must turn to alternative resources. For several decades now nuclear fission reactors have provided a great deal of power for cities. However, fission reactors produce contaminant waste that is difficult to destroy. This is because the radioactive decay time (to a stable non-radioactive state) of the substances being used is of order 100,000 years. One solution to this would be to send all of the radioactive waste into the Sun. However, first we must get that waste into Earth orbit, and the reliability of rocket launches is currently not adequate to take such a risk. The explosion of a rocket during launch containing radioactive waste will spread dangerous particles for many hundreds of kilometers around the globe.

Until we have developed efficient sources of energy we are not ready for the rest of the universe, and more importantly it is not ready for us. This is not such a concern for the time being, because currently, we do not possess the technology to fully exploit the natural resources of the Solar System anyway. However, this situation is changing rapidly, and the technology is being developed with some priority. This is for several reasons. Firstly, there is the belief that global climate change is related to manmade activities of energy generation. Second is the depletion of our natural energy resources. As a consequence of this rapidly moving technological situation, humanity's destiny is either going to go one of two ways. We will either destroy ourselves as our energy reserves dwindle with an increasing population competing for precious resources, or we will be successful, tame our behavior, and move on to other pastures – outer space. Others have argued that all civilized communities must achieve control over their population growth if they are to survive the catastrophe of population explosion and its associated problems of overcrowding and shortage of resources [2]. Whether the universe is ready for us or not, we are likely to be going there soon, and we must prepare ourselves for the exciting journey that awaits future generations. In the words of the Russian physics teacher Konstantin Tsiolkovsky, "Earth is the cradle of mankind, but we cannot live in the cradle forever."

To deal with our future energy needs the physics community has been working since the 1950s on the design of a civilian nuclear fusion reactor. Fusion energy is what powers the Sun, and it has the advantage of being a clean source of energy. However, although significant progress has been made in the development of a fusion reactor, we are not there yet. So if we were to move out into space now, we will be reliant upon twentieth century technology and associated industrial processes, which damage any natural environment as well as waste energy. But the development of fusion energy technology is moving ahead rapidly. It is just a matter of time before we have working nuclear fusion reactors powering the majority of the world's electrical supplies. Once these reactors have been built and become operational, the next generation of reactors will be even more efficient and smaller. A consequence of this is that alternative applications of fusion energy will be considered. A whole new field of science will be born based upon the application of fusion energy and household devices not yet conceived by the human mind will become common. This progress will inevitably have applications to both robotic and human spaceflight, and the development of a space propulsion drive based upon the principle of fusion reactions will become possible. Until that time, we must be content with the progress made by our ambassadors among the stars – robotic probes.

As of the year 2011 we are busy receiving data from robotic probes on many different trajectories throughout the Solar System. Our civilization is still in the information-gathering phase. The question is – when we have got the information we want, will we have the courage to explore the outer Solar System and beyond to finally embrace our destiny to become a space faring civilization in the near future? Humanity is at the brink of an incredible age. In fact, it would not be an overstatement to say that we are on the brink of an evolution in humanity's great journey. It is difficult to predict when we will become a space faring civilization, but the taming of fusion energy on Earth (should we be successful) and for use in space travel must suggest that we are near that point – perhaps only a few decades to a century away.

The closely interconnected world that we live in, as observed from the nightly news on our television screens, is enabled by the communication age. What has driven us to these technological accomplishments? What makes us want to go further, faster and better than before? Is it simply the need for self-entertainment or is it something more profound? We always want to know what is over that next horizon. So we walk long distances, we ride horses, we build cars, and eventually we build machines that fly. What all these modes of transport have in common is they all rely upon some system of propulsion to provide the forward momentum. For the human, it is simply the muscles in his legs and the forward movement of his arms and pelvis. The use of the horse is an example of a paradigm shift in thinking; let the energy be expended by some other means rather than due to the muscles of the person, an improvement in efficiency. But then we realize that horses, too, have a physical limit. They also must stop for rest and food. So we build a mechanical horse, which doesn't need rest – the bicycle, motorcycle and motorcar – except when wear and tear of the components require they all be replaced. But these machines also need fuel, and how much we can give it depends upon how much fuel it can carry and the rate at which it is used. It is a self-limiting technology. So we look skywards, and see the birds that effortlessly glide above our heads surveying the ground to horizons we cannot see. Humans mimic nature, master the principles of flight and fly like the birds – perfection is achieved. We have been able to do this because we have something that the rest of the animal kingdom lacks – a highly developed brain. This allows us to seek novel solutions, which ultimately have the purpose of prolonging life or continuing the survival of the species.

Humankind has evolved over many millions of years to a species that can now boast to have traveled on all of the continents, all of the oceans, walked on the Moon and even sent robotic ambassadors to enter the atmosphere of Venus, Mars, Saturn, Titan, several asteroids and a comet. Our robotic probes have also entered the orbits of many worlds, including Mercury, Uranus, Neptune and several other moons such as Gannymede, Europa, Io and Callisto. But all these accomplishments still do not add up to a spacefaring civilization. We are still children playing with toys at the dawn of our destiny. We are yet to send a probe to our most distant world, Pluto, which resides at 40 astronomical units away (1 AU is the distance from Earth to the Sun), although when this book was written a spacecraft mission called New Horizons was en route. We also are yet to send probes to the outer Kuiper Belt and Oort Cloud, which reside at 40 to 500 AU and 2,000 to 50,000 AU, respectively. The most distance spacecraft we have sent includes Voyagers 1 and 2 along with the Pioneers 10 and 11 spacecraft. The furthest any of these have traveled to date is around 100 AU.

If we have only traveled this far away from Earth, how much further do we need to go to the nearest star? Alpha Centauri is 4.3 light years away. That is, it takes light (the fastest thing in the universe) over 4 years to travel the distance of 40,000 billion km, or 272,000 AU. Our home galaxy, the Milky Way, contains several hundred billion stars spread over 100,000 light years. We have a long way to go before we can truly be considered a spacefaring civilization. We have merely dipped our toes into the vast ocean of space, been frightened by the temperature and not had the courage to dive in far enough. This is the 'failure of nerve' that Arthur C. Clarke frequently discussed in his writing. This lack of confidence in our species to go further despite the risks is like a ball and chain around our species, constantly tethering our species to a single star and its neighboring worlds. We must find the courage to shake off the chains of our home world and seek new resources elsewhere. This is the only way to ensure the permanent survival of our species.

When we examine Low Earth Orbit (LEO) space operations or robotic missions to the planets today, we also notice a large degree of international cooperation. Is this a model for how the first missions to the stars will one day be achieved? Sending spacecraft to other worlds is likely to be expensive, with high risk, so minimizing these factors necessitates international cooperation. This endeavor also has the potential to unite humanity behind a single vision despite its diverse religions, cultures, and laws. In order to make more progress in exploring outer space, we will need a rocket engine – a very powerful one – which is part of the focus of this book. The first steps towards this vision of becoming a spacefaring civilization are the complete robotic exploration of the Solar System and nearby space. This has already started and with sustained effort should continue further within the current century. We can imagine the sight, a large space vehicle being constructed in Earth orbit over a period of years and eventually a crew of a dozen or more astronauts blasting off towards the first stars on a rocket ship powered by some exotic engine, perhaps already invented. Many today believe that interstellar travel is impossible, but it will be shown throughout the chapters of this book that in fact it is perfectly possible and engineering designs already exists for achieving this goal (Fig. 2.1).

2.4   Why We Shouldn't Go Out into Space

Why would anyone want to go into space, let alone to another planetary system? Over a period of billions of years, humans have evolved on this planet to the complex life forms that we are. We have a wonderful biosphere that has been perfected by nature for us as well as the rest of the animal kingdom on Earth. We have a planetary environment rich in plant life, plentiful in oceans of liquid water and overall a moderate climate, all enabled by our gravitational center – the Sun. Any move to another solar system will involve significant hazards, and even when we get there it is highly unlikely that the biosphere of another world will be ideal for us. Firstly, the surface pressure is likely to be either too high or too low, which

Fig. 2.1 Illustration of a planetary rocket scene, the vision of dreams for centuries

means we cannot breathe the air freely but will require some form of respiratory apparatus. Secondly, the surface temperature will be either too cold or too hot. If too hot any equipment may malfunction. If too low, then we may simply freeze to death. Then there are the charged particles that enter the atmosphere from the local solar wind. On Earth, we are protected by a magnetic field. On another planet, it may be insufficiently strong or non-existent, should a liquid iron core not be present. We would require substantial protection from any cosmic radiation.

The atmosphere of another planet may not be compatible with our basic survival requirements of oxygen, carbon dioxide and nitrogen. In all likelihood, any astronauts stepping foot onto this new world will need some form of space suit to enable them to function. Many of these risks can be mitigated by understanding in advance the many worlds and moons of our own Solar System. Then there is the journey itself to get to this new world. This will involve traveling across vast distances of space, not depending on any rescue parties and avoiding asteroids or dust particles as they approach with large kinetic energies. Even a single particle can present a significant collision risk if the vehicle is traveling at high speed. This is due to the fact that kinetic energy is proportional to the velocity squared, so the faster the ship or particle is going, the higher the collision energy involved. There is also the risk from bombardment of cosmic rays en route, which may cause cancer. If an artificial gravity field is not created then astronauts should expect significant calcium loss and bone decay. So given that there are these significant risks, why go? To undertake such a high-risk enterprise and place humans in harm's way, the potential rewards would have to be very high indeed.

There are many educated people who take the view that traveling to the stars is either impossible, a useless expenditure of our energies or both. In particular, they cite many reasons why interstellar travel should not be attempted. The first objection is an obvious one, the stars are too far away and any travel times will be prohibitively long. They also cite massive fuel requirements. Then there is the risk and difficulties with sustaining any crew in space for long periods, as discussed above. One way to offset the presented objections is to launch a mission with minimum mission duration. This would mitigate the risk to any crew and reduce the probability of dust or other impact hazards as well as radiation exposure. The key driver then for any mission has to be, *minimize travel time.*

Many people also take the viewpoint that sending missions to distant destinations is a waste of our useful resources, both scientific and financial, when we should be concentrating on solving the problems back here on Earth – that is, problems such as poverty, war, unemployment, education, health and of course surviving global climate change. The cost of such a mission is considered to be a major factor.

There are two further objections to interstellar travel, which are more technical. The first relates to the fact that technology is continually improving, and so why launch a mission with an engine, when it could be overtaken by a more advanced engine in a later decade. This is the so-called incessant obsolescence postulate, when no matter when a mission is launched eventually another mission will overtake it with faster engines. This postulate may not hold however in the case of identifying an optimum launch opportunity, where launching a mission several years later with faster engines may not get the second spacecraft there sooner than the first. There is also the question of waiting for the technology to come to fruition. This has been discussed in the literature and the author Andrew Kennedy [3] described the problem in terms of the incentive trap to progress. In such a situation a civilization may delay interstellar exploration in the hope that an improved technology, perhaps based on radically new insights, will be ready later on. Kennedy describes the problem thus:

> It is clear that if the time to wait for the development of faster means of travel is far greater than the length of the journey, then the voyagers should go ahead and make the journey. But if the likely future travel time plus the length of wait is equal or less than the current journey time then they should definitely wait.

However, using the equations associated with growth theory, it was shown that a window of opportunity exists where the negative incentive to progress does turn positive and so a spacecraft can arrive at the destination earlier than a later launch. Leaving before the minimum time allows future growth to overtake the voyagers; leaving after the minimum time will mean the voyagers cannot catch up to those who left at the minimum. Similarly, in the event of a propulsion physics breakthrough that does not rely on the conventional methods of crossing space, this postulate would become irrelevant. This relates to the time dilation effect of

Einstein's special relativity, where any crew that accelerates at high fractions of light speed away from home will become time separated from the loved ones they leave behind. Poul Anderson elegantly explored this in the science fiction novel *Tau Zero*.

2.5   Why We Should Go Out into Space

Now that we have discussed some of the objections to interstellar travel, we shall consider the arguments for it. If a calamitous disaster were to occur tomorrow, such as a sufficiently large asteroid, we would be doomed as a species. We have insufficient protection and nowhere else to go. The situation is so serious that we may not even notice the impending threat until it is too late. There is currently no internationally organized, government-supported program to watch the skies for this threat. In effect, Earth is naked in space and the priority of any technologically developed civilization must be to protect it from the hostile elements of space, if we are to expect Earth to continue to look after us. This is not science fiction: the threat is real and we simply have two choices – defend ourselves or run. However, we cannot run because we are not a spacefaring civilization, although our chances for survival would be increased if we had somewhere to go. Therefore all we can do is to defend ourselves where possible.

For some reason, the existence of this threat is controversial in some quarters and still being actively debated, with many not taking the threat seriously, despite being shown evidence of a recent impact of the comet Shoemaker-Levy 9 in 1994 into the atmosphere of Jupiter at a speed of around 60 km/s. A similar event occurred in 2009. There are also internal risks such as natural catastrophes from earthquakes and tsunamis or even rapid global climate change. Then there is the fact that both our Sun and planet have a finite age of around 5 billion years. If our existence is to be permanent, at some point we must grow up and leave this Solar System, finding alternative homes among the stars. Natural disasters originating on Earth and from space do threaten the stable existence of our cherished human civilization, and such threats should be taken as real and present dangers.

The direct impact of successful space missions has a positive effect on society that is difficult to calculate and long lasting. We are still reaping the benefits of the Apollo program today. However, one of the issues we currently face is that most of the tremendous achievements in spaceflight were accomplished in the 1960s and 1970s. The people involved in those programs are getting older, and we may find ourselves in a position where no one living has worked on manned lunar projects, for example. If this were to occur, we could have a failure of confidence or even a failure of capability. This is the reason why we must continue to build upon the achievements of the past in the old tradition of an apprentice. If we do not, we face the prospects of returning ourselves to the Dark Ages and taking centuries to catch back up with ourselves. Hence, bringing all this together we are led to our first of seven motivations for space travel which has to be continued survival. The philosopher and writer Olaf Stapledon understood this well:

> Is it credible that our world should have two futures? I have seen them. Two entirely distinct futures lie before mankind, one dark, one bright; one the defeat of all man's hopes, the betrayal of all his ideals, the other their hard-won triumph.

The second motivation for space travel is related to a balance of resources. There is a limited amount of material on the planet and an exponentially increasing population cannot be sustained indefinitely within those limits. It is simply a question of numbers. To sustain the population growth, resources must be found and better distributed, but once depleted the population must be curtailed or move on. Imagine a hypothetical lift with a maximum capacity of 30 people. Let's assume that each person is a clone and to begin with there are 2 people in the lift. Each clone produces 2 more clones so that there is a clear geometrical progression in the total population. By the second generation there are 6 clones in the room. By the third generation there are 14 clones, and by the fourth generation there are 30 clones. The maximum capacity of the lift has been reached. Now let us go back and assume that by the second generation half of the total population leaves the lift and goes into a different lift, so that there are now only 3 clones per lift. Carrying on the same geometrical progression by the third generation there will be 7 clones per lift and by the fourth generation 15 clones per lift – half the total capacity of the original setup.

Now, with planet Earth's human population it's a bit different, because we are talking in very large numbers, of order 6 billion people growing at a population increase of around 2% per year. With the limited resources of the planet this growth is not permanently sustainable. Indeed, when the population does eventually hit the resource limit this could result in a very unstable world as nations compete for what resources remain to sustain their national populations and economic prosperity.

Now let us pretend that we had built a gigantic housing complex on the Moon. It is not simply a question of just shipping that population off on many thousands of rockets to the new homes. A simple calculation shows that this is impractical. If a rocket were to ship off from Earth carrying 100 people every day of the year for 10 years, this would still only be a total population move of 365,000 people. If this were continued for a century, it would still be much less than 1% of the total population, which was continually increasing anyway. So hoping to ship vast quantities of the population off to other worlds is not the solution to population control. Instead, we must seek alternative mechanisms to ensure a continuing and successful civilization. The first is population growth control. This necessitates extreme and draconian measures such as limiting childbirth or even longevity – clearly not desirable options in any culture that values liberty. The second mechanism is expansion of the population outwards off Earth, so that the population can grow on other worlds and out beyond the Solar System not limited to the resources of one planet.

The third motivation for interstellar travel is creating new economic opportunities, although more directly relevant to interplanetary exploration. For a growing and healthy civilization the creation of new business opportunities is vital to the economic growth of a nation and the world, as well as important for the growth of technological innovation. There are several potential business markets associated with space exploration, from space tourism to the mining of rare resources off world. This could be iron ore from an asteroid or helium-3 from the Moon. This will also slowly bring to fruition a Solar System-wide infrastructure that could be vital to the eventual expansion of our species to the outer Solar System and beyond.

The fourth motivation for interstellar travel is the quest for knowledge. This can be knowledge in the scientific sense, such as geological or astronomical. Alternatively, it may be related to understanding our place in the grand scheme of the universe, a kind of spiritual or metaphysical knowledge. How can anyone seeing an image like that of the Sombrero Galaxy shown in Plate 1 not be moved by the immensity of space? This may be a necessary part of any religious faith that needs to continually re-invent itself in order to offer comfort, guidance and relevance to its members.

Arthur C. Clarke said: [4] "Any sufficiently advanced technology is indistinguishable from magic." Certainly, if humans were to travel back in time and showed Victorian England a laptop computer or a 3D computer game, they would think it was magic. Indeed, it could be argued that this is ultimately the test of any advanced civilization. If a comparison is made of the current technology to a previous era, and there are no products that can be presented as magic, then technological progress is too slow. There is no evidence that we on Earth are reaching that point currently. The last 100 years has seen astonishing advancements in technology, and there are many visionary ideas around today that may become reality tomorrow. However, eventually we may reach a point where the limit of our knowledge dries up, and the only way to replenish it is to go to other places in the universe to learn more about it. Knowledge also brings about other benefits, such as medicines to cure diseases or technologies to ease the burden on our hard existence and increase our overall quality of life. Seeking knowledge about ourselves and our environment is also a fundamental part of our makeup. Speculating further one may go as far as to say that curiosity itself may even be a reflection of a survival instinct.

The fifth motivation for interstellar travel is the spreading of life. Given that there are around 400 billion stars in our galaxy and around 100 billion galaxies observed in the universe, the probability that Earth is the only world with life is very small. Any conservative estimates for parameters such as the lifetime of planets and stars will still lead to a substantial number of possible worlds with life. However, we also have to consider the evidence, and to date with our limited knowledge of space we are not aware of any substantive evidence of life (let alone intelligent life) existing outside of our home planet. If this was the case, then the responsibility of us to spread our life seed to the rest of the universe is very great. Indeed, there can be no nobler cause for the purpose of humanity. Of course, it is most probable that there are many worlds out there with life. Several moons in our own Solar System are good candidates for life of some form and there is evidence that life may once have existed on Mars, as discussed in Chap. 7.

However, intelligent life is another issue altogether and could even be considered a reason to embark on such missions itself – to trade knowledge with another species. If we were to meet such intelligent life then this would have profound implications for humanity and our knowledge of biology and medicine. Meeting intelligent beings from another world will also challenge humanity socially. We have already shown throughout history that we struggle to overcome our racial and cultural tensions, although things are slowly improving. If we as a species react so poorly to someone of a different skin color, how would we react to someone who looked literally alien or even had social values and ethics vastly different from ours? The occasional incident of physical birth defects in a human and how that person is treated by the rest of society also demonstrates that even today we still have deep prejudice and fears that suggest we are not yet ready to meet our interstellar neighbors. We must first become a society where appearance is not a factor for determining social ranking and behaviors.

The sixth motivation for interstellar travel is the freedom factor. The development of human beings to their modern form has been a slow process. Homo sapiens have evolved from hominids (great apes) and in turn from placental mammals over a period of several million years. The oldest discovered human skeleton is around 4 million years old. As humans have developed we have learned to adapt ourselves to the hostile world around us. The application of technology has then enabled us to exploit that world to our advantage. It started with our technological use of stone in the so-called 'Stone Age.' We then progressed into the 'Bronze Age' and eventually into the 'Iron Age.' This technological progression has given us unparalleled power on Earth to control all other life, derived from our intelligence, and helping to place us firmly at the top of the food chain. But as Clarke clearly explains we should not be complacent: [5]

> Most of this planet's life remains to this day trapped in a meaningless cycle of birth and death. Only the creatures who dared to move from the sea to the hostile, alien land were able to develop intelligence. Now that this intelligence is about to face a still greater challenge, it may be that this beautiful Earth of ours is no more than a brief resting place between the sea of salt and the sea of stars. Now we must venture forth.

We are now in a different age, having exploited all metals heavier than iron and those lighter such as aluminum. Do we call this the 'Metal Age'? Or is it the ability of our species to develop a language and record our history by written records that characterize us. This has led to what we may call the 'Information Age,' dominated today by computing technology that is largely based on silicon materials. The development of computers has allowed us to calculate very large sums, with direct application to space travel. However if one reviews the eras of human history, a clear trend emerges: humankind develops technology as a means to dominate the environment to aid in survival. But what happens when we reach the peak of what we can develop within the confines of this planet and its natural resources? This is a difficult question to answer, but one way to avoid this situation is to not reach this point. Hence, in order for us to continue to develop technology we must expand in all degrees of freedom possible. Hence we are led to an interesting question: What is meant by freedom?

We tend to think of freedom as flexibility to move in any particular direction, freedom to go left or right, for example. But there are other types of freedom; such as to go where you want and when you want. This includes a temporal component to freedom. There is also financial freedom to purchase what one requires, political or religious freedom to hold certain views. Speculating, we could postulate that freedom is a manifestation of built-in (evolved) human desires to dominate the environment wherever possible, in order to survive. This partly explains why we explore all environments on our planet, be it sea, air, land or beneath Earth itself. This is a continuing quest, and there is still much to explore, and so what about space? Humankind is fascinated by the endless possibilities of space, the infinity of stars in all directions. We want to go there and our desire to do so may be an expression of our continued freedom.

Today, a select few elite individual governments dominate space travel. All space operations are run from large government organizations, which heavily regulate the activities of private industry. It is not easy today for a private individual to independently raise the capital, build a rocket and launch themselves to the Moon. This will firstly be very difficult to achieve due to the structure of our society. Secondly, some governments may not allow the individuals to fly in the fear that they may perish in the attempt. They may have the intellectual skills to build the rocket, and courage to attempt the journey, but they won't be allowed to fly (incidentally this situation is now being directly challenged by the emergence of the space tourism market). This is a form of constrained freedom. They are prevented from directly expressing their evolutionary desires. These desires are what conquered the American frontier, to explore the boundaries of known reality, to know what is over the other side and exploit the resources for self-benefit and hopefully for the benefit of those around you.

The U.S. Declaration of Independence contains the inalienable rights of man, namely life, liberty and the pursuit of happiness. There are many people not just in the United States but also throughout the world who desire space travel but are prevented from doing so, largely due to cost. This is a constraint on their individual freedom, their liberty and their right to pursue self-happiness for the benefit of all humankind. Space travel must be opened up to the world at large and humanity allowed to continue to explore the cosmos. In his autobiography *Disturbing the Universe* physicist Freeman Dyson says: [6]

> Space travel must become cheap before it could have a liberating influence upon human affairs. So long as it costs hundreds of millions of dollars to send three men to the Moon, space travel will be a luxury that only big governments can afford. And high costs make it almost impossible to innovate, to modify the propulsion system, or to adapt it to a variety of purposes.

This is absolutely right, and this must change if we are to become a spacefaring civilization. It is worth adding that the recent X-Prize competition developments and the successful flight of SpaceShipOne illustrate that opportunities for space travel are expanding to the general public.

Other than the rights of individuals to pursue their chosen happiness, the situation could actually be made worse if an oppressive government dominated the world and prevented space travel in its entirety [7]. This could be due to a cultural belief that such exploration is either immoral or not the best choice of funds, which could be directed to less nobler pursuits. Alternatively, such oppression could actually stimulate space exploration as civilized humans attempt to find a safer place to exist. In the circumstances that an oppressive government had reach throughout the entire Solar System, then the only route of escape for the human colony would be towards more distant locations – planets around other stars. Global war could also force people to seek other places to inhabit, not confident that governments have the capability to ensure their continued safety.

Finally, there is a possible seventh reason for interstellar travel that is very profound if true but also highly speculative. It relates to an unconscious reaction to return to the stars. We already know that the atoms that make up our body derive from the fusion cores of stars and indeed this is true of everything around us. All matter on Earth is made up of atoms that were built within the core of a star. It is quite possible that there is a mechanism in our build up that is driving us into space, a 'starstuff driver' – a concept closely related to evolution. This is an idea similar to 'why do birds sing?' and it is plausible that the answer is hidden within our genes. Others have discussed this idea, such as the physicist Greg Matloff, who states: [8]

> In one of his books, the late Carl Sagan states that we are all 'starstuff,' since the atoms that compose us were mostly generated in the explosive demise of a super giant star. To take the poetic analogy further, it is interesting to note that interstellar flight requires us to either behave like a star or fly very close to one. So starstuff can use the stars to visit them.

Many a species of bird will return to its original nesting site from which it was born to begin its own generation, and then they too will feel compelled to do the same. It is a possibility, although highly speculative, that in our quest to understand ourselves better, the intelligence level of humans has led to an unconscious desire to return to the source from which we are fundamentally derived – the stars themselves.

To summarize, we have identified seven key motivations for interstellar space travel: continued survival; balance of resources; economic opportunities; quest for knowledge; spreading of life; freedom factor; and starstuff driver. Many other motivations can be thought of, but ultimately, they can be reduced to a subset of seven reasons. Sending human beings to worlds around other stars is the ultimate expression of a noble civilization that wishes to know more about itself and the apparent reality that it inhabits throughout its physical existence – called by us the universe.

Finally, in this section it is worthwhile quoting Arthur C. Clarke from his book *Profiles of the Future* to show clearly where his opinion was on the so-called impossibility of interstellar travel:

> Many conservative scientists appalled by these cosmic gulfs have denied that they can ever be crossed. . .And again they will be wrong, for they have failed to grasp the lesson of our age – that if something is possible in theory, and no fundamental scientific laws oppose its realization, then sooner or later it will be achieved.

The idea of humankind going to the stars, of being a spacefaring civilization, is what can be termed Clarke's Vision, although many others have shared in the origin of this idea. To this author's mind this vision is an optimistic future, which fully embraces technological progress for the use of a compassionate and forward looking society in search of like-minded races. Those today who pursue that vision with the same passion are in essence children of Clarke, ambassadors of a great dream that humanity's destiny lies not just here on Earth but also on another Earth orbiting another star. To encourage others in successive generations to steer our civilization towards that goal is the greatest legacy Clarke and his literature left behind.

2.6   Practice Exercises

2.1. Make your own list of reasons why space exploration should be continued, listing both the benefits and the costs to society. List the main scientific discoveries that could be made with the sending of only robotic probes throughout our Solar System. Think of the economic, political and cultural implications to our species of a world determined to explore space. What effects would this have on our society?

2.2. Think about the infrastructure requirements for a Solar System-wide economy that includes regular manned space missions out as far as Pluto. Describe this future, the heavy industry background as well as the transport requirements for Earth to orbit and from orbit to the planets. Using your derived model extrapolate from current technology to the future you have envisioned and estimate a time period for when you think the first unmanned and manned interstellar missions would occur.

2.3. Using the future you have described in the last problem think about how we go from our current technological-political world to that future. What key technological, social, political and economic steps are required to make that future happen? Describe your own roadmap.

References

1. Kardashev, NS (1964), Transmission of Information by Extraterrestrial Civilizations. Soviet Astronomy-AJ. Vol.8, No.2.
2. Hoerner, S von (1975) Population Explosion and Interstellar Expansion, JBIS, 28, 691.
3. Kennedy, A, (2006) Interstellar Travel; The Wait Calculation and the Incentive Trap of Progress, JBIS, Vol.59.
4. Clarke, AC (1999) Profiles of the Future, Indigo.
5. Clarke, AC (1999) Greetings, Carbon-Based Bipeds! A Vision of the 20th Century as it Happened, Essay on Rocket to the Renaissance, HarperCollinsPublishers.
6. Dyson, F, (1979) Disturbing the Universe, Basic Books.
7. Molton, PM (1978) On the Likelihood of a Human Interstellar Civilization, JBIS, 31, 6.
8. Matloff, G (2010) Personal Letter to this Author.

Chapter 3
Fundamental Limitations to Achieving Interstellar Flight

From an economic point of view, the navigation of interplanetary space must be effected to insure the continuance of the race; and if we feel that evolution has, through the ages, reached its highest point in man, the continuance of life and progress must be the highest end and aim of humanity, and its cessation the greatest possible calamity.

Robert H. Goddard

3.1 Introduction

Many have argued that even if you could build a probe to travel to the nearby star systems, it would take so long and require so much energy that this would be an obstacle for going in the first place. In this chapter we address this question head on and consider what the velocity and energy requirements are for sending a robotic probe across the vastness of space. This analysis is necessary so that we may understand the challenge of interstellar flight and how it measures up technically compared to an interplanetary mission. Before a spacecraft can embark on such a journey however, it first must reach Low Earth Orbit.

3.2 How Much Speed Is Required to Reach the Stars?

In recent decades, we have witnessed the achievement of the successful lunar landings, launched using technology similar to that developed for the German V2 rocket (and by the same scientists). We have watched the gradual construction of orbital platforms such as the Russian Mir Space Station and the International Space Station (ISS). Space-based telescopes and satellite communication has been enabled; we live in a telecommunications age. For years, many of these achievements have been driven and owned by national governments. But this is

K.F. Long, *Deep Space Propulsion: A Roadmap to Interstellar Flight*, DOI 10.1007/978-1-4614-0607-5_3, © Springer Science+Business Media, LLC 2012 now all changing, and the challenge of the new century is the conquering of space by the private industrialist; economic returns await those who try. The X-Prize competition is a leading example where the space tourism market is now opening up. There is every reason to be positive about the coming decades and optimism regarding the eventual colonization of the Moon, Mars and asteroids by humankind in the decades ahead. All such missions depend mainly upon chemical propulsion technology, so we need only build upon the achievements of the past.

Before we consider the requirements for achieving interstellar flight, let us understand what is required just to get into space. This is necessary so as to understand the context upon which such a mission is to be attempted as well as to develop some fundamental theory. First, we must achieve Earth orbit and escape from the bounds of its gravitational influence. We must consider what velocities are required in order for a vehicle to merely escape from the gravitational field of Earth. This can be simply estimated by equating the equation for centripetal force and the equation for gravitational force, which depend on the central mass M, orbiting mass m, orbital radius r, velocity of rotation v and the gravitation physics constant G. This gives the following:

$$\frac{mv^2}{r} = \frac{GMm}{r^2} \tag{3.1}$$

If we rearrange (3.1) we obtain a relation which describes the required velocity to reach circular orbital velocity:

$$v_{circ} = \sqrt{\frac{(GM)}{r}} \tag{3.2}$$

On Earth a rocket must accelerate to a velocity of 7.9 km/s (26,000 ft/s) in order to achieve circular orbit. Similarly, to fully escape from Earth's orbit, a rocket must accelerate to a velocity of 11.2 km/s (36,700 ft/s), which is $\sqrt{2} \times$ circular velocity. This is the equivalent of the kinetic energy attained at burnout, which is equal to or greater than the potential energy. Equating the kinetic and potential energies one easily finds the relation:

$$\frac{mv^2}{2} = \frac{GMm}{r} \tag{3.3}$$

Then re-arranging for the velocity we obtain the following what is equivalent to the escape velocity:

$$v_{esc} = \sqrt{\frac{(2GM)}{r}} \tag{3.4}$$

Table 3.1 shows the orbital and escape velocity required for the different planetary bodies in the Solar System. Clearly, the more massive an object the larger

Table 3.1 Properties for various Solar System bodies that show the circular orbital velocity v_{circ} and escape velocity v_{esc} for a satellite of mass m in orbital around a larger mass M

Object	Mass (Earth = 1)	Mass (kg)	Equatorial radius (km)	v_{circ} (km/s)	v_{esc} (km/s)
Sun	3.3×10^5	1.989×10^{30}	6.959×10^5	437	618
Mercury	0.055	3.302×10^{23}	2,439	3	4.3
Venus	0.815	4.869×10^{24}	6,052	7.3	10.4
Earth	1.0	5.974×10^{24}	6,378	7.9	11.2
Moon	0.012	7.348×10^{22}	1,738	1.7	2.4
Mars	0.107	6.419×10^{23}	3,397	3,6	5
Jupiter	317.83	1.899×10^{27}	71,492	42.1	59.5
Saturn	95.16	5.685×10^{26}	60,268	25.1	35.5
Uranus	14.5	8.662×10^{25}	25,559	15.0	21.3
Neptune	17.204	1.028×10^{26}	24,764	16.6	23.5
Pluto	0.002	1.3×10^{22}	1,150	0.87	1.3

the escape velocity required. Hence to escape Earth requires a velocity of 11.2 km/s, but to escape the gravitational influence of the Sun the spacecraft must exceed a whopping 618 km/s, although in the reference frame of Earth this is reduced to around 42.1 km/s due to the fact that Earth is already in motion around the Sun at an enormous velocity. So this is realistically what any spacecraft launched from Earth must overcome. It is also worthwhile noting at this point that the escape velocity of the planet Mars is substantially less than that of Earth. Any permanent settlement on Mars may therefore be a better launching post for future missions to the outer planets.

For any object (artificial or natural satellite) in orbit about another (larger) gravitational mass, the motion will be governed by three fundamental laws. These were first described by the seventeenth century German astronomer Johannes Kepler. These are thus named Kepler's laws of planetary motion and are as follows:

1. A satellite describes an elliptical path around its center of attraction.
2. In equal times, the areas swept out by the radius vector of a satellite are the same.
3. The period of a satellite orbit is proportional to the cube of the semi-major axis where the semi-major axis is the length of longest diameter in an ellipse. This is described mathematically by the following equation, where a is the semi-major axis of the object being orbited and t is the orbital period:

$$t^2 = \frac{4\pi^2}{GM} a^3 \tag{3.5}$$

The first law is contrary to the popular belief that planetary orbits are circles with the Sun at the center. This is not the case; planets will describe elliptical orbits with the Sun at one of two foci of an ellipse. This is the same for any satellite and means that the object in orbit will have a point of closest approach and one of furthest approach. For the Sun–Earth system these are called the perigee and apogee. The second law essentially says that when the object is near to the closest point of its gravitational center of attraction it will move faster. The third law simply states that for objects further away from the Sun, they will take longer to complete one orbit.

For example, Pluto is around 40 times the distance of Earth from the Sun and takes around 240 years to go around it. These laws are fundamental to any consideration of satellite or spacecraft orbits around any central mass. Even an interstellar probe decelerating into a distant planetary system must take account of these laws – otherwise known as celestial mechanics.

When World War II came to an end, Arthur C. Clarke didn't waste any time in making his vision come true. In October 1945 he published an article in an obscure magazine called *Wireless World* [1]. This article is considered to be the first technical paper describing the geostationary satellite. The dynamics of a geostationary satellite can easily be demonstrated by considering the centripetal acceleration a_n of an object in orbit around Earth with radius R and height above Earth's surface h as described by:

$$a_n = \frac{v^2}{R + h} \tag{3.6}$$

The acceleration of any object under gravity is given by:

$$a_n = \frac{GM}{(R + h)^2} \tag{3.7}$$

If we balance these two equations and rearrange for the velocity we get a variant on (3.2):

$$v_{circ} = \left(\frac{GM}{R + h} \right)^{1/2} \tag{3.8}$$

The angular velocity v describes the motion of an object about a central point. This can be defined as a function of angular frequency ω as follows:

$$v = \omega(R + h) \tag{3.9}$$

Combining this with (3.8) we then obtain the angular frequency:

$$\omega = \frac{(GM)^{1/2}}{(R + h)^{3/2}} \tag{3.10}$$

The mass of Earth is around 6×10^{24} kg and the radius 6.4×10^6 m with an angular frequency of around 7.3×10^{-5} s^{-1}. We also need the Gravitational constant, which is 6.7×10^{-11} Nm^2 kg^{-2}. If we rearrange (3.10) we find the expression for the height of a satellite above Earth's surface to be:

$$h = \frac{(GM)^{1/3}}{\omega^{2/3}} - R \approx 35,800 \; km \tag{3.11}$$

This is the height of a geostationary satellite, which stays in a fixed position relative to Earth. Clarke estimated that only three such satellites would be required separated by 90° to maintain continuing global coverage of Earth.

Once a spacecraft can use the Hohmann transfer orbit. This basically involves moving the spacecraft from one lower circular orbit to another higher orbit by firing the engines in two pulses to produce velocity changes. The spacecraft then completes half an ellipse during the maneuver to enter into the higher orbit. The engine is fired in the direction opposite to that which the spacecraft wishes to go. To do the opposite would result in the spacecraft orbit dropping to a lower one. The application of a Hohmann transfer is also useful in planning deep space missions where velocity from the gravitational field of planetary objects can be added to the spacecrafts velocity. Similarly, adding velocity to the planetary body and subtracting it from the spacecraft can slow a spacecraft down. When the spacecraft decelerates and slows down into the planetary orbit, it's said to be captured, and this process is a gravitational slingshot – one method of adding velocity to a spacecraft in order to send it on a deep space trajectory. Essentially any spacecraft performing a slingshot maneuver with a planetary object will be exchanging momentum with that object. If a probe gathers momentum as it passes Jupiter for example, then Jupiter has lost some of its momentum, and vice versa.

Once a spacecraft achieves orbital velocity, it may then need to change its orbit, such as to dock with a space station or satellite. To go further however, to explore other worlds and new frontiers is an entirely different problem. No surprise then when we find out that it comes down to one big issue: propulsion. To attain maximum velocity any rocket engine must either burn at high thrust (high acceleration) for a short period or burn at low thrust (low acceleration) for a long period until a high cruise velocity is reached. However, any engine that has high thrust for a short period will also experience high temperatures, which is the fundamental constraint on any thrust produced, due to the melting properties of materials. Similarly, any engine that has thrust for a long period will require substantial power to continuously deliver the thrust. So we arrive at the two limiting factors to any propulsion technology: temperature and power.

The other relevant factor is mass; any material that is launched into Earth orbit will cost typically around $50,000/kg. The reason a launch vehicle requires so much cost/mass to get into orbit is because Earth's gravity is trying to pull everything back down. Then once in space and heading off on a predefined trajectory, any spacecraft will also encounter the gravitational fields of other planets in the Solar System and this will involve trajectory maneuvers, which also require fuel. In principle, the lighter the vehicle, the easier and cheaper this will be. However, in order to keep the vehicle light one must minimize fuel mass. But then if one has minimal fuel mass, this limits the maximum velocity and therefore distance attained for a given fuel. When applying these considerations about long-term deep space missions, we are really presented with four possible mission solutions as shown in Table 3.2.

Table 3.2 Speed-duration routes for space missions

Method	Description
Standard route	To cover a large distance in space the engine must carry lots of fuel, which means more total mass (structural + propellant).
Efficient route	Alternatively, the engine may carry a more efficient fuel to minimize mass for the same journey.
Rapid-efficient route	To cover the same distance in a quicker time, the engine must not only carry a more efficient fuel, but also achieve a higher exhaust velocity early on (which implies high thrust) for the acceleration phase.
Gradual-quick route	To cover the same distance in a quicker time, the engine carries a more efficient fuel, but achieves a higher exhaust velocity gradually over time (implies low thrust but gradual velocity increase) for the acceleration phase.

This simple analysis can be seen by considering the definition of velocity v, and distance S moved in a given direction and by the duration or time t as defined by simple linear theory:

$$S = vt \qquad (3.12)$$

For a specific mission, the distance is fixed (neglecting variations on the trajectory for now) so the only parameters that are variable are the velocity and time taken over the journey. An increased velocity will result in short mission durations.

Supposing we have the technology available, how fast do we need to go? For simplicity, we can consider the speed requirements for a linear distance profile to an effective Alpha Centauri distance of 4.3 ly or 272,000 AU, where a 1 light year $= 9.46 \times 10^{15}$ m $= 63,240$ AU and an astronomical unit (AU) is the distance between Earth and the Sun, where 1 AU $= 1.496 \times 10^{11}$ m. We ignore the fact that Alpha Centauri is out of the ecliptic plane. We also ignore acceleration requirements. Table 3.3 shows typical journey times to reach this distance for given constant velocities. The data clearly shows that to reach the nearest star in a time frame of order a century or less, a vehicle must travel at a cruise velocity of >10,000 km/s, which equates to >3% of light speed. We note that for mission durations of ~50 years the spacecraft must reach ~1/10 of light speed and for mission durations of ~ a century or so the spacecraft must reach ~1/100 of light speed. For much faster missions of order ~ a decade or so, the vehicle must reach a speed of order 1/3 of light speed.

To put these speed requirements into perspective, we can compare this to the fastest spacecraft that we have so far sent out into deep space. These are the Pioneer and Voyager spacecraft. Pioneer 10 was launched in March 1972 and is currently traveling at around 13 km/s or 2.6 AU/year. Pioneer 11 was launched in April 1973 and is currently traveling at ~12 km/s or around 2.4 AU/year. Voyager 1 was launched in August 1977 and Voyager 2 in September 1977 and both are traveling at around 17 km/s or 3.6 AU/year. In January 2006 NASA also launched the New Horizons mission, which will visit Pluto and move on to the Kuiper Belt. It is currently traveling at around 18 km/s or 3.8 AU/year. To reach Alpha Centauri

Table 3.3 Linear velocity scale to Alpha Centauri (neglecting relativistic effects)

Velocity (km/s)	% Light speed	Time to effective α-(cent)
1	0.0003	1.3 million years
10	0.003	130,000 years
100	0.03	13,000 years
1,000	0.3	1300 years
10,000	3	130 years
25,000	8	50 years
100,000	33	13 years
200,000	66	6 years
300,000 = c	100	4 years

we must cross a vast distance of 272,000 AU. At current speeds, most of these spacecraft would reach there nearest line of sight star in many tens of thousands of years.

If a spacecraft could attain a sufficient cruise velocity to reach the nearest star, Alpha Centauri, within a reasonable timeframe, what sort of mission options would there be? Designers can play around with various mission profiles by varying the parameters of acceleration, velocity and mission duration, which is measured by the time for data return to Earth (mission duration to destination +4.3 years for signal transmission at speed c). We can continue to assume simple linear mission analysis by using equations of motion which as well as the terms described above includes the initial velocity u, constant acceleration a and final velocity v as follows:

$$v = u + at = at_{(u=0)} \tag{3.13}$$

$$v^2 = u^2 + 2aS = 2aS_{(u=0)} \tag{3.14}$$

$$s = ut + \frac{1}{2}at^2 = \frac{1}{2}at^2_{(u=0)} \tag{3.15}$$

This is a crude analysis but is a good starting point to begin considering the requirements to an Alpha Centauri distance of 4.3 light years. In particular, most space missions will involve large parabolic trajectories and several phases of acceleration and deceleration by using gravitational slingshots. A parabolic trajectory is an escape orbit with an eccentricity equal to unity. The eccentricity is the amount by which the orbit deviates from a perfect circle, defined with an eccentricity equal to zero. An elliptical orbit would have an eccentricity less than unity. For a changing acceleration one must turn to the use of calculus equations of motion, which we do not cover in this text.

Also, there are very few astronomical targets that lay in the ecliptic plane of the Solar System. But for our simple analysis, we can ignore gravitational slingshots and deviations from the ecliptic plane. We can also ignore a deceleration phase for the analysis and assume a flyby-only trajectory. We can assume constant acceleration for an initial period of time. Table 3.4 shows the results of several

Table 3.4 Linear mission
analysis to effective
α-Centauri

Acceleration phase	Cruise phase	Minimum data return from effective α-(cent)
0.01 g for 1 year	0.01c for 430 years	~436 years
0.01 g for 5 years	0.05c for 84 years	~94 years
0.01 g for 10 years	0.1c for 42 years	~57 years
0.1 g for 1 year	0.1c for 42 years	~48 years
0.1 g for 5 years	0.5c for 8 years	~18 years
0.5 g for 1 year	0.5c for 8 years	~14 years
1 g for 1 year	~1c for 4 years	~10 years

hypothetical mission profiles. For comparison, the Daedalus Project (see Chap. 11) had a mission profile that involved two acceleration phases, the first at 0.03 g to 0.071c followed by 0.01 g up to a cruise speed of 0.12c to get to Barnard's Star 5.9 ly away in around 46 years.

One is quickly led to some simple conclusions about practical requirements for acceleration (0.01–1 g), mission velocity (0.1–0.5c) and mission duration (10–100 years). The lower limit on the acceleration will result in prolonged duration missions, and acceleration >1 g (e.g. 10 g) would both (a) not impact the mission duration due to the speed of light limit and (b) give rise to uncomfortable accelerations for any crew on board for a manned mission. Also, mission durations of a century of more would be outside the working lifetime of a designer or any crew (which may not be desirable) as well as place stringent environmental pressures on the technology. An ideal mission profile would be one that employed 0.1 g acceleration for a few years up to 0.3c resulting in total mission duration of ~50 years. Conventional thinking about future interstellar missions is that they are likely to be one of two types:

- Type I: A short ~50-year mission using high exhaust velocity engines to accelerate to a moderate fraction of the speed of light, 0.1–0.3c, completing the mission within the lifetime of designers.
- Type II: A long ~100–1,000 year mission using low exhaust velocity engines, completing the mission duration over several generations of designers.

It is generally believed that a Type I mission would require a large technology jump, but a Type II mission would require only a moderate jump, except perhaps with the environmental lifetime requirements. Interstellar travel is just one form of space travel as defined by typical distance scales in the cosmos. Because scales in the universe are very large (astronomical), we must turn to definitions used in astronomy to comprehend the vast distance scales involved. Different forms of space travel can then be categorized as shown in Table 3.5. In this book we are concerned only with the problem of interstellar flight, but clearly if we ever hope to go further the challenges mount up in orders of magnitude based purely on the significant distance scales involved. The nearest galaxy, Andromeda, is at a distance of around 2.5 million light years away. This means that when we observe this galaxy through a telescope today we are seeing it 2.5 million years ago, because even light takes that long to cross the void that separates the 'islands' in our universe.

Table 3.5 Cosmological distance scales

Type	Distance
Near-Earth travel	0.003 AU, 10^{-8} light years (ISS-shuttle, satellites, lunar exploration)
Interplanetary travel	40 AU, 0.001 light years (Mercury–Pluto)
Extraplanetary travel	40–500 AU, 0.001–0.01 light years (Kuiper Belt) 500–50,000 AU, 0.01–1 light years (Oort Cloud)
Interstellar travel	50,000–271,932 AU, 1–4.3 light years (Nearest star Cent A)
Intergalactic travel	10^8–10^{10} AU, 2000–160,000 light years (Milky Way)
Extragalactic travel	10^{11} AU, 10^6 light years (Nearest galaxy M31 Andromeda)

Now all the above analysis is based upon linear theory, but in reality rockets are governed by the ideal rocket equation (discussed in Chap. 5) which is logarithmic. When the analysis is done correctly this leads to a modified boost equation [2] for the distance attained S_b during the burn time t_b using an engine exhaust velocity v_e, as a function of the vehicle initial mass M_i and final mass M_f (after propellant burn) as follows:

$$s_b = v_e t_b \left[1 - \frac{Ln\left(M_i/M_f\right)}{\left(\frac{M_i}{M_f}\right) - 1} \right]$$

$$(3.16)$$

This equation ignores special relativistic effects, which is considered appropriate for low fractions of the speed of light such as between 10% and 30%. We can compute this equation for the two-stage Daedalus concept (discussed in Chaps. 5 and 11). This would result in two boost phases — the first of 2.05 years duration to a distance of 0.048 light years to reach a staged velocity of around 7% of light speed, followed by 1.76 years duration to a distance of 0.224 light years to reach a further staged velocity of around 5% of light speed – adding up to a total velocity of 12% of light speed, or around 36,000 km/s. This is assuming an exhaust velocity of around 10,000 km/s for both stages and a mass ratio M_i/M_f of around 7.9 for the first stage and 5.1 for the second stage. The total boost would be covered in 3.81 years and then the spacecraft would coast for a further 46 years to Barnard's Star mission target at 5.9 light years away. The total mission duration would be around 50 years.

3.3   How Much Energy Is Required to Reach the Stars?

We can begin to consider the energy and power requirements for an interstellar mission by simply thinking about the energy required to impart to a vehicle to produce kinetic energy for forward momentum, assuming 100% conversion efficiency.

Table 3.6 Minimum energy and power requirements to Alpha Centauri

Mass (tons)	1	10	100	1,000	10,000	100,000
Energy (J)	10^{18}	10^{19}	10^{20}	10^{21}	10^{22}	10^{23}
Power (GW)	50	500	5,000	50,000	500,000	5,000,000

Table 3.7 Approximate mission profiles for a 10-ton flyby interstellar probe to Alpha Centauri 4.3 light years distant

Description	Slow mission	Medium mission	Fast mission
Initial acceleration (g)	0.001	0.01	0.1
Cruise velocity (km/s)	1,547 (0.51%c)	9,281 (3.1%c)	30,937 (10.3%c)
Fraction light speed	~1/200th	~1/30th	~1/10
Boost duration (years)	5	3	1
Boost distance (light years)	0.013	0.046	0.051
Cruise duration (years)	840	137	41
Minimum energy (J)	1.2×10^{16}	4.3×10^{17}	4.8×10^{18}
Minimum power (GW)	0.1	5	150

Note that for the fast mission scenario the balance between acceleration and boost duration is limited to the speed of light

This will give us our minimum requirements. The kinetic energy locked up in a given mass can be huge. For example a 1 g particle moving at only one tenth of one percent of light speed (300 km/s) will have a kinetic energy of around 45 million Joules or the equivalent of around 10 kg TNT. We can consider a situation where a vehicle accelerates for 0.1 g up to 0.3c. We can then calculate the required energy input from the kinetic energy, which is proportional to the particle mass m and velocity v, as follows:

$$E = \frac{1}{2}mv^2 \tag{3.17}$$

This leads directly to an approximate power requirement by dividing by the number of seconds during the boost phase t_b if we assume that the propellant mass dominates the total mass of the vehicle:

$$P = \frac{E}{t_b} \tag{3.18}$$

It is easily shown that the minimum power to push a 1-ton vehicle to 1/3 of light speed over a period of 3 years is around 50 GW. For the same speed a 100,000-ton vehicle would require around 5 PW of power. Table 3.6 shows the results of several vehicle masses (where 1 metric ton = 1,000 kg) and the associated energy and power requirements.

From this analysis we can produce several possible mission profiles to be considered so as to assess the suitability of various propulsion schemes (Table 3.7). There are other limitations to launching deep space missions, other than the challenges derived simply from physical laws, particularly for missions to the nearest stars. The first is technological maturity; there is an enormous difference between a speculative idea and a designed engineering machine. The range of interstellar propulsion concepts proposed in the literature varies from the fantastic to the highly credible. Other reasons are the political or financial restrictions. To launch an interstellar mission a strong case has to be presented, so as to secure both political backing and the necessary funds, a topic discussed further in Chap. 15.

This ends our brief introduction to the technical challenges of achieving Earth orbit and the required energies to move off on an interstellar mission. Over the centuries many people have voiced the opinion that interstellar flight is impossible. But those who are familiar with the technical challenges have a different opinion. One of the experts in the field of interstellar propulsion was the physicist Robert Forward, who said [3]:

> Travel to the stars will be difficult and expensive. It will take decades of time, GW of power, kg of mass-energy and trillions of dollars...interstellar travel will always be difficult and expensive, but it can no longer be considered impossible.

The simple calculations conducted in this chapter illustrate the challenge of the problem that indeed GW of power is required. The challenge of this book is to demonstrate that there are credible schemes by which this can be someday accomplished.

3.4   Practice Exercises

3.1. Using (3.4) compute the escape speed necessary to leave the galaxy. Assume that the mass of the galaxy is around 2×10^{11} solar masses and that the launch point is 30,000 light years form the galactic center. Convert all the terms to consistent units. Use of this equation for the entire galaxy makes an assumption about the gravitational influence of the surrounding stars to be negligible and instead the rocket is essentially flying through an inertial frame. Also, the model assumes a spherically symmetric mass distribution of the galaxy. Discuss these two approximations. Assuming you could leave the galaxy at the calculated speed, how long would it take the spacecraft to reach the edge of the nearest galaxy, Andromeda, located around 2.5 million light years away?

3.2. A 5 ton spacecraft starts at the orbit of Earth and you want it to travel to the distance of Pluto at around 40 AU away from the Sun. Using the data and equations given in this chapter and assuming a linear mission profile (equivalent to a static solar system assumption with no gravitational influences other than from the Sun), calculate the cruise velocity of the spacecraft after it has undergone two boost periods for 2 days and then for 1 days at 0.0001g and 0.0002g respectively. How long will it take the spacecraft to reach Pluto's orbit and what will its final cruise speed be assuming flyby only with no deceleration? Repeat these calculations using the logarithmic boost equation (3.16) assuming a mass ratio of 2 and 3.1 for the first and second stage respectively, a 100 km/s exhaust velocity for both. Compare and discuss the different approaches.

3.3. Light, the fastest thing in the Universe, moves at a speed of around 300,000 km/s. What constant acceleration is required to reach this speed (assuming you could and ignoring relativistic effects) if you accelerate for 1 year? Do you notice anything special about this acceleration, and what does it suggest for manned interstellar missions? Show that light moves at a speed of around 30 cm/ns. Next, convert the units of light speed from km/s to 'light years' and 'years' and see if you get a number of unity. Show that it is given approximately by one Earth gravity assuming acceleration in the same units. Finally, show that it takes a light signal sent from Earth less than 2 s to reflect and return from the Moon.

References

1. Clarke, AC (1945) Extra-Terrestrial Relays, Wireless world.
2. Bond, A & AR Martin (1978) Project Daedalus: The Mission Profile, JBIS, S43.33–S.
3. Forward, RL, (1996) Ad Astra, JBIS, 49, pp.23–32.

Chapter 4
Aviation: The Pursuit of Speed, Distance, Altitude, and Height

How many more years I shall be able to work on the problem I do not know; I hope, as long as I live. There can be no thought of finishing, for aiming at the stars, both literally and figuratively, is a problem to occupy generations, so that no matter how much progress one makes, there is always the thrill of just beginning.

Robert H. Goddard

4.1 Introduction

All spacecraft engines work on the principle of generating thrust from a reaction. The same principle applies to aircraft, and the pursuit of space would not have been possible without these aviation developments. In this chapter we briefly review some of the essential elements behind propulsion theory that apply to conventional engines. This is necessary so as to place the later rocket developments into proper context as well as build from an initial foundation for how thrust is generated. By attempting the problems at the end of this chapter we learn about the limits of jet propulsion technology in reaching high altitudes, and the requirement for rocket engines will be much clearer.

4.2 The History of Propulsion in Aviation

In order to understand the challenges of interstellar flight, we first must realize the challenges and achievements of the past – how humankind went from being a land-based hunter-farmer race to a ship building-airplane flying trading race throughout the globe.

The dream of flight started a long time ago, perhaps before recorded history. But it was in 1852 that the world was changed forever. In this year the Frenchmen Henri

K.F. Long, *Deep Space Propulsion: A Roadmap to Interstellar Flight*, DOI 10.1007/978-1-4614-0607-5_4, © Springer Science+Business Media, LLC 2012

Giffard's airship flew over Paris at a speed of over 2 km/s. This airship was very special for one particular reason; its horizontal movement through the air was driven by a steam engine, which was connected to a propeller, thus providing thrust for the vehicle. This was the first time in history that humankind had developed a flying machine that embodied powered flight and began the long program of work for humankind to finally conquer the air and mimic the flight of the birds.

It was another half century later before similar technology was first used in the successful demonstration of powered flight using a winged vehicle. Orville and Wilbur Wright had investigated many different types of engines for use in an airplane. This included using engines from automobiles and marine craft. To their disappointment no suitable engine was available, so they had to design and build one themselves. They did so by studying the aerodynamic designs of propellers using home built wind tunnels. The Wright brothers did not receive a high school diploma, but were fortunate enough to come from a well-educated family. Their mother had a degree in mathematics and their father was a church Bishop with considerable practical skills (he had designed a version of a typewriter), similar skills of which he presumably passed on to his two sons. Eventually, in 1892 the two brothers set up a bicycle manufacturing, repair and sales business. The knowledge gained on gears and mechanical movement must have been highly useful in their later attempts at building mechanism for the purpose of manned flight.

In order for a vehicle to plane through the air (airplane), two forces are required. Firstly, the vehicle requires lift to offset gravity, so that it does not fall. This is largely provided by the pressure difference over the wings of the structure. Secondly, the vehicle requires forward movement, as generated by some propulsive mechanism – an engine. The Wright brothers understood this and had conducted substantial research into both topics, which eventually led to their place in history. They studied the wings of birds and were led to the conclusion that continuous stability of a flying wing was enabled by use of torque forces, so that any gusts of wind would be quickly adjusted for.

The success of their propeller design, along with their excellent control system allowed them to eventually demonstrate the first powered flight of an airplane in December 1903 at Kitty Hawk in North Carolina. It was a calm weather morning and several witnesses were assembled to observe the Wright brothers latest attempt at powered flight. Orville was chosen to be the pilot on this occasion and his airplane was to be powered by a 12 Horse Power engine mounted on the top surface of the bottom wing. During the ground take-off run Wilbur ran along side holding the tip of the wing to prevent it from dragging in the sand. Eventually, the airplane lifted off from the ground flying an erratic set of dive and climb maneuvers at around 3 m (10 ft) altitude, until finally coming to rest having gone a total distance of 37 m (120 ft). Not very far some might say, but it was the furthest any man made winged air vehicle had previously gone under its own power. It was one of the most significant moments in aviation history.

The achievement of the Wright brothers was not built in isolation. Others had been experimenting with the idea of powered flight for sometime, including George Cayley in England and Otto Lilienthal in Germany. Some people believe that the

Italian Renaissance man Leonardo Da Vinci should be given due credit for his studies of flight in the sixteenth century. Although his airplane design may not have worked it certainly laid the groundwork in studies of wing design and lift generation.

In recent years one of the new fields that have opened up in aviation is the development of aircraft that can touch the boundary of space by reaching suborbital altitudes. By taking inspiration from the 'Spirit of St. Louis' Atlantic crossing of 1927 by Charles Lindbergh the current developments are being led by the X-Prize competition, which was won in 2008 by Burt Rutan and his SpaceshipOne. The X-Prize is primarily concerned with space tourism, and perhaps one day may lead directly to vehicles that can reach low Earth orbit so that tourists may eventually stay on an orbiting space station for several weeks at a time.

If a full space market does open up, then this could lead to spinoffs in the aviation industry so that flights from New York to Sidney will be possible within a matter of hours. The British-French Concorde was one of only two civil airliners ever to carry passengers at supersonic speeds, the other being the Soviet Tupolev Tu-144. These passengers were moving more horizontally than vertically up into space. However, if passengers are capable of undergoing supersonic transport, then why not space-flight, too? When the Wright brothers first flew their airplane at Kitty Hawk, it is doubtful that they considered it would lead to a worldwide aviation industry and eventually flights into space. The world has truly changed in just a century.

Since the first Wright brothers flight, we have seen the first powered lift of a helicopter (1907), the first airplane crossing over the Atlantic (1927), and the first aircraft powered by a jet engine (1939). Looking back over the developments in aviation we have learned that successful sustained flight through any medium (including a vacuum) requires power. Now all airplanes derive their forward momentum from thrust produced by some sort of engine. The same is true of rockets. But before we begin our exploration of how rockets work, it is first useful to discuss some of the engineering and performance elements of various types of engines that have been produced for the sole purposes of propelling objects through the air. We briefly discuss the development of four types of mechanisms; the propeller, the reciprocating engine, the jet engine and later variants.

The most basic form of thrust generation is by use of the propeller. This is a bit like an airplane wing. It is designed to generate a lifting force that provides forward momentum. The propeller airfoil is essentially a small wing, but it has a twisted geometry so that the chord line, the distance between the trailing edge and the point on the leading edge where it intersects the leading edge, changes throughout its length. The wind seen by the propeller is due to the airplane's motion as well as due to the rotation of the propeller section, generating the net thrust for forward motion, with the propeller diameter being the key parameter in the thrust determination. The rotation of the propeller causes a slipstream of air that exhibits both rotational and translational motion and acts as a form of power loss. Because of these losses and others such as skin friction drag and boundary-layer separation at the tip, the propeller is never completely efficient and becomes less so the faster the propellers rotate, which is why they are only used in the propulsion of aircraft that travel at subsonic speeds.

During the 1930s developments in propeller design led to the variable pitch propeller, which maintains maximum efficiency across the airfoil, and the constant speed propeller, which allows the pitch angle of the airfoil to be varied continuously so that the engine revolutions per minute are constant over a variation in flight speeds. Ultimately, the speed of any vehicle that relies upon propeller technology is limited to speeds less than that of sound (340 m/s at sea level). Alternative propulsion technology is therefore required if you want to go faster.

The reciprocating (piston) engine was first developed in the eighteenth and nineteenth centuries. It works on the principle of burning gasoline fuel and a piston that reciprocates back and forwards inside a cylinder, so that at each stage of the reciprocation valves are opened or closed to allow air to enter and replenish the mixture and release any burned gases. The reciprocating motion of the piston is converted into rotational motion of a shaft, via a connecting rod. The air is ultimately pushed backwards in the opposite direction to the propeller thrust, so that a change in momentum is imparted, due to the mass flow rate of air moving from the front to the back of the vehicle. The basis of a reciprocating engine is the four-stroke cycle as follows:

1. *Intake stroke*; piston moves downwards at constant pressure, intake valve open to take in air and fuel mixture.
2. *Compression stroke*; compress the constant mass of gas and air to higher pressure, at start of stroke ignite mixture with electric spark producing combustion at constant volume. High temperature drives piston downwards.
3. *Power stroke*; gas expands isentropically, without any change in the thermodynamic state (entropy), and exhaust valve opened.
4. *Exhaust stroke*; burned gases pushed out of cylinder. Engines exist that have more than four strokes per cycle but will not be discussed here.

The number of cycles induced in the engine per second will translate directly to more work done and therefore more power to propel the aircraft. The power is also dependent upon the intake air pressure and the displacement, which is the volume swept out by the propeller. Because a reciprocating engine controls the power by changing the revolutions per second it is a significant technological advancement from basic propeller driven aircraft that do not employ a piston driven principle. However, because of various energy loss mechanisms, speeds below that of sound are also only possible with this technology. To go faster still, we must find alternative ways to generate more thrust.

Therefore, we arrive finally at the development of the jet engine, and it is from here onwards that we must begin to examine some of the mathematics behind propulsion. As we will soon see, this is because the principle of jet propulsion is very similar to that for rockets. Many people know the story of Sir Frank Whittle and his quest in the 1930s to get jet technology accepted. As an officer in the British Royal Air Force, it was during his time at Cranwell Technical College that Whittle first published his doctoral dissertation "Future developments in aircraft design" in 1928. This paper laid down the foundations of jet propulsion, but because reciprocating engines so dominated airplane propulsion his ideas were not considered seriously. This then formed the basis of a patent entitled, "A reaction motor suitable for Aircraft Propulsion," granted in 1931. Unfortunately, the British Air Ministry largely ignored Whittle's ideas for a jet engine, and it wasn't until 1939 (at the outbreak of World War II) that he was finally taken seriously and awarded a development contract for his engine design.

It was in 1935 that Whittle first started to develop a practical jet engine. However, although Whittle had built and tested a jet engine on the ground, glory would be stolen from him by developments in Germany. On April 12, 1937, the German Hans von Ohain designed and built a gas turbine engine, strapped it to an airplane and the HE 178 became the first jet-propelled airplane in the world, generating 3,727 N of thrust (or 838 lb where one pound = 4.45 N) and moving at 194 m/s. Because of these rapid developments in Germany, Whittle soon got his support from the Air Ministry and eventually developed the Gloster E.28 jet aircraft, which flew on May 15, 1941, generating 3,825 N of thrust and moving at 151 m/s. In the next section we consider the mechanics of the jet engine in more detail.

4.3 The Physics of Aviation Propulsion

So how does the jet engine work? Essentially, a jet engine is just a duct, where air is passed through, heated by combustion and expelled rearwards at higher velocity. So the free stream airflow, which enters the duct at the speed of the aircraft, is accelerated to a much higher velocity, and it is because of this increase that the vehicle is able to go faster. This is an example of Newton's action-reaction principle. The concept of thrust is fundamental to propulsion theory, and it can be interpreted to mean the rate of change of mass flow ejected per unit of time. Since mass multiplied by velocity is momentum, thrust can also be interpreted to mean the rate of change of momentum of the gas and therefore also for the vehicle it reacts against.

Another way to think about the meaning of thrust in a jet engine is to think of it in terms of a net pressure difference between the internal and external surface. The net internal force acting on a gas within an engine derives from three sources – the force in, the force from the engine surface and the force that comes out. Relating force as the rate of change of momentum (mass multiplied by velocity) from Newton's second law of motion and denoting the mass flow rate of the gas $\dot{m}$ through an exit area A_e, it is possible to describe the changes in the momentum at the entrance and exit of an engine as a function the internal pressure p_{in}, exit pressure p_e, internal velocity v_{in} and exit velocity v_e. This leads to the fundamental equation of thrust for a jet engine as follows [1]:

$$T = (\dot{m}_{air} + \dot{m}_{fuel})v_e - \dot{m}_{air}v_{in} + (p_e - p_{in})A_e \qquad (4.1)$$

This equation has several important results, one of which is that the thrust is proportional to the mass flow rate of the combustion products ejected from the engine, and indeed what velocity those products are ejected at. The expression for thrust can also be used to calculate the available power requirements. Power has the units of energy per unit time, which is equivalent to thrust multiplied by distance per unit time, hence the power is simply the product of the thrust and exit velocity of the combustion products with an efficiency factor η to take account of energy loss mechanisms:

$$P = \eta T v_e \tag{4.2}$$

The technical name for a jet engine is a turbojet, and it has several parts. The first is the air intake where subsonic air is passed through a divergent diffuser and reduces the air to a slower velocity. After this the processes of compression and combustion take place. The compressor takes in the air, compresses it to high pressure using a series of blades and allows fuel to be injected and burned in the combustion chamber at constant pressure with dramatically increased temperature. On combining with the fuel the hot fluid expands outwards and through a turbine section that consists of a series of blades used to transmit the gases through a shaft and eventually through the nozzle, which accelerates the fluid to high velocity. Turbojets are not generally used in modern commercial aircraft due to their low efficiency at low speed and have a typically poor specific fuel consumption.

The type of engine used today in both modern commercial aircraft as well as most military jets is the turbofan (Fig. 4.1), which is more efficient at low speeds than the turbojet and is also much quieter. A turbofan stands out from a turbojet in having a large ducted fan mounted at the front end of the shaft. The turbine drives this, and the ducted fan accelerates a large mass of air between the outer and inner shrouds and the unburned portion of which then mixes with the exhaust gases at the nozzle end of the engine, increasing the efficiency of the performance. The thrust is quite large because it is derived from the exhaust nozzle as well as from the fan blades. The turbofan engine is really quite ingenious in that it combines a turbojet engine with that of a propeller type without the disadvantages of both. The turbojet is inefficient due to the large exhaust losses, while producing a large thrust. The propeller driven engine is quite efficient but will only produce a small amount of thrust.

The ramjet engine is essentially like the conventional jet engines discussed above but with all of the fans, the turbine and the propellers removed (Fig. 4.2). Air enters through the intake duct at supersonic speeds and is then decelerated by compression to subsonic speeds by a diffuser. Fuel is then injected and burned with the air in the combustion chamber to high pressure and temperature. The combustion gases then move rearwards passing into the nozzle area, where it is expanded and accelerated by a divergent nozzle section to supersonic exhaust speeds. Ramjets are very simple compared to the other engines discussed above, and they can produce a high thrust. However, because they require a supersonic air intake, they are of little use to current commercial airliners, unless they intend to move at several times the speed of sound such as in a hypersonic airplane. Ramjets are most efficient at around three times the speed of sound and can operate up to about five or six times the speed of sound. Indeed they require a supersonic gas in order to

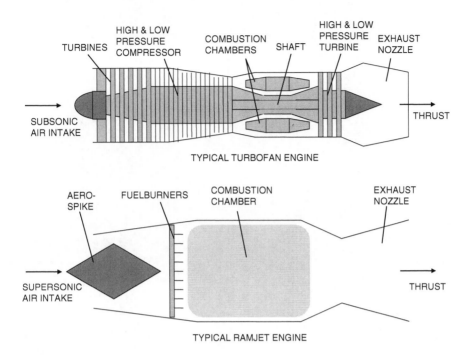

Fig. 4.1 Basic air breathing engine schematics

get a good compression ratio in the diffuser, which is dependent upon the speed of the gas. This may be seen as a disadvantage to this sort of engine type. What limits the maximum speed of a ramjet is the temperature of the combustion, which is limited to the temperature at which the material structure making up the engine shell begins to fail.

The British-based company Reaction Engines Ltd is developing one design for a type of engine that incorporates ramjet technology. The Synergic Air BReathing Engine (SABRE) [2] is a pre-cooled hypersonic hydrogen-fueled air breathing rocket engine. It is designed to power the Skylon launch vehicle, a concept for access to low Earth orbit. The engine has the interesting feature of an air pre-cooler that cools the hot ram compressed air and so allows high thrust to high efficiency at high speeds throughout the entire flight to high altitude. Because the air is cooled to a relatively low temperature, the material structural requirements are not so harsh, so the engine can be designed using lightweight materials, reducing the overall mass of the vehicle. The Skylon vehicle would use the engine to go to a speed of well over five times the speed of sound and then use a rocket engine to reach Earth orbit. This is discussed further in Chap. 5.

A supersonic combustion ramjet, otherwise known as a *scramjet* engine, may be able to go at speeds approaching 24 times the speed of sound, which exceeds orbital velocity of 7.9 km/s. If this is proven to be so, then an aircraft fitted with a scramjet engine may ultimately lead to the first spaceplane and orbital flights for passengers

Fig. 4.2 Illustration for the concept of an Earth to Orbit Spaceplane

(Fig. 4.2). It requires a minimum speed of around eight times the speed of sound, which may mean that a rocket engine is required first to accelerate the vehicle to high velocity. The scramjet is just like a ramjet, with the exception that the fluid flow is still supersonic when entering the combustion chamber. This means that when fuel is added to the mixture, combustion takes place supersonically. Because the airflow entering the diffuser is not slowed to subsonic speeds in a scramjet, no shocks are generated and this avoids a reduction in efficiency. Also, because the combustion is supersonic, the static temperature is relatively low, presenting little risk of structural material failure.

One proposal for a scramjet-driven spaceplane was the National Aero-Space Plane (NASP), otherwise known as the X-30. This was being managed by several companies led by Rockwell International, in response to a directive from the then U. S. President Ronald Reagan to produce an 'Orient Express' for such a purpose. This was to be a single stage to orbit vehicle capable of hypersonic speeds to Earth orbit, with a specific impulse of 1550 s generating a liftoff thrust of around 1.4 MN.

The vehicle was aesthetically very beautiful and was based upon a wave rider type design. The lift was to be generated mainly from the fuselage using a principle known as compression lift, where the lift is generated from the vehicle's own supersonic shock waves. There were also two small wings, used to provide the required control and trim. Research into the X-30 ceased in 1993, but it is quite possible that the X-30 was a highly visionary piece of aeronautical technology, which may be the template for human access to space in the coming decades. Some research has continued under the guise of the X-43 program.

An interesting question then is to ask if it is possible to apply these different classes of engines used in aviation for transport in deep space. The answer is simply no, and you have probably guessed the reason – space is a vacuum and there is no air to combust with a fuel, although the equations used to describe the performances are very similar, and this is the reason why jet propulsion has been discussed in this book, to provide a proper context from which the reader can think about propulsion for the application of space transport.

In fact, this leads us directly to the first requirement of a rocket engine discussed in the next section, that of an oxidant. Of course, if a spacecraft was to enter the atmosphere of another planet that contained a gas such as air, then in theory these types of engines could once again be used. However, most engine designs are highly optimized for specific atmospheric conditions. Without knowing in detail the atmospheric conditions of an unknown world (pressure, temperature, molecular content) any vehicle that attempted to fly through it would likely be inefficient and aerodynamically unstable. This would be one reason why a parachute descent of any atmospheric probes would be preferred, but even that has issues that are beyond the scope of this book.

4.4 Practice Exercises

4.1. Visit a local air museum and look at different aircraft engines. Print off some diagrams from the web and try to familiarize yourself with the different parts. Understand where the air flow comes in and what happens to it as it passes through the different parts of the engine and how it is exhausted out the back. At what point is the fuel fed into the engine and where is it combusted? Next go and find some rocket engines, compare and contrast the differences and similarities.

4.2. Using (4.1) compute the thrust for a jet engine flying at around 9 km (30,000 ft) altitude and has the following parameters. $A_e = 0.5$ m^2; $p_e = 3100$ kg/m^2; $p_{in} = 3,000$ kg/m^2; $v_{in} = 200$ m/s; $v_e = 500$ m/s. You will first need to compute the mass flow rate of air through the engine by multiplying the air density (assume 0.018 kg/m^3) by the free stream velocity v_e and by the exit area A_e. You should assume that the mass flow rate of fuel is negligible compared to that of air.

4.3. The aircraft climbs to several points of higher altitude corresponding to 12,200 m (40,000 ft), 15,200 m (50,000 ft), 18,300 m (60,000 ft) where the air density lowers to 0.004 kg/m^3, 0.001 kg/m^3 and 0.0004 kg/m^3 respectively. Calculate the thrust at these new altitudes assuming the same air speed and free stream properties defined in the previous problem. Given the thrust levels drop with these increasing altitudes, what else can be done by using (4.1) to maintain the thrust level equivalent to an altitude of 9 km? What does this suggest about the limits of jet propulsion technology in terms of high altitude flight and access to space?

References

1. Anderson, JD, Jr (1989) Introduction to Flight, Third Edition, McGraw Hill International Editions.
2. Varvill, R & A Bond (2004), The SKYLON Spaceplane, JBIS, Vol.57, No.1/2.

Chapter 5
Astronautics: The Development and Science of Rockets

What is it that makes a man willing to sit up on top of an enormous Roman candle, such as a Redstone, Atlas, Titan or Saturn rocket, and wait for someone to light the fuse?

Thomas K. Wolfe, Jr.

5.1   Introduction

In this chapter we learn about some of the history of rocketry and the fundamental principles that underline the subject. Crucial to any rocket calculations is the ability to determine its velocity increment under multiple staging conditions. We explore this by using example calculations and learn that there is a close connection between the mass of a rocket and the final velocity that it may attain. Using this knowledge we briefly review some of the rockets used in historical space programs and discuss some ideas for the near future.

5.2   The History of Propulsion in Rocketry

In the last chapter, we discussed in some detail the development of a propulsive engine for use in thrust generation of vehicles moving through the air. We now must turn our attention to the much higher regions of space and the requirements for a different type of engine. However, we should note at the outset that the mechanics of the operation is the same, that of an action-reaction principle according to Newton's laws of motion.

Over the centuries many have thought about the prospects of space travel. In 1657 Cyrano de Bergerac wrote a story called "Voyage dans la Lune" where a machine flies to the Moon using a principle of rocket staging (discussed below). This was a remarkable concept and way ahead of its time. Then the science fiction writer Jules Verne wrote a wonderful story in 1867 called From the Earth to the Moon, which also discussed a human mission to the Moon. In the story Verne's rocket works on the principle of an enormous bullet. The shell would be fired from a large cannon with sufficient escape velocity to reach lunar orbit. Such a sudden acceleration would of course likely kill any human crew, but the principle of propulsion by a reaction mechanism was at least correct. In 1901 HG Wells published his romantic story "The First Men in the Moon," which negated gravity altogether using an exotic substance called Cavorite – if only things were this easy!

The Russian schoolteacher Konstantin Tsiolkovsky first properly described the fundamental principles of the rocket engine when in 1903 he published his paper, "The Exploration of Cosmic Space By Means of Reaction Devices," which first discussed reactive propulsion and interplanetary travel using liquid hydrogen and oxygen as a fuel. He had understood that in order to produce thrust one must bring together an oxidant (air) and a fuel (liquid hydrogen) that could be combusted together, the reaction products of which would be directed through a nozzle chamber at supersonic speeds out of the rear of the rocket, hence giving thrust for forward motion.

In 1919 the American schoolteacher Robert Goddard published, "A Method of Reaching Extreme Altitudes," and then went on to eventually demonstrate his ideas practically by launching a liquid fuel rocket in March 1926. This rocket was around 3 m long, reached an altitude of around 56 m and attained a maximum speed of 27 m/s. The first practical rocket (capable of carrying a payload) was the German V2 launched on London in 1944, during World War Two. This horrifying yet amazing machine was the first to exit outside of Earth's sensible atmosphere at an altitude of 80 km (262,000 ft) over a range of 322 km (one million ft), achieving a speed of 1.6 km/s, or nearly five times the speed of sound. In essence, it was the world's first space rocket. It was powered by liquid oxygen and alcohol fuel, was around 14 m long, 1.6 m in diameter and had a mass of around 12 tons. The German-born engineer Wernher von Braun played an important role in the design of the German V2 rocket and also later the U.S. Saturn V, the rocket that eventually took men to the Moon.

While German V2 missiles were falling on London, members of The British Interplanetary Society (BIS) [1] were meeting in London pubs discussing the future of space travel and the possibility of 'flying' to other worlds. During this time they were presented with the surprise arrival of the German rockets. Most in London saw them as a great threat to civilization as we know it, but members of the BIS nonetheless saw the potential of future rocketry in the German weapons, which had arrived much sooner than anyone had dared imagine. These were people such as Arthur C. Clarke, the science fiction writer and inventor of the communications satellite, Val Cleaver, the rocket engineer who supervised the design effort to build the British Blue Streak missile (discussed later), Archibold Low, the father of radio guidance systems and Philip Cleator, who along with Low was the founder of the BIS.

Between the years 1948 and 1951 members of the BIS undertook a study for a vehicle that would launch from Earth and deliver a lunar lander to the Moon [2]. The design was quite sophisticated and included elements such as three-axis gyro control,

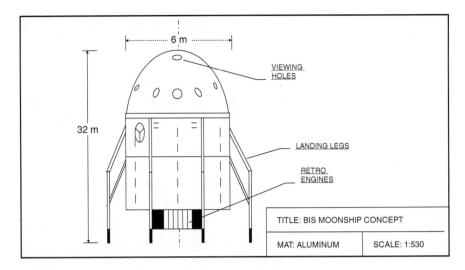

Fig. 5.1 BIS spaceship design

the use of steerable exhaust vanes and the ability to gimbal the second stage. At liftoff the engine would generate around 125 tons of thrust. Although the vehicle was never constructed it is these sorts of studies that influenced the design of rocket vehicles in later years. The hemispherical segment of the lander spacecraft was known as the 'Life Ship' or 'Moon Ship,' and it is this section that was to be carried to the Moon as shown in Fig. 5.1. The entire mass of the 'Moon Ship' was said to be around 1 ton (including a crew of three), although in hindsight this seems very optimistic and perhaps several tons would be more likely. For comparison the two-man crewed Apollo 11 lunar module, which touched down on the Moon in 1969, was around 15 tons, including the descent and ascent stage.

The designers recognized that enormous amounts of propellant would be required. They designed the total vehicle in a cellular manner so that parts could be ejected throughout the journey once the propellant and motor had been used. This massively increased the efficiency of the design over historical ones. The technique was said to reduce the propellant requirements for a mission to the Moon from millions of tons to thousands of tons. Air resistance was identified as having a negligible effect on the performance, which was mostly dominated by the large power to lift the vehicle out of Earth's gravity. At the front of the lander was located a parabolidal section constructed out of reinforced ceramic and capable of withstanding temperatures of 1,500°C due to frictional heating during the ascent through the atmosphere. This would be ejected once the vessel had reached space. In space the entire vessel would rotate at one revolution in over 3 s to provide artificial gravity, which would have been uncomfortable for the crew having to endure the constant spinning. This rotation would be halted during the descent down to the surface of the Moon. The BIS spaceship design was a fascinating idea and was later updated to a nuclear-powered version. It addressed many issues that

would eventually become part of real rocket design. It is an understatement to say that this work was pioneering and influential in the later Apollo spacecraft designs.

The BIS still exists today and is a wonderful group of the best of British (and non-British) eccentrics from a diversity of backgrounds, all sharing a common passion for spaceflight. This organization is a private charitable foundation, which despite having little influence on government space policy has always been at the forefront of developments. In the 1930s one of its members, Ralph Smith, helped to produce a wonderful set of drawings for the future lunar lander design [3], as shown in Fig. 5.2.

A comparison to the actual Apollo lander, which touched down on the Sea of Tranquility in 1969, shows many similarities; in particular, the presence of four legs for the uncertain lunar soil. The renowned space artist David Hardy recently produced a rendition of the BIS Moonship, which he called the BIS Retrorocket (Plate 2), painting the lunar landscape in the tradition of what it was thought to be like all those decades ago as visualized by the American artist Chesley Bonestell. The lunar surface was painted as having very tall mountains, highly jagged in appearance, an exciting landscape. In reality, the Apollo missions showed that the lunar surface is a highly barren place, with no exciting mountains silhouetting the horizon, although admittedly the astronauts were targeted to landing at relatively flat locations for mission safety. In his book *Visions of Space* Hardy points to the fact that many artists thereafter painted the lunar surface in the same style as Bonestell (including the famous *Tin Tin* cartoon stories created by the Belgian artist Georges Rémi) and has speculated that this could be one of the factors that later led to disillusionment with the Apollo program when the photos first appeared. This is an intriguing suggestion and highlights the important role that space art plays in inspiring the public.

In meetings after the war Clarke and Cleaver met up with the famous writers J. R. R. Tolkien and C. S. Lewis to discuss the moral issues of space travel and science fiction upon which there were large disagreements. In the end, they couldn't agree, so instead they all got thoroughly drunk [4]. Even today members of the BIS have a slightly eccentric reputation, largely due to their ability to speculate about fantastic worlds and ways of getting there since their founding in 1933. However, this is not just a British phenomenon. In the United States the American Interplanetary Society (AIS) was formed in 1930 by David Lasser, who wrote an exciting book on space travel entitled *The Conquest of Space* [5] which Arthur C Clarke was to see in the window of a WH Smith store in Minehead, England, in 1931 when he was only 14. Clarke says that this book had a profound effect on him and that it literally changed his life. The AIS was later to become the American Rocket Society and then the society was professionalized with the Institute of American Sciences to become the American Institute of Aeronautics and Astronautics. But while rocket societies were stirring on British and American soil, there was the formation of the German society for space travel founded in 1927, and this was the first space society formed in history. Hermann Oberth, who wrote about his theories of rocket propulsion, was the founder of the society. He conducted extensive rocket research, which culminated in his publication "The Rocket into Planetary Space" in 1923. This ultimately led to the development of the V2 rocket.

When we look back over the historical developments of flight and modern rocketry, it is quite remarkable that the final achievements were built upon individual efforts from several continents. People from the former Soviet Union, Germany, England, China and the United States have built upon these foundations. China had originally developed rockets for use against the Mongol hordes. Based upon black powder these solid rockets were also used in fireworks displays. It is a shame that over the centuries China lost this knowledge, but it is ever the more remarkable that in this new century China is only just now entering space and even have their aspirations set on future Moon landings.

It is an interesting fact that Tsiolkovsky, Goddard and Oberth were all teachers of some form. But it is from these three people and their ideas that all of modern rocketry derives. This includes most of the rockets developed by the United States and the former Soviet Union during the 1960s–1970s space race. Gradually, more was learned, and as a species we finally utilized the sky above our heads. Essentially, what came out of all historical rocket research are two types of rocket designs. The first is called a solid fueled rocket, which uses solid propellants for both the fuel and oxidizer. The most basic example is a gunpowder fueled rocket, which is what China was using in the thirteenth century. Indeed, model rockets used by children and hobbyists are still powered by solid fuel propellants. They are also used as booster rockets such as for the space shuttle's Solid Rocket Boosters (SRB). An important aspect of solid fuel propellants is the shape of the charge, which burns at different rates as the shape changes. However, a far more efficient type of rocket engine emerged in the twentieth century known as a liquid fueled rocket, which uses a combination of a fuel and oxidant to produce energy from their combustion. Robert Goddard first developed the liquid fueled rocket. He used propellants in liquid form such as hydrogen- and oxygen-based fuels. A solid fueled rocket is one that uses a solid propellant such as an ammonium nitrate composite. Let us explore how rockets work a bit further.

5.3   The Physics of Rocket Propulsion

The solid and liquid fueled rockets work on the same principle of ejecting a mass rearward from a body, so that by Newton's laws of motion a reactionary force causes the body to move in the opposite direction. This is a simple principle and is the mechanism for most propulsion concepts, applied even to interstellar missions. This simple principle can be shown by expressing the product of a mass m under acceleration a by a force F given by:

$$F = ma \tag{5.1}$$

A typical rocket will work on the principle of combustion and thrusting. Taking the fuel and oxidant and then combining them both energetically will enable this. Once ignited the hot particle products are directed towards the rocket nozzle and expelled rearwards.

By Newton's action-reaction principle this generates thrust to move the vehicle forward. A combination of the mass flow rate $\dot{m}$ (kg/s) of the expelled product, the exhaust velocity of those products and the pressure difference between the combustion chamber and the ambient medium give the thrust of a chemical rocket. The area of the exit nozzle is also important in determining the optimum thrust generation. In Chap. 4 we discussed the thrust equation for a jet engine. Now we shall see that it is a simple extension of this earlier derivation to produce the equation for the thrust of a rocket engine. This is achieved by removing the term for the mass flow rate of air (i.e., there is no air in a vacuum) and the total mass flow rate $\dot{m}$ is actually the combination of that from the fuel and oxidant. This then leads us to the rocket thrust equation as follows:

$$T = \dot{m}V_e + (P_e - P_{in})A_e \tag{5.2}$$

The exhaust velocity has a dependence upon a quantity called the heat capacity ratio γ, which measures the heat capacity at constant pressure to the heat capacity at constant volume. The exhaust velocity also depends on a specific gas constant R and the combustion temperature T_o. These can all then be related to derive a relation for the exhaust velocity of the engine, the full derivation of which is beyond the scope of this book.

$$V_e = \left\{ \frac{2\gamma R T_o}{\gamma - 1} \left[1 - \left(\frac{P_e}{P_o} \right)^{\gamma - 1/\gamma} \right] \right\}^{1/2} \tag{5.3}$$

An easier way to examine the performance of an engine is to look at specific impulse. This is an efficiency measure for the propellant burn. An impulse has the units of force multiplied by time and so a specific impulse (meaning divide by mass) will have the units of Ns/kg = m/s or exhaust velocity. If we then divide this by the value of gravity at sea level on Earth (9.81 ms^{-2}) we get the units of seconds, which is a measure of the effective exhaust velocity. Hence we can define the specific impulse as the thrust per unit mass flow at sea level as follows [6]:

$$I_{sp} = \frac{T}{g_o \dot{m}} = \frac{\dot{m} V_e}{g_o \dot{m}} = \frac{V_e}{g_o} \tag{5.4}$$

The most important thing to know about these relations is the dependence of the rocket performance (thrust or exhaust velocity) on two factors: (1) the specific gas constant $R = R'/M$, where R' is the universal gas constant and M is the molecular weight of the fuel. For maximum performance, fuels are required that minimize the molecular weight. A hydrogen-oxygen combination is lighter than a kerosene-oxygen combination and results in a larger I_{sp} (2) the combustion temperature T_o, which depends upon the choice of chemical fuels. Fuels are required that maximize the combustion temperature by having associated high heat of reactions. Chemical propulsion systems are energy limited because of the amount of energy per unit mass attainable from the reactants, which is around 13 MJ/kg and known as the specific enthalpy. The enthalpy is defined to be a measure of the amount of thermodynamic energy inside a substance, and a specific enthalpy just means divided by the mass of that substance. We can then work out the thermal exhaust velocity V_e by inverting the equation for kinetic energy:

$$V_e = \left(\frac{2 E_{kin}}{m} \right)^{1/2} \tag{5.5}$$

Then because we have a specific enthalpy in units of one other mass the exhaust velocity is given by the square root of twice the specific enthalpy, which is ~5 km/s. So this is the maximum amount attainable theoretically from a chemical-based propellant. We will see in later chapters that fission and fusion release energy orders of magnitude above chemical systems and thereby give a much greater exhaust velocity. This limits the exhaust velocity attainable and thereby the specific impulse, although high values of thrust and power can be attained with chemical propulsion systems.

In the discussion on rockets above we learned that the performance of an engine is dependent upon the combustion temperature and fuel molecular weight. What

Table 5.1 Comparison of different rocket fuels

Fuel combination (molecular weight combustion products)	Combustion temperature (K)	Specific impulse (s)
Hydrogen-oxygen	3,144	240
Hydrogen-oxygen	3,517	360
Hydrogen-fluorine	4,756	390
Kerosene (RP-1)-oxygen	3,670	353

should also be realized is that there is a physical limit to how high the combustion temperature can be raised; it must not exceed the melting temperature of the materials used to contain it. Clever schemes using magnetic fields can be employed to mitigate this. But ultimately, chemical fuels have a practical physical performance limit, which equates to a typical specific impulse of around 500 s, although in theory could be as high as 1000 s. Table 5.1 shows the specific impulse for various chemical fuels.

For rockets launching from Earth, the problem with any single rocket is that it can only achieve a velocity of around 4 km/s. However, there is a clever solution to this, which is to use the concept of staging. A staged rocket is a launch vehicle that is constructed of several burn stages. After each propellant section has been used, the stage is thrown away. This removes any unwanted mass from the vehicle and gives the launch vehicle a lower mass ratio. Typically, a multistage rocket will consist of two or three stages, but some can have more. An optimized staged rocket will have the largest propellant section at the bottom and the smallest at the top (serial staging); hence these are known as the first stage and last stage, with all others listed sequentially in between. The Saturn V rocket had three stages, and this is discussed briefly below. Other multistage systems exist, such as on the space shuttle, which has two solid rocket boosters that are burned and jettisoned before the external fuel tank (parallel staging).

Two fundamental concepts in rocketry are that of mass ratio and mass fraction. These are two closely related concepts but with a subtle difference. Mass ratio is simply the ratio of initial launch mass M_i (fuel + spacecraft + payload, otherwise known as the wet mass) to final mass M_f when the fuel has been fully used up (known as dry mass), such as at the destination when the vehicle has reached orbit. Ideally any rocket would have a lower mass ratio and so have less propellant, although for a given engine a higher mass ratio will allow a higher value of the velocity increment or delta-V to be reached. The mass fraction is expressed between 0 and 1 and is determined as follows:

$$m = 1 - \left(\frac{M_f}{M_i}\right) \tag{5.6}$$

A vehicle that had an initial take-off mass of one million tons and a dry mass (after fuel burn) of 200,000 tons would have mass fraction of $1 - 0.2 = 0.8$. This would correspond to a mass ratio of 5. Ideally, any rocket design will have a low mass ratio, as this allows for a greater amount of payload to be delivered into orbit for a given propellant. So mass fraction and mass ratio are a measure of how good the rocket design is. Often in this book a related quantity called the payload mass fraction will also be mentioned, which refers to the ratio of payload mass to dry spacecraft mass (no propellant).

The mass ratio is also incorporated neatly into the ideal rocket equation, first described by the Russian Konstantin Tsiolkovsky. This is obtained by first writing down the expression for the rocket thrust, which is simply equal to the mass times the rate of change of velocity, assuming that aerodynamic drag losses are negligible:

$$T = m\frac{dv}{dt} \tag{5.7}$$

In (5.7) dv/dt is the rate of change of velocity with respect to time, known as a derivative in calculus. We then use the fact that thrust is related to the specific impulse via the equation:

$$T = g_o I_{sp}\dot{m} = -g_o I_{sp}\frac{dm}{dt} \tag{5.8}$$

Here $\dot{m}$ is the mass flow rate of the propellant and is written as a derivative with respect to time dm/dt. The negative is inserted due to the fact that the propellant is reducing with time as it is being used up. We then balance these two thrust equations to give:

$$m\frac{dv}{dt} = -g_o I_{sp}\frac{dm}{dt} \tag{5.9}$$

Re-arranging and canceling out terms we can then use calculus to integrate both sides of the equation between an initial mass m_i and final mass m_f up to the burn velocity v_b:

$$\frac{1}{g_o I_{sp}}\int_0^{v_b} dv = -\frac{1}{m}\int_{m_i}^{m_f} dm = \frac{1}{m}\int_{m_f}^{m_i} dm \tag{5.10}$$

And we finally get the ideal rocket equation to describe the final velocity achieved once the propellant has been completely burned up:

$$v_b = g_o I_{sp} Ln\left(\frac{m_i}{m_f}\right) \tag{5.11}$$

Often v_b is also written as a velocity increment Δv. This relates the burnout velocity of a rocket to its specific impulse and mass ratio. The mass ratio increases exponentially for any defined increase in velocity increment, and inverting the equation can show this:

$$\left(\frac{M_i}{M_f}\right) = e^{\Delta v/v_{ex}} \tag{5.12}$$

where the exhaust velocity is given by:

$$v_{ex} = g_o I_{sp} \tag{5.13}$$

Equation (5.12) tells us what mass ratio is required to achieve a given burn velocity with respect to the spacecraft (not to distant space). The mass ratio is equal to the exponential of the velocity ratio $e^{\Delta v/v_{ex}}$ so if $\Delta v = v_{ex}$ then we have e^1 which is 2.718; in other words the wet mass at launch must be 2.718 multiplied by the dry mass. This doesn't seem too stringent, but the situation rapidly changes, as a higher Δv is required. For example, if $\Delta v = 2v_{ex}$ then we have e^2, which is 7.389. Similarly, $e^3 = 20.085$, $e^4 = 54.598$ and $e^5 = 148.413$. What we are seeing is a rapidly increasing mass ratio, so that to achieve an increasingly higher velocity increment requires massive amounts of propellant.

For a slow interstellar velocity of say 1/10th of 1% of light speed, ~300 km/s, using chemical rocket fuels with a maximum exhaust velocity of 5 km/s would require a mass ratio of ~10^{26}. In other words, the fuel mass would have to be around 100 times the mass of Earth and still take over 4,000 years to get to, say, Alpha Centauri. Because of the exponential mass ratio dependence on the velocity ratio, the velocity increment attainable is usually limited to around 2–3 times the exhaust velocity. For optimum efficiency $\Delta v \sim v_{ex}$ so there is not a large difference between them. If $v_{ex} \gg \Delta v$ during the early thrust period then the energy will be in the exhaust and not in the spacecraft, but if $v_{ex} \ll \Delta v$ during the late thrust period, this is equivalent to wasting propellant [7].

The higher the velocity increment required, the larger the mass ratio for a given exhaust velocity. To lower the mass ratio, one must find ways to increase the exhaust velocity. This requires alternative propulsion schemes using fuels that are more energetic. It is of fundamental importance to rocketry to understand that the final velocity increment Δv can be much higher than the exhaust velocity. This is because the acceleration of the vehicle is not provided by the expansion of the exhaust products into the ambient medium but by the reaction of the combustion products against the reaction chamber walls. Note also that for any mission that required an equal deceleration phase on the trip the mass ratio would be squared. Similarly, for a mission that accelerated, cruised, decelerated and then came back on the same profile decelerating back into orbit, the mass ratio would be to the fourth power.

An example will demonstrate how this can be applied to both a single and multi-stage rocket design. For this purpose we use the fusion-powered Project Daedalus concept discussed in Chap. 11 [8]. This is a two-stage design with each stage having an exhaust velocity of approximately 10,000 km/s. The two-stage Daedalus design had a structural mass of 1,690 tons (first stage) and 980 tons (second stage). Similarly, it had a propellant mass of 46,000 tons (first stage) and 4,000 tons (second stage). The second stage mass includes a 450 tons payload. Let us firstly assume that the Daedalus vehicle is a single stage design, so that:

$$M_{structure} = 2,670 \text{ tons}$$

$$M_{propellant} = 50,000 \text{ tons}$$

We then apply the ideal rocket equation to find the single-stage velocity increment ΔV_o as follows:

$$\Delta V_o = 10{,}000 \text{ km s}^{-1} \left(\frac{50{,}000 + 2{,}670}{2{,}670} \right) = 10{,}000 \text{ km s}^{-1} \text{ Ln } (19.7)$$

$$\Delta V_o = 29{,}819 \text{ km s}^{-1} (9.9\%c)$$

Now let us compare this value to a calculation that splits the Daedalus design into two separate stages:

$$\Delta V_1 = 10{,}000 \text{ km s}^{-1} \left(\frac{50{,}000 + 2{,}670}{4{,}000 + 2{,}670} \right) = 10{,}000 \text{ km s}^{-1} \text{ Ln } (7.9)$$

$$\Delta V_1 = 20{,}664 \text{ km s}^{-1} (6.9\%c)$$

$$\Delta V_2 = 10{,}000 \text{ km s}^{-1} \left(\frac{4{,}000 + 980}{980} \right) = 10{,}000 \text{ km s}^{-1} \text{ Ln } (5.1)$$

$$\Delta V_2 = 16{,}256 \text{ km s}^{-1} (5.4\%c)$$

The total velocity increment for the vehicle at the end of the boost phase is then simply the sum of the velocity increment from each stage as follows:

$$\Delta V_{tot} = \Delta V_1 + \Delta V_2 = 20{,}664 + 16{,}256 = 36{,}920 \text{ km s}^{-1} (12.3\%c)$$

This shows that the total velocity increment from having multiple stages is greater than for a single stage because $\Delta V_{tot} > \Delta V_o$. We can also see then that the total velocity increment is given by the sum of the velocity increments from each engine stage:

$$\Delta V_{tot} = \sum_1^n \Delta V_n \tag{5.14}$$

Further multi-staging of the Daedalus design is discussed in Chap. 11. But it is useful to note that the amount of staging does hit a physical limit and it is usually around four to five stages where obtaining additional velocity increment becomes unpractical. Let us apply this understanding of multi-staging to the Saturn V rocket that took men to the Moon [9]. This was a three-stage rocket with the approximate staged masses shown in Table 5.2. This was a massive rocket standing at nearly 111 m tall and with a main tube diameter of nearly 11 m (Fig. 5.3).

Table 5.2 Saturn V stages

First stage S-1C	Second stage S-11	Third stage S-1VB
LOX/RP-1 propellant	LOX/LH$_2$ propellant	LOX/LH$_2$ propellant
$M_{structure} = 131$ tons	$M_{structure} = 36$ tons	$M_{structure} = 11$ tons
$M_{propellant} = 2152$ tons	$M_{propellant} = 445$ tons	$M_{propellant} = 107$ tons
$M_{total} = 2283$ tons	$M_{total} = 481$ tons	$M_{total} = 118$ tons
$v_{ex} = 2.9$ km/s @66 km altitude	$v_{ex} = 4.2$ km/s @186 km altitude	$v_{ex} = 4.2$ km/s @323 km altitude
$I_{sp} = 265$ s	$I_{sp} = 424$ s	$I_{sp} = 434$ s

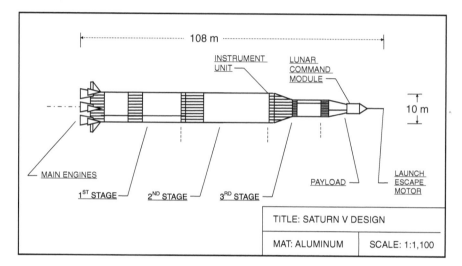

Fig. 5.3 Saturn V rocket

We also must take account of the approximately 45 tons payload, which included the Lunar Command Module and re-entry vehicle. Calculating the different staged velocity increments:

$$\Delta V_1 = 2.9 \text{ km s}^{-1} \left(\frac{2{,}704 + 178 + 45}{552 + 178 + 45} \right) = 2.9 \text{ km s}^{-1} \text{ Ln } (3.78)$$

$$\Delta V_1 = 3.85 \text{ km s}^{-1}$$

$$\Delta V_2 = 4.2 \text{ km s}^{-1} \left(\frac{552 + 47 + 45}{107 + 47 + 45} \right) = 4.2 \text{ km s}^{-1} \text{ Ln } (3.24)$$

$$\Delta V_2 = 4.93 \text{ km s}^{-1}$$

Now the third stage calculation is a bit more complicated because it was not separated until after a couple of Earth orbits and when the vehicle had entered the trajectory for trans-lunar injection. So we keep the third stage structural mass for this calculation:

$$\Delta V_3 = 4.2 \text{ km s}^{-1} \left(\frac{107 + 47 + 45}{47 + 45} \right) = 4.2 \text{ km s}^{-1} \text{ Ln } (2.16)$$

$$\Delta V_3 = 3.24 \text{ km s}^{-1}$$

$$\Delta V_{tot} = \Delta V_1 + \Delta V_2 + \Delta V_3 = 12 \text{ km s}^{-1}$$

This shows the velocity of the Saturn V on its way to the Moon and is the typical speed required that should at least exceed 11.2 km/s, although in this calculation we have made some approximations and in reality the vehicle performance would have been slightly greater to take into account gravity and aerodynamic losses that came to a deduction of around 1.5 km/s.

5.4 Rockets for Space

Throughout the last century there have been many types of rockets, and it is worthwhile spending some time discussing just a few of these so as to provide the reader with some background information for the types, sizes and masses of historical, current and some future launch vehicles. It is outside the scope of this book to review all launch vehicles tested historically, as there are too many even to name. However, we briefly discuss a select few to illustrate some variations in rocket designs and performance (see Table 5.3). Most of these rockets grew out of Intercontinental Ballistic Missile (ICBM) programs. The launch vehicle can be classed depending upon the payload capacity to low Earth orbit. A small launch vehicle would carry a payload of less than 2 tons, whereas a medium launch vehicle would carry a payload of up to 10 tons. A mid-heavy and heavy would carry a payload of up to 20 tons and 50 tons, respectively. Finally, any launch vehicle capable of carrying a payload of 50 tons or greater is classed as a Super-Heavy type.

The German V2 or "Vengeance" weapon was the first ballistic missile to go outside of the sensible atmosphere and was in use between the years 1944 and 1952, although initial development started in 1942. V2's were used towards the end of World War II and were fired on various parts of Europe, with over 3,000 being launched on London, leading to the deaths of several thousands. Some have speculated that elements of the V2 design were based on ideas from the American rocket pioneer Robert Goddard – this seems credible [10]. It is also possible that the German effort to develop this rocket was a useful distraction from the perspective of the Allies, diverting important resources away from the German air force and army who wanted to build more planes and tanks. The V2 was a single-stage rocket using a fuel combination of liquid oxygen and alcohol-water. Hydrogen peroxide with

Table 5.3 Approximate layout for different launch configurations where the payload mass range depends upon the mission to LEO, geostationary or heliocentric orbit or trans-lunar injection

Rocket	Number stages	Fuel	Length (m)	Wet mass (t)	Payload mass (t)
V2	1	LOX/alcohol	14	12.5	0.98
Black Arrow	3	RP-1/Hydrogen Peroxide	13	18.1	0.14
Mercury/Redstone	1	Ethyl alcohol/LOX/Hydrogen Peroxide	21.1	28	3
Atlas	2	LOX/RP-1	58.3	335	13–30
Sputnik	1	LOX/Kerosene	30	267	1.3
Soyuz	2	LOX/RP-1	46	308	6.4
Proton	4	LOX/RP-1 and N2O4 UDMH	53	694	5–22
Ariane V	2	LOX/LH2 + SRB	54	700	6.8
Delta IV	2	LOX/H2	63–72	250–733	3.9–22.7
STS	3	LOX/H2 + SRB	54	2,000	29.5
Saturn V	3	LOX/H2/RP-1	108	3,039	45–119
Behemoth	3	Nitric Acid/Hydrazine	80	6,500	35
Titan IV	3–5	Hydrazine/Nitrogen Tetroxide	44	943	6–22

a potassium permanganate catalyst was also used to produce steam to power the fuel pumps. The V2 was designed to reach 80 km (264,000 ft) altitude before the engine shut off, traveling at supersonic speeds of 1.6 km/s (5,200 ft/s) to a range of 320 km. Compared to modern rockets, the V2 was not particularly tall, standing at 14 m (45.9 ft) with a diameter of 1.65 m (5.4 ft). The total mass of the rocket at liftoff was 12,500 kg (28,000 lb), and it carried a 980 kg (2,200 lb) warhead. It is without doubt that the research conducted to develop the V2 set the stage for the next half century of developments in rockets for space, in both the East and the West.

Towards the end of the war the Germans were developing an even more capable weapon called the A9, which was intended to reach the United States. It was successfully tested, reaching an altitude of 90 km and a top speed of 1.2 km/s, or well over three times the speed of sound. After the war the Americans experimented with many V2 launches and one of them called Bumper was launched in 1949 from the White Sands proving ground in New Mexico, reaching an altitude of nearly 400 km and a top speed of 2.5 km/s, or over seven times the speed of sound.

It is worth mentioning that the German rocket scientist Wernher von Braun developed rocket designs that were even larger than the Saturn V. von Braun was always interested in space travel, and this appears to have been the underlying motivation for much that he did before he was working for the Nazis during World War II. His 1934 doctoral dissertation was entitled "Construction, Theoretical, and Experimental Solution to the Problem of the Liquid Propellant Rocket," and he pursued his passion obsessively. His collaboration with both the German and U.S. military rocket programs seems to have always gone in parallel with his personal vision of manned space travel. Launching rockets to the Moon was certainly made possible by his efforts, although his vision was much wider – deep space and Mars.

In 1951 one of his designs for a large rocket began to appear in *Colliers* magazine. The Behemoth was a monster of a rocket, standing at over 80 m (265 ft) tall and with a total launch mass of 6500 tons (14,000,000 lb). The three-stage rocket used a fuel combination of nitric acid and hydrazine. The first stage produced a thrust of 124.5 MN with a specific impulse of 230 s and a burn time of 84 s. The second stage produced a thrust of 15.6 MN with a specific impulse of 286 s and a burn time of 124 s. The third stage produced a thrust of 1.9 MN with a specific impulse of 286 s and a burn time of 84 s. This was an impressive rocket, although it had a takeoff mass nearly three times greater than the Saturn V or the space shuttle. In 1949 von Braun wrote a story called "Project Mars: a Technical Tale" (not published until 2006) [11] in which he described the launch of the Sirius rocket to Mars. It is clear that from the early days he was fascinated by the possibility of sending people to Mars, and it must have been a grave disappointment for him not to see the fulfillment of that vision. Although seeing men walking on the surface of the Moon must have been very exciting for him, turning away from the Moon to concentrate on LEO was not compatible with his vision of manned space travel. It is sad to think that in the year 2010 we are still stuck in LEO.

On October 4, 1957 the Soviet Union surprised the world by the launch of the first artificial satellite, called Sputnik 1. History records this event to be largely responsible for initiating the 'space race' between the Soviet Union and the United States. The satellite orbited Earth every 96 min emitting a radio signal detectable throughout the world for several weeks. The satellite was placed into orbit using the single-stage Sputnik rocket used on three occasions between the years 1957 and 1958 and which was derived from the R-7 Semyorka rocket. Standing at 30 m (98 ft) in length, it was much larger than the German V2, with a diameter of 3 m (9.8 ft) and a total launch mass of 267 metric tons (590,000 lb). The rocket used a fuel combination of liquid oxygen and kerosene. The four strap-on booster engines were designed to produce 970 kN of thrust with a specific impulse of 306 s, burning for around 120 s. The first stage of the rocket produced around 912 kN of thrust, with a specific impulse of 308 s, burning for 330 s. The fuel used in both the boosters and the first stage was a combination of liquid oxygen and rocket propellant that consisted of refined kerosene. The total thrust at liftoff was around 3.8 MN, carrying a payload mass of 1.3 tons. The rocket was subsequently used as a strategic intercontinental ballistic missile and was in use between the years 1959 and 1968. Derivatives of the original R-7 Semyorka include the Vostok and Soyuz rockets, all-important in the Soviet space program.

After the Americans captured a lot of the German V2 technology and scientists as part of Operation Paperclip, in the 1950s they constructed and launched the Redstone missile, which eventually became the Redstone rocket. The Mercury-Redstone was used for some of the Mercury flights to LEO, and this included the first and second flight of an American in space by Alan Shepard in Freedom 7 in May 1961 and Gus Grissom in Liberty Bell 7 in July 1961. The rocket was 21.1 m tall (69.3 ft) with a tube diameter of 1.8 m (5.83 ft), and at liftoff it would have a total mass of nearly 28 tons with a payload capacity of nearly 3 tons. It could ascend to an altitude of around 95 km (312,000 ft). Ethyl alcohol and liquid oxygen with hydrogen peroxide powered the rocket.

The experiences with the Redstone rocket led to the Jupiter rocket designs. A class of Jupiter rockets has recently been considered as one option for replacing the aging space shuttle fleet. The two proposals are the Jupiter-130 and the Jupiter-246 rockets, and both are based around the current space shuttle external tank design along with a set of solid rocket boosters. The Jupiter-130 rocket would be capable of lifting around 60 tons into LEO with a specific impulse of up to 269 s and a thrust of 6,550 kN. The Jupiter-246 rocket would be capable of lifting around 90 tons into LEO. This would be the same core first stage as the Jupiter-130 but would have an extended upper stage design, and it would be capable of missions to the Moon. This would have a specific impulse of up to 460 s and a thrust of up to 8,734 kN. The Jupiter-246 would also use an upper stage giving an additional 661 kN of thrust.

The Atlas family of rockets has been around since the 1950s, mainly for military use, and has been a highly successful series. An exception to this military application however was the launch of the astronaut John Glenn in Friendship 7, who in February 1962 became the first American to orbit Earth and was launched using an Atlas rocket, as well as the other manned orbital flights during the Mercury program. One proposal for replacing the space shuttle is to build the Atlas V rocket, which has been successfully launched in several flights since 2002. The two-stage Atlas V rocket is 58.3 m (191.2 ft) tall and has a tube diameter of around 3.8 m (12.5 ft). It has a liftoff mass of around 335 tons and can carry a payload of up to 30 tons to LEO or 13 tons to a geostationary transfer orbit. It has launched many successful missions including New Horizons, which is currently on its way to the dwarf planet Pluto. One proposal for a Heavy-Lift Vehicle (HLV) variant would be a three-stage rocket capable of lifting around 25 tons to LEO. The heavy life version of an Atlas V has a specific impulse of up to 311 s and a thrust of 4,152 kN.

All of the Gemini flights were launched using a Titan rocket in the 1960s. These multistage rockets used a toxic propellant combination of hydrazine and nitrogen tetroxide. One of the biggest in the Titan family was the Titan IV, which was developed as an alternative for delivering 22 tons payloads to LEO or a nearly 6 tons of payload to geostationary transfer orbit. It was 44 m (144 ft) tall with a tube diameter of around 3 m (10 ft) and it had a total liftoff mass of around 943 tons. During the 1980s the Titan IV was considered as part of space shuttle architecture for launching a lunar module to land on the Moon, but this was abandoned. Since the 1960s the United States has also been launching the Delta IV rockets, considered one of the most successful workhorses of the U.S. launcher program. The latest in the series was the two-stage Delta IV rocket, which can lift a 22.7 tons payload into LEO. A heavy lift version was also developed that included an additional upper stage. During the 1980s a very heavy lift version was also designed for the "Star Wars" program using additional solid rocket boosters from the space shuttle. It was to lift a total payload of around 45.5 tons consisting of huge laser platforms. The design was never built.

Little known to most people even in Britain was the Blue Streak rocket program conducted in the 1970s. The vehicle was intended for the launching of ICBMs and eventually led to the development of the Black Arrow rocket, for the peaceful uses of space. Throughout its development however Black Arrow suffered continuing funding problems and uncertainty over its future. The Black Arrow rocket was in use between the years 1969 and 1971 and was based upon its earlier Black Knight design. All of the Black Arrow rockets were built by the Saunders Roe Company under contract with the Royal Aircraft Establishment, on the Isle of Wight, a small island off the southern coast of England. Static firing tests were conducted at the High Down site on the island although the actual launches took place in Australia.

The Black Arrow was a three-stage rocket, with a 135 kg payload located within the third stage for deployment at LEO. It used a fuel combination of refined kerosene (RP-1) and concentrated hydrogen peroxide as the oxidizer component. The rocket was 13 m (43 ft) tall with a diameter of 2 m (6 ft 7 in), and it had a total mass of around 18 tons (40,000 lb). The first stage produced a thrust of 255 kN with a specific impulse of 265 s, burning for 131 s. The second stage produced a thrust of 68 kN with a specific impulse of 265 s, burning for 116 s. The third and final stage was used as a solid propellant engine to produce a thrust of 27.3 kN with a specific impulse of 278 s and burning for 55 s. There were a total of four launches of the Black Arrow rocket.

The first R0 was in June 1969, but due to a control malfunction had to be destroyed after launch. R1 was set to launch in March 1970 but was also destroyed during the launch due to a fuel leak. R2 launched in September 1970, but due to a technical failure could not launch the satellite. Then finally on October 28, 1971, the first British 66 kg satellite X3 Prospero was placed into a near perfect orbit, carrying a micrometeorite experiment and solar cell test. This made Britain the sixth nation to put its own satellite into orbit with its own launch rocket. The satellite is still there today, although not being used. A fifth rocket named R4 was also built but was never launched due to the cancellation of the program. Now sitting in the London Science Museum, it is a testament to a great achievement, but also to a missed opportunity.

The satellite launch industry was to become a multi-billion dollar sector. Britain played a fundamental role in the early formation of the European Space Agency (and then later withdrew) as well as in the creation of the first Ariane rockets. The main European launcher today is the Ariane V, which first flew in 1997 and is operated by the successful company Arianespace. It is a 58 m tall rocket and at liftoff it has a mass of around 700 tons carrying a 6.8 tons payload into orbit, which is frequently made up of two satellites. The main Vulcain engine stage uses a propellant of LOX and LH_2, supplemented by two solid booster rockets made up of ammonium perchlorate, aluminum and polybutadience, which provide around 90% of the total 12,500 kN thrust at liftoff. Each booster has a propellant load of around 237 tons. When the Ariane V was first designed it was with the intention to locate a small manned space shuttle on top called Hermes. This was a novel proposal that never left the drawing board, probably due to financial reasons and shifting political backing.

One of the most successful launch vehicles has been the U.S. space shuttle, illustrated in Fig. 5.4 and shown in Plate 3. It had six orbiters in the fleet at its peak, called Enterprise (experimental), Challenger, Discovery, Atlantis, Columbia, and Endeavour. These vehicles began launching missions into LEO in 1981. The space shuttle has had two serious fatal accidents in 130 flights, these being STS-51-L

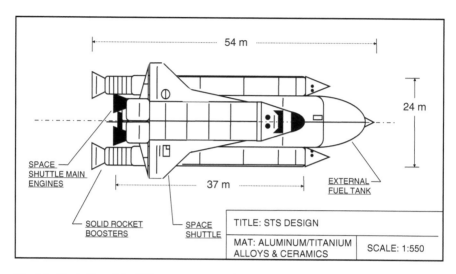

Fig. 5.4 Illustration of the U.S. space transportation system

(Challenger in January 1986) and STS-107 (Columbia in February 2003). This is a reliability of 98.5% or one mission loss in every 65 flights. The original design aim of the space shuttle was that it would be highly reusable, launching around a 100 times per year, but in reality the launch rate has been around 10% of that figure. Although the orbiter itself and the solid rocket boosters are reused, the external tank is not but is discarded in orbit to burn up on re-entry. Considering the cost of getting into orbit an ideal situation would have been to place those external tanks into a safe geosynchronous orbit so that they could become a highly useful material resource in future decades.

The other complication with the space shuttle was the thermal protection system, consisting of something like 30,000 heat tiles, which have to be carefully checked after each flight. In theory, the space shuttle associated technology is adaptable in the form of a Heavy Launch Vehicle proposal already discussed above. The Soviets also built a space shuttle known as Buran, which was successfully tested in one unmanned flight. The design was a replica of U.S. technology, and the large launch costs associated with each flight is possibly what killed the Soviet space shuttle program.

In the 1990s NASA was considering options to replace the aging space shuttle. One such option was the delta wing shaped Venture Star. It would take off from a vertically positioned launch and then land horizontally on a runway like an ordinary airplane. It was to be 38 m in length with a total mass of around 12 tons. The unique feature of the Venture Star was the rear-mounted linear aerospike engine. The Venture Star used a composite liquid hydrogen fuel tank, necessary to demonstrate lightweight technology for single stage to orbit missions, but the failure of the tank led to the project's cancellation. The contractor Lockheed Martin was to build the Venture Star and commercial missions were to begin in 2004 at a rate of

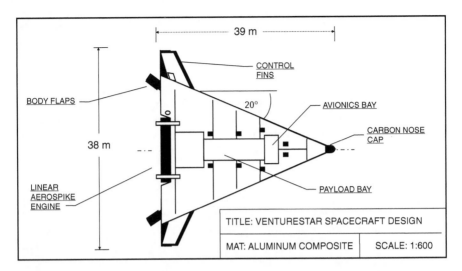

Fig. 5.5 Illustration of venture star spaceplane

around 20 per year. A half-scale model or advanced technology demonstrator of the vehicle was built called the X-33, which first flew successfully in 1999. This is illustrated in Fig. 5.5 and shown in Plate 4. The design for this vehicle was elegant, and it is a shame it did not enter into operational service. However, with advances in material science there is always the possibility of restarting what was a very credible program.

With the impending replacement of the U.S. space shuttle, government scientists have been speculating on future launch system possibilities. Several candidates have emerged. The first was that selected after the initial U.S. President Bush presented his vision for space exploration. The Administration had decided to develop a new launch capability based upon the Ares rocket designs. Ares V was designed to lift the Altair lunar vehicle into Earth orbit. Ares I would be launched separately, carrying the Orion spacecraft and crew. The Orion and Altair capsules would then rendezvous in orbit and perform a lunar orbit insertion.

Considerable effort was used in developing the Ares launch vehicles, which would be capable of lifting 25 tons into LEO. This was all part of the NASA Constellation program aimed towards sending people to the Moon and eventually to Mars. However, some questioned the Ares approach and said that it was just like the Apollo program architecture with very similar rockets (although a good engineer might say if the idea isn't broken then don't fix it). The Ares rockets were expected to have high reliability but not necessarily high frequency of flights. Another proposal was the Heavy Launch Vehicle, which would lift a massive 72 tons into LEO and would use the rockets already on the existing space shuttles, while replacing the orbiter with a lighter vehicle. Several independent scientists had proposed a variant on the Jupiter family of rockets, which would lift between 39 and 96 tons into LEO. It would use rocket technology similar to the space shuttle, but because it was much more massive it could also be used for missions to the Moon.

During the summer of 2009 the newly formed Administration of President Barak Obama ordered a review of the American human spaceflight program. The review committee was led by Norm Augustine and included former astronaut Sally Ride, the first American female in space. They reported their findings in September of 2009. They concluded that the Vision for Space Exploration originally mandated by former President Bush, known as the Constellation program, was unsustainable with current funding levels. Instead the committee suggested three alternative options known as pathways. Pathway 1 was a direct attempt to land a man on the Moon, although it might require a test mission to the Moon first. Pathway 2 was to concentrate on the establishment of a lunar base with the ultimate aim being to explore Mars. Pathway 3 was to concentrate on flexible mission scenarios such as the exploration of the local inner Solar System. This would include missions to the Lagrange points, near-Earth objects or even landings on the moons of Mars. Exploration of the Moon and Mars would be an end goal for this exploration program.

The third pathway was an interesting suggestion, with recommendations to develop a commercial launch capability along with greater international collaboration on space exploration. This third way seemed to make a lot of sense. National space agencies should be passing on knowledge transfers of their space exploration experience and eventually permit commercial companies to operate missions to LEO and ultimately the space station. Instead, governments should concentrate on pioneering new frontiers in space, and the third pathway seemed to allow this opportunity, but with the ultimate near-term goal being the establishment of a lunar and Martian colony. In the end, however, none of the above scenarios will be possible unless national space agencies are allocated funding levels appropriate for what they are being asked to do. So in early 2010 President Obama canceled the Constellation program and set in place instead a cheaper path to space, which relied upon the private commercialization of the launch vehicle industry. As of the time of writing this book, no firm decisions have yet been made on a clear mission target or how this is to be enabled, although it is likely that low gravity wells (moons, Lagrange points) are to feature as the main mission targets.

When the space shuttle fleet is finally retired there will be no U.S. launcher capability for taking astronauts up to the International Space Station. One option for NASA is to rely on a Russian rocket during this time. One of the most reliable rockets in the world is the Russian Proton series, which was originally designed to go around the Moon. A Proton rocket was used to launch many missions to the planets, including the Phobos mission to Mars and the Luna series to the Moon. In recent times they have been used to lift sections up to the ISS for full assembly. The first launch of a Proton rocket took place in 1965 and dozens have been launched since. It is a four-stage rocket with a total height of around 50 m and a tube diameter varying between 4 and 7 m. For the first, second and third stages, it uses a dinitrogen tetroxide (N_2O_4) oxidizer with unsymmetrical dimethylhydrazine (UDMH) propellant, but for the fourth stage it uses LOX/RP-1 propellant. The Proton can deliver a 22 tons payload to LEO or a 2.5 tons payload to lunar orbit. At liftoff a Proton rocket will generate a thrust of around 9,000 kN.

Another Russian option is the Soyuz rockets. The name Soyuz actually refers to the booster rocket and the crew capsule that sits on top of it. Used as early as 1960 it has played an important role in getting Cosmonauts into LEO ever since. Indeed, in the wake of the Space Shuttle Columbia accident the US also had to rely on Soyuz rockets for human Earth to orbit transport while the investigation got under way. The Soyuz capsule also remained on the station as a vital two or three-person rescue lifeboat should anything go wrong in orbit, with each capsule replaced as the crew is replaced. Despite a few tragic accidents over the years it has been quite a reliable piece of space hardware. However, many within the United States are unhappy with relying on a foreign launch provider.

Let us consider for a moment the ideal requirements for an Earth to orbit transport system. Firstly, there is *reliability* so that we can launch a payload into orbit successfully with a minimal failure rate. Next there is *reusability*, where 90% of the carrier is reused for future missions and a high quantity of them, such as at least 100. Then there is *frequency* so that the launch vehicle can be used a 100 times in a year if required. This would require a quick turnaround time for the vehicle – within a few days at most. *Adaptability* is also important so as to carry a variety of payloads and also to allow an extension of the carrier for an assortment of missions – to Earth orbit, the Moon and beyond. Finally, any such launch system should be *economical* so as to be as cheap as possible in terms of cost per unit mass access to space. All of the launch vehicles discussed above are the ballistic rocket type, with emphasis on massive thrusting of an aerodynamic projectile to high altitude.

However, there is another way to get into orbit and promote space travel at the same time and this is the concept of a winged spaceplane – essentially a vehicle that takes off from a conventional runway and heads straight up into orbit. A vehicle that could do this would go from London to Tokyo in around 3 hours compared to the 10 hours flight expected with conventional commercial aircraft such as a Boeing 747.

The dream of a spaceplane has been around for decades, but what really made this seem possible were the experimental flights of the NASA X-15 in the late 1950s and early 1960s, which reached a top speed of just under seven times the speed of sound. Neil Armstrong, the first man on the Moon, was one of the test pilots for the X-15 program. Flying on board a very fast aircraft would not be a new experience for some passengers who may have flown across the Atlantic on board the British and French Concorde at over two times the speed of sound in the latter part of the last century.

But what are the technical challenges towards the development of such a spaceplane? The main problem is velocity – to reach orbit. The requirement to achieve orbit is a velocity of 7.9 km/s, which equates to around 23 times the speed of sound (340 m/s at sea level). To achieve velocity you need energy, and this is proportional to the velocity squared. In other words, if you want to go twice as fast, you need four times the amount of energy. The key to this is of course (1) the fuel (2) the engine design. This is why there are now designers all over the world experimenting with new engine designs that can give performance advantages and make the cost to orbit just that little bit cheaper. Many of these engine designs

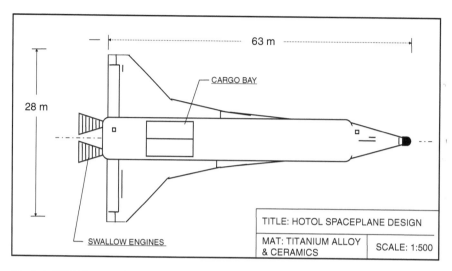

Fig. 5.6 HOTOL spaceplane

are based upon supersonic combustion technology or a scramjet, using the atmosphere to supply the oxidant at high speed while transferring the main thrust generation to a rocket engine when the atmosphere becomes too thin and the vehicle is essentially entering the near vacuum of space.

In the 1980s two companies in Britain got together to create a concept for a single stage to orbit Horizontal Takeoff and Landing Spaceplane called HOTOL, which is illustrated in Fig. 5.6 [12]. The two companies were British Aerospace (now called BAE Systems) and Rolls Royce. The vehicle would have been around 63 m × 12.8 m in height with a wingspan of 28.3 m. It would take off from a rocket sled to assist the initial launch and then use jet engines to accelerate to a speed of seven times that of sound, and then once it reached around 30 km altitude the rocket engine would kick in. The unique engine design, called the Swallow, was a liquid hydrogen and liquid oxygen fueled concept but with a varying propellant feed as the vehicle climbed to higher altitudes. The payload capacity of HOTOL was around 7 tons to an altitude of 300 km in its 4.6 m (15 ft) diameter cargo bay. The initial design was for an unmanned vehicle, although later designs did exist for a manned concept. At the time the ESA choice for the future was the French proposed Hermes mini-shuttle topped Ariane V, HOTOL or just Ariane V. Unfortunately, funding for the HOTOL concept was withdrawn in 1988, and the ESA choose Ariane V as its main launcher of satellites into LEO. One of the things criticized in the HOTOL design was the thermal loading protection upon re-entry. Another issue was the position of the engine, forcing the center of gravity to be rearwards of the vehicle, which may have presented stability issues in flight.

Many of the designers who worked on HOTOL are now involved with its successor, called Skylon and illustrated in Fig. 5.7. This design aims to solve most of the technical problems associated with HOTOL [13]. At a length of 82 m

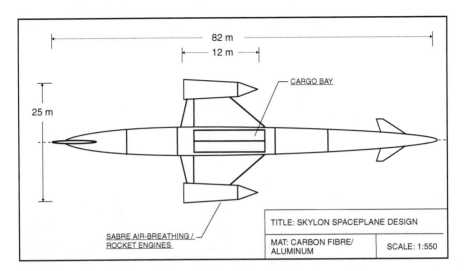

Fig. 5.7 Skylon spaceplane

it is not a small vehicle and is around twice the length of the U.S. space shuttle. It has a liftoff mass of over 300 tons and a dry mass of around 41 tons. Designers are investigating the use of heat exchanger technology as well as an oxidizer-cooled combustion chamber and a specially adaptive nozzle all to be included in the Skylon design. The twin SABRE engine is a pre-cooled hybrid air-breathing rocket engine, running a combined cycle along with the rocket engine. What has made Skylon so attractive compared to HOTOL is the lightweight heat exchanger in the engine. The performance of the engine is expected to be at a specific impulse of around 2,000–2,800 s (due to its air-breathing engines) and is much more efficient in its use of propellant compared to conventional launch vehicles. Because the engines are positioned on the wings there are no center of gravity issues like those associated with HOTOL. The fuselage is constructed of a lightweight carbon fiber structure that is sufficiently strong to support the mass of the aluminum fuel tanks.

It is envisaged that Skylon would be reusable at a turnaround time of 2 days for around 200 flights without the need for conventional rocket staging. It would cost something like $3,000/kg for each payload in a single flight and would be a substantial reduction on costs associated with the space shuttle. The typical mission has the vehicle taking off from a conventional runway like an ordinary aircraft and then accelerating to over five times the speed of sound. At an altitude of around 26 km the rocket engines would be switched on to place a 12 tons payload into LEO or 300 km or a 10.5 tons payload to 460 km. If heading directly towards the International Space Station it could deliver a 9.5 tons payload or up to 13 astronauts for a single flight. Although the payload capacity is around half of the space shuttle it is an impressive amount considering the nature of the vehicle design. Extensions of the design exist that would allow up to 60 people to be carried into space for the emerging market of space tourism. It is estimated that the cost of the Skylon research and development program to flight is around $10 billion. Reaction Engines Ltd, has received substantial funding for the Skylon project, including well over a million dollars of European ESA funding that includes UK government support. Overall, the Skylon concept has great potential for expanding access to space for many people, and not just trained astronauts, while at the same time massively reducing costs to deliver satellites to LEO. It is a project that deserves political and financial support.

One design developed by Orbital Sciences Corporation is the Pegasus rocket. This concept was first tested successfully in 1990 and has since flown on numerous occasions, including launching the Interstellar Boundary Explorer (IBEX) probe discussed in Chap. 8. It works on the principle of giving the rocket a headstart by an air launch high up into the atmosphere. A large carrier aircraft, Lockheed L-1011, takes the solid rocket to an altitude of around 11 km from which it is released, free to ascend under its own thrust into LEO. The 17 m length rocket stage also has its own wings spanning 6.7 m to assist it through the upper part of the atmosphere during the early stages of rocket burn. The rocket section has a mass of 24 tons not including the payload. When it reaches an altitude of 88 km the second rocket stage kicks in and boosts it to over 200 km. A third stage then takes it further to an altitude of over 700 km. The beauty of this concept is that it can in principle be launched from any conventional runway, and the operators of Orbital Sciences Corporation claim that it halves the cost of putting a satellite into orbit using a conventional launcher. The disadvantage of this system however is that the payload capacity is limited to around 500 kg for LEO or for an escape mission to the planets a mass of around 135 kg. But the simplicity, reusability and cost effectiveness of this design in a market where satellites are becoming more compact points the way to the future of launch systems.

Another concept was developed by Burt Rutan for his company Scaled Composites (SpaceShipOne was his answer to the X-Prize competition) is to launch a vehicle to 100 km (328,000 ft) altitude, the edge of space – although too low to launch a satellite. It would be carried by its mothership, White Knight, and then released an at altitude of 21 km. Then, using its 74 kN thrust hybrid solid rocket engine, it would boost to a speed of around 1 km/s or over three times the speed of sound, reaching a maximum altitude of around 112 km. Tourists on board the next generation design, SpaceShipTwo, will be able to experience several minutes of weightlessness and terrific panoramic views of Earth. The successful completion of the X-Prize objective on December 17, 2003 (100 years after the first Wright brothers' flight) marked a major milestone in aviation history. SpaceShipOne has the capacity to carry a payload of 3.6 tons to an altitude of perhaps as high as 150 km. Although it doesn't reach a high enough altitude to release a satellite, combining the technology of SpaceShipOne with that of the Pegasus concept could allow cheap access to Earth orbit and a significant reduction in the cost of launching satellites.

The Falcon 9 rocket is another proposed private launch vehicle for placing cargo and astronauts into space. It's built by a company called Space Exploration Technologies (SpaceX) and in July 2010 had a successful launch of the rocket from Cape Canaveral in Florida. The two-stage rocket design uses liquid oxygen and kerosene (RP-1) fuel and has a spacecraft sitting on top called 'Dragon,' which is where any crew or a payload of up to 10 tons total mass would be located. When on the launch platform it is 54 m in height (178 ft) and has a liftoff mass of 333 tons. Rockets like Falcon 9 may lead the way in demonstrating low cost access to space, although there are other approaches. The full commercialization of access to space will take the funding away from the limited resources of government-funded space programs.

For interstellar missions, chemical fuels are clearly inadequate. An interstellar mission is likely to require a specific impulse exceeding one million seconds in order to reach a third of light speed accelerating at several g (suitable for a robotic mission). The space shuttle and Saturn V rockets are fantastic achievements for near Earth and lunar operations, but clearly fall short of this longer distance goal. From an examination of these historical rocket designs it can clearly be seen that for all of their monstrous power they are totally inadequate for more ambitious missions such as going to the stars. The thrust may be large, but it is too short-lived to result in the very high specific impulse and thereby velocity increments required. Hence we must invent new methods of propelling a vehicle through space.

In this chapter and the one preceding it we have learned about different types of propulsion systems such as the jet engine for air travel and the rocket engine for space travel. In particular, we have learned about the key mechanism responsible for the thrust generation and more importantly how to calculate the thrust for both air-breathing engines and rockets. In later chapters of this book we shall examine a variety of propulsion systems for space travel, which are based upon non-chemical reactions. We shall also learn how to calculate the thrust and other important performance quantities for some of these different concepts and eventually we shall examine some historical propulsion concepts for interstellar flight, learning what performance levels are possible. From this extensive knowledge, it is the intention of this book that the reader will then have developed a theoretical mental process that they can apply to any new propulsion design for space. The challenge then will be to invent new concepts using this tool, to be able to confidently and capably calculate the likely performance.

Finally, it is important to note that although chemical rockets will not form the basis of an interstellar probe, they will still be very important in the overall mission architecture. Any probe built on Earth must then be transferred into LEO for final assembly and launch. For a large interstellar probe this may necessitate considerable space infrastructure such as the use of orbital docking ports and crew stations – all delivered to orbit using chemical propulsion systems. There are alternative mission architecture paths such as the use of a space elevator, but if assembled from a top down (orbit to surface) approach this would likely require material from the Moon and the use of electromagnetic mass drivers to accelerate any material to Earth orbit – all implying that the first launch would be several centuries away at the earliest. To launch the first unmanned interstellar probe mission by the year 2100 chemical rockets and SSTO vehicles are likely to play an important role in the early phase of the mission.

5.5 Practice Exercises

5.1. A rocket engine has a specific impulse of 50 s and a constant mass flow rate of 10 kg/s. What is the exhaust velocity and thrust of this engine design? A new rocket engine is tried out which has a variable mass flow rate and follows a $\dot{m} = 10t^{-1}$ trend so that the mass flow rate starts at 10 kg/s, where t is the burn time that we assume here to be equal to the specific impulse. What is the thrust of the engine after 50 s? Assuming that the thrust then remains constant after 50 s at this final level, what specific impulse would be required in order to generate a equivalent thrust to the original design?

5.2. A single stage rocket is designed using chemical propellant that has a maximum performance of 500 s specific impulse. It has a mass ratio of 3. The total structure and payload mass is 10 tons. What is the mass of propellant? What is the final velocity increment after complete propellant burn and assuming no structural mass drop? The structure is redesigned so that the mass ratio is increased to 3.5. What is the new velocity increment? Finally, a decision is made to split the design into two separate stages with a structural mass of 6 tons (first stage) and 4 tons (second stage), where only the first stage structural mass is dropped after its propellant is burned. What is the final velocity increment?

5.3. The space shuttle has an external tank of liquid hydrogen and liquid oxygen with a total gross mass of 756 tons and an empty mass of 26.5 tons. The external tank has a specific impulse of 455 s and supplies the fuel for the three main engines. It also has two solid rocket boosters (SRBs), each of which has a gross mass of 571 tons and an empty mass of 68 tons. These have a specific impulse of 269 s and supply around 83% of the total thrust at liftoff and are jettisoned at an altitude of 46 km. At liftoff the space shuttle burns both the main engines and the SRBs. The sequence of the mission is to burn both the external tank and SRBs. When the SRBs are jettisoned the ET continues to burn until it is used up and then it is jettisoned. Finally, the main engines burn with a specific impulse of 316 s to maintain altitude while accelerating towards its orbital altitude of 380 km, where the International Space Station awaits. The complete system has a gross liftoff mass of around 2,000 tons. Using this information and the equations given in this chapter, compute the velocity increment of each stage of the mission and the final velocity as it nears its orbital altitude.

References

1. Parkinson, B (2008) Interplanetary A History of The British Interplanetary Society, A BIS publication.
2. Ross, HE (1939) The BIS Space-Ship, JBIS, 5.
3. Parkinson, B (1979) High Road to the Moon, BIS Publication.
4. Spufford, F (2003) Backroom Boys The Secret Return of the British Bofin, Faber & Faber Ltd.
5. Lasser, D (1931) The Conquest of Space, Apogee Books & Amelia Lasser.
6. Anderson, JD, Jr (1989) Introduction to flight, third edition, McGraw Hill International editions.
7. Forward, RL (1976) A programme for interstellar exploration, JBIS, Vol.29.
8. Bond, A et al., (1978) Project Daedalus – the final report on the BIS Starship study, JBIS.
9. Godwin, R (Ed) (1999) Apollo 11 The NASA Mission Reports Volume One, Apogee Books.
10. Cleary, DA (2003) Rocket Man: Robert H Goddard and the Birth of the Space Age, Hyperion.
11. Von Braun, W (2006) Project Mars a technical tale, Apogee.
12. Parkinson, B (1987) Citizens of The Sky, 2100 Ltd.
13. Varvill, R & A Bond (2004), The SKYLON Spaceplane, JBIS, Vol.57, No.1/2.

Chapter 6
Exploring the Solar System and Beyond

While the insights of modern physics permit us to dissect the anatomy of interstellar flight, we must forego rash conclusions that any such flights are imminent or feasible. We cannot yet even define an adequate power source. If we had it, many problems of using it would be beyond us. Other obstacles may be even more formidable. For instance, what would happen to an interstellar rocket that hit even a small meteoroid, if the collision were at nearly the speed of light? . . .with our present knowledge, we can respond to the challenge of stellar space flight solely with intellectual concepts and purely hypothetical analysis. Hardware solutions are still entirely beyond our reach and far, far away.

Wernher von Braun

6.1 Introduction

In this chapter we learn about the various objects in and surrounding our Solar System, from the planet Mercury and the dwarf planet Pluto to the long-period comets that originate from the outer parts of the Oort Cloud. In the last 50 years or so since space travel began we have learned much about our neighboring worlds and the reasons why we might want to go there. The day that spacecraft start arriving in the Kuiper Belt and beyond will represent the true moment when humanity has begun to leave the safe cradle of our Sun and venture towards the distant stars.

K.F. Long, *Deep Space Propulsion: A Roadmap to Interstellar Flight*,
DOI 10.1007/978-1-4614-0607-5_6, © Springer Science+Business Media, LLC 2012

6.2 Near Earth

Our own planet Earth is to us a model for a life-bearing world. To us it seems quite large at 12,756 km in diameter and a mass of around 6×10^{24} kg. The atmosphere is perfectly suited to us and is largely made up of 78% nitrogen, 21% oxygen, 0.04% carbon dioxide and water vapor along with other gases. Our climate varies between $-89°C$ and $+60°C$ depending on the seasons and location on the planet. The surface pressure at sea level is what we term one atmosphere, which is equal to around one bar. Earth is located at a distance of around 1.5×10^8 km from the Sun, which we term one astronomical unit (AU). It has an orbital period of around 365 days and of course each day lasts around 24 hours. Our world has a vast ocean of liquid water and as far as we know is the only planet in the Solar System confirmed to harbor life. Life has existed on this world for around 3.5 billion years of its 4.5 billion year history. Earth has a solid core of nickel and iron, surrounded by a liquid molten outer core, mantle and crust. The inner liquid core rotates, giving rise to our magnetic field. Since humans first entered the domain of space they have considered the possibility of staying there. But to do that requires an artificial replication of the environment from which we have evolved, and so enter the requirement for a space station.

The Russians built the first manned space stations. The very first was Salyut 1, which operated from April to October 1971 and eventually had a crew on board for over 20 days, although sadly the crew died on reentry due to a pressure valve opening prematurely. Salyut 4 operated between December 1974 and February 1977 and hosted two separate crews, one for over 60 days. Salyut 5 operated between June 1976 and August 1977, hosting three crews on board, for a total of 67 days. Salyut 6 operated between September 1977 and July 1982 and hosted five different crews, the longest staying for 185 days. Salyut 7 operated between April 1982 and February 1991 and hosted ten crews during its decade-long life. Eventually, the Russians developed the Mir space station, which operated between the years 1986 and 2001, a record 15 years. Mir still holds the record for the longest continuous human presence in a structure in space for nearly 10 years with a maximum of six humans occupying it at any one time. Mir was finally de-orbited in 2001 during a controlled re-entry. There is no doubt that the early Russian experience in manned space stations is unique and a valuable contribution to our understanding about how long humans can remain in the space environment.

The United States eventually caught up with the Russians and placed its first space station Skylab into orbit in 1973, which was in use until 1974. During its time three different crews lived on board. Today, the international community is focused on the final construction phase of the International Space Station (ISS). This is a joint enterprise with the United States, Russia, Canada, Japan and several European nations and has been permanently occupied since October 2000. This orbital platform stands as one of humanity's greatest engineering accomplishments. It ranks up there alongside the so-called Seven Wonders of the World. The ISS is the first manmade wonder of the extraterrestrial type. The ISS is an achievement built upon past successes. Time will tell if international cooperation in the construction of this platform, visible to all humans from the surface of Earth, will help to bring about peaceful relations among nations.

Moving on from Earth orbit we come to our single satellite – the Moon. At first sight it appears to be an uninteresting place, having no atmosphere of its own and a surface temperature that varies between $+138°C$ and $-180°C$, depending upon whether the Sun shines on any part of the surface. The Moon is of course very important for Earth, because it is what causes our ocean tides. It also affects Earth's spin, and a typical day has gone from 6 hours when Earth was first formed to the 24 hours day we have now. In billions of years from now one Earth day is expected to be much longer than 24 hours. It is believed that the Moon has acted to stabilize Earth's axial tilt over its history and so may have been important in creating a stable climate for the evolution of life. The Moon is of course less than a third the size of Earth, being 3,476 km in diameter and with a mass around 80 times smaller. It is thought that the Moon has an iron-rich core surrounded by a mantle and a crust and around 3,000 moonquakes occur per year, measuring between 0.5 and 1.5 on the Richter scale, due to the seismic activity hundreds of kilometers below the lunar surface. Over time, hot lava from the liquid magma interior has created the ancient lava plains visible from Earth. It is not fully understood how the Moon formed, but was probably a result of an impact between the young Earth and another planetisimal-sized body.

One of the greatest moments in the history of our species was initiated by a speech from the U.S. President John F. Kennedy on May 25, 1961. In this speech he made the commitment to send a man to the Moon before the decade was out and return him safely to Earth. This was finally achieved on July 20, 1969, when two crewmembers from Apollo 11, Neil Armstrong and Edwin (Buzz) Aldrin stepped out onto the surface of another world. History records that Neil first spoke those wonderful words:

That's one small step for a man, one giant leap for mankind.

Without doubt, mankind's first steps onto another 'heavenly body' will be remembered for 1,000 of years hence. It is quite likely that even today we do not fully appreciate the significance of this achievement on our human psyche. During the Apollo program in total 12 human beings walked on the lunar surface, between the years 1969 and 1972. They returned some 380 kg of Moon rock back to Earth, clear evidence that men really did go there.

The first landings were over 40 years ago now, and many people have been disappointed with the lack of progress in continuing the Apollo journey. To address this, and the apparent lack of direction with NASA, the U.S. government launched the 'vision for space exploration,' an initiative under the leadership of former President Bush. This plan set out a 2020 timetable for the return of humans back to the Moon, and by 2024 the plan was to have a permanent human settlement with crews rotated every 180 days or so. This is the spirit of previous exploration programs such as the exploration of the Antarctic or the manning of the International Space Station. Unfortunately, economic considerations put a swift end to this plan.

Without a doubt, very exciting times do lay ahead for the next generation. But what are the science exploration drivers for a return to the Moon if we were to go?

First, there is the establishment of a lunar colony. This has been a long time in the waiting since the first landing in 1969. A colony will help us to learn how to live off world and continue to develop the space-based capability of our planet, should we forget. This is a necessary condition if we are to hope to one day travel much further. We must learn to walk before we can run. The use of a lunar base will also provide an opportunity to study the ancient geology of the Solar System and help us to understand more about the evolution of our own planet. If positioned right, the base could provide an excellent opportunity for a surface telescope, giving much clearer views than are obtained from the surface of Earth or even in Earth orbit.

One of the exciting prospects for the Moon is the potential for frozen water, otherwise known as 'lunar water.' This is located within the lunar rocks around the poles, and in 2009 the LCROSS spacecraft identified around 100 kg of lunar ice in a dark crater of the southern pole. Since then a concentration of 10–1,000 parts per million has been identified over the lunar surface. Some hailed this as exciting news, with increased chances for any future lunar settlement. Others were disappointed at the small quantity discovered. There is also the potential to 'mine' lunar helium-3, which could be used in nuclear reactor technology or a space propulsion system such as Project Icarus (discussed in Chap. 17), although some estimates suggest that a large amount of energy expenditure would be required in advance to perform these operations, offsetting any gain. Whatever the truth, we certainly have several reasons to return back to the Moon today and unlock its potential for the use of the human race.

The Solar System is very old, around 5 billion years in fact. How the Solar System came to form is a story that has been pieced together over centuries, although even today we still do not have the complete picture. The best model that we have is known as a proto-planetary formation model. Our Solar System started off as a rotating flattened disc nebula of gas and dust surrounding the young Sun. Over time, grains of dust fell together under the influence of gravity and formed clumps. Eventually these grew in size and grew by a process of coagulation as all of these clumps settled towards the mid-plane of the disc. This allowed them to grow even larger – into 1–10 km sized planetisimals. Eventually, much larger bodies were formed the size of Moons. These then collided, forming even larger objects, which became the rocky cores of the planets.

For a rotating gas cloud as massive as the Sun, this whole process takes around 2–6 million years, producing a protostellar disc of around 500 AU in size. How we have gone from this process to the formation of life and eventually a civilization of humans is a discussion for another text. However, it is a remarkable fact that the substance that makes up our biological structure (mainly carbon) is born from the embers of evolved stars (especially heavy elements), and so it is destined to return. If we are able to successfully discover life on any of the other planetary bodies within our Solar System, this will shed enormous insights into our evolution and potentially unravel part of our common purpose. And so it is that the discovery of life elsewhere must be a major scientific driver for future exploration missions. To understand the potential for such a discovery, it is worthwhile discussing the other planets in our Solar System, their structure and the opportunities for future exploration missions.

6.3   The Colonization of Mars

Mars has been fascinating to human beings for centuries, and there has been much speculation about the possibility of alien Martians. Mars is slightly smaller than Earth, being 6,786 km in diameter with a mass around one tenth that of Earth. The surface temperature varies between $-140°C$ and $+20°C$ and it has a very low surface pressure of around 100th the Earth. It is much further from the Sun than Earth is, at around 1.5 times the distance, and it has a much longer orbital period of 687 days, or around 1.88 Earth years. Its rotation period is similar to Earth, which means that one day is around 24 hours. Mars has its own atmosphere, consisting of 90% carbon dioxide along with nitrogen, oxygen, argon and other elements. Mars is a dusty, barren planet with many rocks that have a high content of iron, giving rise to the familiar orange-red color. When the Sun drops below the horizon twilight can last for up to two hours as the light is continually scattered by the dust and ice particles that reside high up in the atmosphere.

The polar caps of Mars are believed to contain water and carbon dioxide that is covered by a haze of cloud. During the summer months, the carbon dioxide part of the ice caps can shrink to a size only several hundred km across, leaving a residual cap consisting mainly of water. It is thought that in the past Mars was a wetter and icier planet. With gradual warming, however, the poles remain the only place where ice is stable. The planet is still warming up today, and evidence of this can be seen by the presence of pits within the frozen carbon dioxide at the southern pole. Images from orbital spacecraft have shown the presence of flow deposits known as 'water flow,' which is like mud, due to groundwater seeping up through cracks in the surface then freezing on exposure to the low pressure atmosphere. This creates a sort of dam around the crack, which eventually bursts as the pressure builds below it. Water in the form that we know it would not stay in that stable state upon exposure to the atmosphere. Instead, it would boil at a very low temperature and evaporate away. We do not know if water resides upon the Martian surface today, although we are confident that it did in the past. If spacecraft can discover the presence of accessible water on Mars, this will make future exploration missions so much easier.

Mars is an important target for future space missions because of all the planets in the Solar System it appears to be the one most similar to Earth. It has both carbon and nitrogen in the atmosphere and water frozen into the soil. It may have interesting subsurface hydrothermal systems. Its 24-hour day means that astronauts will not require significant adjustments to the Martian time cycles. Mars itself has a low gravity of 3.72 m/s^2 compared to Earth at 9.81 m/s^2. This is due to the difference in mass of 6.486×10^{23} kg (Mars) and 5.974×10^{24} kg (Earth). Consequently, launching a mission from Mars would be a lot cheaper than launching one from Earth, provided the raw materials and manufacturing were available locally. Mars has a complicated geothermal history with the potential for significant human mining. It even appears to have all of the necessary elements for creating industrial products from materials such as glass and metals. Eventually, oxygen and energy can be produced from the water in the soil and atmosphere. The most exciting possibility for Mars is the prospect of bacterial life (discussed later in this chapter) or even plant life in the past. If life of any form can be shown to emerge on another world independent of Earth, then this is strongly suggestive that we are not alone in the universe and life emerges purely as a function of chemistry, given the right conditions.

Mars has a large canyon system, which is caused by internal stresses within the planet. One of these is the Valles Marineris, which measures around 4,000 km with a variable width of 150–700 km. There are also many smaller valleys up to a 1,000 km across that may have been formed by running water. In its past, Mars is thought to have had many floods, glaciers and volcanic eruptions that have shaped the surface the way we see it today. Mars can stand on record as having the tallest volcano in the Solar System known as Olympus Mons. It is around 24 km (80,000 ft) high and three times bigger than Mount Everest on Earth. Mars also has tremendous dust storms that contain charged particles and speed across the planet between 50 and 100 m/s. Most of these are quite small and harmless in the low-pressure environment. These are so called 'dust devils,' which appear as the planet's surface warms up in the morning and disperse in the evening as the surface cools. Some of these can be quite fierce, though, reaching up to 10 km high and several 100 m across. They can cover the planet for months at a time, presenting a real problem for surface exploration missions.

Mars has two moons, known as Phobos and Deimos. Deimos is quite small, at only $15 \times 12 \times 11$ km in size, and it orbits Mars every 30 hours. It is a heavily cratered moon but contains no meteorite ejecta deposits on its surface because its gravity is too weak to prevent any surface material from escaping. Phobos is much bigger, at $27 \times 22 \times 18$ km in diameter, and it orbits Mars around three times a day. Its most famous landmark is the Stickney Crater. The interesting fact about Phobos for astronomers is that the moon is moving closer to Mars at a rate of just under 2 m per century so that eventually in around 50 million years it will crash into the surface. Alternatively, it may break up and form a Martian ring. Whatever scenario comes true, Phobos may be a problem for far future Mars colonies and orbital space stations.

Any spacecraft traveling from Mars to the outer planets will have to pass through the Asteroid Belt, which is between 2 and 3.5 AU and consists of 1,000 of objects ranging between 1 and 1,000 km in diameter. Because Mars is less massive, the escape velocity for any rocket is also much less, at around 5 km/s compared to Earth's 11.2 km/s. In the last century many spacecraft visited Mars, including Vikings 1 and 2 in 1976, Pathfinder and Global Surveyor in 1997 and the European Mars Express from 2003.

In 1949 the German rocket scientist Wernher von Braun wrote a science fiction novel called *Project Mars: A Technical Tale* (finally published in 2006), which described a possible mission to Mars in the future. He said that [1]: "Despite our preoccupation with the problems of today, we must not neglect those of the morrow {tomorrow}. It is the vision of tomorrow which breeds the power of action."

Indeed, the planet Mars provides a powerful and attainable vision for humanity to aspire to and to begin to initiate wider long term exploration plans today. The exploration and colonization of Mars is our future. It is not a question of '*if*' but one of '*when.*' The technology exists today to undertake such bold missions and move us away from our fixation with orbital operations. We must learn to have courage in the face of great risk, for the rewards of succeeding are many.

In 1989 the U.S. space agency NASA published a plan to go to Mars [2]. It was based upon a 90-day study that involved tripling the space station size to add vehicle assembly hangers and constructing vehicles for the Moon and Mars; any eventual mission would take around 18 months, with only a 3 day stay on the Red Planet. It was a mission that would have cost of order \$450 billion and was destined for rejection by the then political leaders. To move out into the Solar System we need to develop a strategy for space exploration that is sustained by economic and practical self-sufficiency. This means using the resources of Solar System bodies to provide the consumable propellants and life support materials to reduce the costs and increase the performance of space missions. The objectives of the missions would be defined by scientific and economic returns. This necessitates a '*live off the land*' approach to human planetary exploration where technology and materials are recycled to minimize waste. We must plan for a permanent presence rather than just a quick visit. To achieve these things, we need to take measured risks in human space exploration using innovative technology that leads to high gains in the short term. Current space exploration plans are limited to a handful of highly qualified individuals, and this wrongly constrains human access to space. Private industry must be allowed to pioneer the way, but motivated by government support and strong leadership.

In an attempt to demonstrate that missions to Mars could be achieved for a more moderate cost, the American engineer Robert Zubrin proposed the Mars Direct plan, which was estimated to cost \$55 billion over 10 years [3]. The plan was to launch a heavy lift booster carrying the Earth Return Vehicle (ERV) that included a chemical plant, a nuclear reactor and a quantity of hydrogen fuel, because the hydrogen abundance is only about 5% in the rocks on Mars and is difficult to extract. It would arrive at Mars after a 6-month journey using a conjunction trajectory mission that is characterized by long surface times, short in-space durations and minor propulsion requirements.

Once on Mars, the ERV would use the nuclear reactor to react the hydrogen (H_2) with the atmospheric carbon dioxide (CO_2) at elevated temperatures and pressures to produce methane (CH_4) and water (H_2O) via the reaction $CO_2 + 4H_2 \rightarrow CH_4 + 2H_2O$ that creates methane and water. Oxygen is also extracted from the water by electrolysis over a period of 10 months, which can be used for rocket fuel. Then, 26 months after the launch of the ERV, the Mars Habitat Module (MHM) is launched from Earth carrying a crew of 4 and arrives 6 months later. It uses the spent upper stage along with the MHM to generate artificial gravity by a tether. Eventually the MHM reaches Mars, lands and the crew stays for a period of 18 months, conducting scientific surveys of the Martian surface, thereby maximizing Mars surface time. When completed, they return to Earth in the ERV that is by then fueled up and ready to go. The total mission duration would be around 910 days or 30 months. Eventually, missions to Mars will result in crew stay durations of order 10–12 years. The Mars Direct Plan is a bold and well thought through proposal, and international space agencies should be finding ways to plan for Mars Direct-type missions in future program plans. Indeed, the NASA vision for space exploration does state that Mars is a secondary goal after a human return and settlement of the Moon. Ideally, any lunar mission scenarios should optimize Mars exploration ties to prepare us for the greater challenges ahead.

In 2006 members of the British Interplanetary Society, led by the scientist Charles Cockell, published an extensive report on the design of a base located at the Martian geographic North Pole [4]. This was Project Boreas, a study that ran from 2003 and was an international project involving over 25 scientists and engineers. Its primary aim was to design a station to carry out science and exploration in the Martian polar region. The crew would be up to around ten people to allow for flexibility in exploration objectives in the fields of geology, geophysics, astronomy, climatology and astrobiology. The station was designed with present-day technology and considered all aspects of the station such as the power requirements, thermal control, science laboratories, human habitation and life support systems. Other aspects to the mission were also considered, such as surface drilling and surface transportation. The proposed mission date for such a station was 2038 with a crew staying for the duration of the mission, lasting three summers and two winters, and then returning to Earth in 2042, several years later.

Exploration-based missions like that proposed for Project Boreas will make eventual human colonization of Mars possible. Raymond Halyard has described such a scenario [5]. He looked at the establishment of three self-sufficient permanent outposts, all within transport range of each other, by the end of the twenty-first century. With regular supplies delivered from Earth the expansion of the small colony into a much larger colony will become possible. The first colonies would be a set of three manned by 18 people in each one, with each station capable of manufacturing some of the equipment and structures for the next colony station – seen as a necessary condition to ensure colony growth. With an assumed 2% per year contribution of equipment and structures towards the next settlement the colony would double approximately every half century. Earth would act as the support structure using nuclear-thermal propulsion technology for supply missions, as the colony gradually builds in size, solving many technical challenges along the way, perhaps many resulting in loss of life as humans exist at the frontier of survival conditions.

Let's consider for a moment, how a human settlement of Mars would work in practice and the sort of timescales involved in future landings. For the purposes of this exercise let's use the definitions: *Before Colonization Time* (BCT) and *After Colonization Time* (ACT). Here we define a colony to be a self-sustainable community of ~1,000 people not dependent on Earth for any resources (including water, food, oxygen), other than medical supplies, generating its own trade economy, satellite governance, and with Earth as a customer base. At the beginning of Mars colonization, missions will be launched (the first landings) from Earth to Mars at a steadily increasing rate. This will be the start of the period BCT. This actually began in 1965 with the first space probe to visit the Martian system, Mariner 4, launched by the United States. Subsequently the first spacecraft on the Martian surface was Mars 3 in 1971 launched by the Soviets. At the same time, missions will be launched from Mars back to Earth as crew and soil/rock samples, for example, are sent home for analysis. After a significant number of Mars missions, a considerable habitat base will be built up. Perhaps over time, several such habitat bases will be constructed located at different positions on the Martian surface. Eventually, maybe after a few decades, sufficient infrastructure will have been built up on Mars so that the Earth to Mars missions will settle down to a steady rate and then drop off to a lower number of missions as only replacement crew are sent, along with essential components that need replacing. By this time, the infrastructure built up on Mars will be sufficient to enable some crew to remain on the surface for very long periods of time, and eventually settle. This will also see the first human Martians being born on the planet.

For all intents and purposes, the human Martian settlement will be considered a permanent colony, which will then begin producing its own trade economy and independent self-governance, given the large human population. Having sufficiently explored the Martian surface, the need will arise to explore the surrounding space around Mars, including the moons Phobos and Deimos as well as the many asteroids nearby. Mining operations will then begin for materials such as iron-nickel ore to boost Earth production levels. The continued build up of infrastructure and space launch capability from Mars will see a dramatic rise in the number of missions launched from Mars into space to explore the rest of the Solar System and beyond.

The first space race was to the Moon between the United States and the former Soviet Union. The second space race is the private commercialization of Earth orbit and lunar travel for tourism. The third space race will be the full private commercialization of space, driven by economic business returns. The exploration of Mars will be the true beginning of the third space race. At some point, the industrial scale missions will increase and the number of missions launched from Mars into space will exceed the number of missions launched from Earth into space. This will occur naturally, due to the closer proximity of Mars to the outer planets and Asteroid Belt, as well as the lower Martian gravity making launches from Mars more cost effective. By this time, the colony on Mars will be substantial, perhaps numbering ~10,000s humans. This will be the start of the period ACT, and this will be the turning point in human exploration of outer space.

Both robotic and human missions will then be launched at an ever-increasing rate from the Martian surface into the outer Solar System, accelerating human ambitions in space travel. This will initiate the true era of human interplanetary travel. Such missions will become normal and will also result in an increase in the technological capability (e.g., propulsion) as more ambitious missions are attempted. This may see the advent of human exploratory missions to the Kuiper Belt and eventually the Oort Cloud. Missions to the nearest stars will then be a logical continuation of this effort, and the initiation of human interstellar travel will finally begin. In the end, the colonization of Mars will bring about a paradigm shift in thinking about the wider exploration opportunities for space travel. Traveling to other stars will be seen as a natural technological progression of humanity's expansion into the cosmos, bringing about advances in new space propulsion systems as we reach for more ambitious missions (Fig. 6.1).

Fig. 6.1 The exploration of Mars from orbit with a space shuttle in near orbit, although not a very realistic scenario

One of the best candidates today for long distance space missions is fusion propulsion, discussed in Chap. 11. However, fusion reactor technology has not yet reached maturity and is still the subject of prototype demonstrators on Earth. To prepare for the day when fusion propulsion will become available, plasma propulsion offers a near technology demonstrator as well as providing an efficient propulsion system for Mars missions. An engine of this sort has been proposed in recent years known as the Variable Specific Impulse Magnetoplasma Rocket, or VASIMR [6]. This is a highly developed technology that can be scaled for fusion designs, in theory. It bridges the gap between high thrust-low specific impulse engines and low thrust-high specific impulse engines and can perform in either mode. Exhaust velocities of ~300–500 km/s may be possible. The engine is ideal for Earth-Moon or Earth-Mars missions if a small fission reactor can supply the electricity. Using a 12 MW reactor, it can get a crew to the vicinity of Mars within 115 days. The technology is reaching maturity, having been successfully ground tested with its full rated power of 39 kW using argon propellant. Current research is aimed at testing the second stage of the design, by ion-cyclotron boosting the plasma stream to 200 kW. Investment in this technology now will be essential if we are to achieve human missions to Mars within decades and develop fusion-based technology for longer term goals. So this technology is a clear demonstration of how a focus on a single technology for a Mars-based mission can bring missions to the stars that little bit closer.

The eventual colonization of Mars will be a turning point in the human exploration of space. In particular, it will be marked by a radical change in attitudes towards space travel. Prior to this period, attitudes will be along the lines of: space travel is too expensive; too difficult; mission targets are too far; what is the justification for going? Viewpoints will persist that claim we are fundamentally limited by our technology and we should be focusing inwards on our problems on Earth and remain at home. These attitudes will only lead to one result – stagnation and regression of the human race. Once we have achieved the colonization of Mars, attitudes will change and a renewed confidence about human capabilities in space travel will be born. Attitudes will be along the lines of: Space travel can be done relatively cheaply; there are many reasons why we must go into space; we are capable of achieving anything; we must focus outwards and think about the long-term future of humanity and we must reach the nearest stars and colonize other worlds. The human race will have moved from a negative attitude to a positive one, with eyes open to the full possibilities that space travel offers our species. When the human colonization of Mars finally gets underway one can envisage six key phases of exploration as follows:

1. Initial landings and establishment of exploration base.
2. Permanent presence, self-sustaining energy reserves (fuel, food, oxygen).
3. Exploration of many sites, establishment of multiple communities.
4. Establishment of first Martian cities.
5. Exploration of near Mars orbit including the moons.
6. Exploration beyond Mars.

Alternatively, one may consider the negative prospects for future space travel. This could be along the line of a stagnation scenario, where humans never settle Mars but instead retreat back to Earth. There is also a divergence of interest scenario, between two self-governed powers on both Earth and Mars. This could be a divergence of interests due to political, economic, territorial, technological or even religious lines that lead to interplanetary tension between people of both worlds. Then there is the extraterrestrial scenario, where intelligent beings arrive in our Solar System, but seeing that Earth is already inhabited, decide to colonize Mars before us. This is assuming they are not malevolent and chose to terminate our species. Finally, we must consider a failed colonization scenario, where we try to colonize Mars but fail, due to the challenges of day to day living on the Martian surface. It may just be too difficult a challenge for us. Similarly, we may attempt to explore the remainder of the Solar System and fail. This again would lead to stagnation of our race. But if we do not try, we will never know.

However, with the renewed confidence of having colonized Mars, should we be successful, we will then see that nothing is beyond our potential. One consideration will be the actual terraforming of the Martian environment in the distant future, to make it adapt to our human physiological needs. Such ideas have been widely explored in the literature and would likely involve a large settlement with an industrial base. Mars was once warm and wet and with time it can be made to be so again by trapping sunlight within the Martian atmosphere, perhaps assisted by ozone production, so that it can slowly heat up over many years and release some of the water frozen in the surface [7]; Mars is the symbol of hope and endless possibilities.

6.4   Other Planetary Objects

Being the planet closest to the Sun, Mercury is often not visible in the night sky. Perhaps for this reason, it is not given much attention. Most astronomical interest has focused on Mars or the outer planets. But Mercury has its own fascination, and it is worth considering the possibilities for future exploration of this innermost of worlds. Mercury does not have any moons and is a small world in comparison to the other inner planets, being around 4,879 km in diameter and around 1/20 the mass of Earth. It is very close to the Sun at only 57.9 million km or just over 2/3 the Earth's distance and experiences extremes of temperature: $-170°C$ to $+430°C$, a staggering temperature range of $600°C$.

Despite its proximity to the Sun, Mercury does have its own atmosphere (although very thin) consisting of 42% oxygen, 29% sodium, 22% hydrogen, 6% helium, 0.5% potassium, as well as some argon, carbon dioxide, nitrogen and water vapor. The presence of sodium and potassium in the atmosphere makes Mercury an inhospitable planet. Mercury has an orbital period of around 88 days and a rotation period of around 59 days, which means that the planet makes three spins on its axis for every two complete orbits, and one Mercurial day lasts for two of its years.

One of the most interesting features of Mercury is the presence of a magnetic field due to a large iron core (presumably in a liquid state), which makes up around 75% of the planet's diameter or 40% of its volume. This compares with Earth, which also has an iron core that makes up around 55% of its diameter or 17% of its volume. Mercury may be an attractive proposition therefore for future explorers looking for large quantities of iron, Mercury having more than any other body in the Solar System. Despite its massive iron core Mercury has a very weak magnetic field that is only around 1% of Earth's magnetic field. This is a big mystery for future space missions to investigate. Mercury is a heavily scarred planet, with one particular 'scarp' being 300 km long extending across the surface, as a result of the planet contracting in size and cooling due to compressive forces. It has been speculated that Mercury may have large quantities of ice at the poles deposited by passing comets, particularly at the south pole, where little sunlight reaches the shadows of the craters. Several spacecraft have visited the orbit of Mercury, including Marina 10 in 1974, Messenger in 2010 and the planned BepiColumbo in 2014. The Mercury Messenger mission was launched in 2004 and went into orbit around Mercury in 2011.

The planet Venus is known affectionately as the sister planet of Earth. It is named for the Roman god of love, which is ironic considering the planet's huge $+480°C$ surface temperature, which is hot enough to melt lead and even hotter than the surface of Mercury despite being twice the distance from the Sun. The temperature is so high on Venus because short wavelength solar energy in the form of visible light waves penetrates the thick atmosphere but then the heated planet re-emits light at a much longer wavelength that becomes absorbed within the clouds. It becomes trapped, maintaining a high surface temperature. This is the greenhouse effect. Venus doesn't have any moons of its own and is a planet similar in size to Earth, around 12,104 km in diameter and about 15% less massive than

Earth. The crushing atmospheric pressure is equivalent to nearly 1 km down in the oceans of Earth and the atmosphere consists of 96.5% carbon dioxide, 3.5% nitrogen and sulfur. It is located at around three quarters Earth's distance from the Sun, has an orbital period of around 225 Earth days and a rotation period of around 243 days, so that a day on Venus is longer than 1 year. Venus has the peculiar behavior of rotating on its axis in the direction opposite to its orbital motion (retrograde motion). The planet has very quick winds that are able to move around the complete planet in about four Earth days.

Several spacecraft have visited Venus, including Mariner 2 in 1962, Mariner 10 in 1974, Pioneer Venus Orbiter in 1978, Magellan in 1990 and Venera 7 in 1970 (one of ten Venera probes to successfully land on the surface). When spacecraft land on the planet, they do not last for very long due to the dense corrosive atmosphere. Spacecraft have to descend through several cloud layers of sulphuric acid. If the acid droplets could reach the surface in the form of rain this would be deadly to any astronaut colony. However, because of the high pressures the acid simply evaporates at much higher altitudes. But the atmosphere still creates massive chemical reaction problems such as corrosion for any spacecraft that land there. For the same reasons, most asteroids entering the atmosphere do not generally make it to the surface, but are destroyed on entry and so Venus has few impact craters.

Venus is the most volcanic planet in the Solar System, and it is the volcanism that creates the thick carbon dioxide atmosphere, which prevents heat energy from escaping. It has been speculated that in the distant past Venus may have had an ocean and may have had several continents, consisting of flat volcanic lava plains that are constantly shifting and reforming the surface. It has many volcanoes, and the largest is Theia Mons, which is around 6 km (20,000 ft) high. Venus has some very strange-looking dome shaped hills probably formed by successive lava eruptions. These are around 25 km in diameter and 700 m in height. The internal makeup of Venus is thought to be similar to Earth, with a solid core of nickel and iron surrounded by a mantle and crust. The planet does have a magnetic field, but it is miniscule compared to Earth's.

Jupiter is the largest and most massive planet in our Solar System. For these two reasons it holds much fascination. It has also played an important role in history, in focusing the debate over the role of science in human civilization. This was played out in dramatic events recorded between the Italian astronomer Galileo Galilei and the Catholic Church. Galileo was put on trial in 1633 for heresy because he claimed that Earth orbited the Sun. This idea was suggested earlier by the astronomer Nicolaus Copernicus in 1514 and so is called the Copernican model. We now know that Galileo was correct.

Jupiter is a colossal planet of around 142,984 km in diameter with a mass over 300 times that of Earth. Its atmosphere consists of 80% hydrogen, 13.6% helium along with methane and many other gases. It even has hydrogen cyanide, which is believed to be formed by the interaction of ultraviolet light and lightning discharges in the atmosphere. The presence of ammonia crystals in the upper cloud layers is thought to be responsible for the white cloud layers; these are then colored by the presence of phosphorous and sulphur compounds, giving Jupiter its notable appearance. It has a surface temperature of around $-110°C$ and an atmospheric cloud pressure of between 0.2 and 2 atmospheres. Jupiter also has an extensive magnetosphere. Scientists are still unsure about the internal structure. It may have a solid rocky core 20 times more massive than Earth, which makes up around 4% of the mass of the whole planet.

Some of the most incredible sights on Jupiter are the tremendous storms driven by 0.2 km/s (700 km/h) winds. One of these is an enormous spinning wheel-like storm around 20,000 km long and 10,000 km wide known as 'the Great Red Spot,' where internal winds spin for around 6 days in a counter clockwise motion. This is usually visible on the surface through any moderate-sized telescope, and the storm is so big that Earth could fit inside of it. The atmosphere itself moves fastest at the equator and slowest at the poles; this is known as differential rotation. It is located at 778 million km from the Sun, which is way over five times the distance of Earth. It has an orbital period of around 12 years and a rotation period of just less than 10 hours. Because it is so massive it has a whopping escape speed of 59.5 km/s, and the large gravity well is also useful for exchanging momentum in spacecraft 'slingshot' maneuvers, for the use of either acceleration or deceleration. Several spacecraft have visited the Jovian system, including Pioneer 10 in 1973, Pioneer 11 in 1974, Voyager 1 and 2 in 1978 and Galileo in 1995.

People of Earth got a wakeup call from Jupiter in July 1994 when 23 separate fragments from Comet Shoemaker-Levy 9 crashed into the atmosphere at speeds of around 60 km/s, colliding with the equivalent energy release of millions of metric tons of TNT and sending ejector up to 3,000 km out above the atmosphere. Since 1994 there have been several other impact events with Jupiter. If such a collision were to occur on Earth, we simply would not survive it. Jupiter may make an excellent location for mining of helium-3 fuel for nuclear fusion reactors or propulsion power systems. This material would be mined from the atmosphere and then delivered to Earth on container ships. Jupiter could become a major re-fueling station in the centuries ahead.

To date, Jupiter has an astonishing 62 moons, and a new one seems to be discovered every few years. One of those moons, called Io, is known to be highly volcanic as a result of tidal resonances with the big planet. Images of the surface of Io clearly show dark spots flagged by bright red pyroclastic deposits resulting from explosive debris ejected from the surface. The sources of these ejecta are known as volcanic calderas. The surface is covered in sulfur compounds giving rise to its riotous color and silicate rocks surround the lava flows. Images also show 121 km (400,000 ft) high plumes coming out of the calderas, a warning to any orbiting spacecraft to keep a safe distance. The lack of water in Io probably makes it unlikely that this world harbors life. Some of the ejecta from Io result in the deposit of charged particles into Jupiter's magnetosphere, giving rise to auroral displays.

The moon Europa has to rank as one of the most interesting moons in the Solar System. The surface shows icy plains with long dark ridges and fractures in the crust up to 3,000 km in length. The surface ice thickness is unknown, but opinions vary between tens of meters to tens of kilometers. It is believed that beneath this ice sheet lays a subsurface ocean, perhaps up to 60 km deep. The main evidence for the subsurface ocean is the identification of hydrated mineral salts.

The saline ocean, which is full of electrolytes, has electric currents produced in it by induction from Europa's motion through Jupiter's magnetic field. These currents then generate Europa's magnetic field. If Europa is similar to Io then volcanism may be heating the moon internally through a process of tidal resonance and the seabed may be littered with volcanic vents like those found on Earth. It is quite possible that these are surrounded by life forms living off the internal heat flow, protected by the presence of a magnetic field around the moon.

The planet Saturn has to stand out as the most beautiful of all the planets. Saturn is also the second largest and second most massive planet in our Solar System, next to Jupiter. It is 120,536 km in diameter and around 95 times the mass of Earth, 26% of which is located in the core. It has a surface temperature of $-180°C$ and its surface pressure is larger than 1,000 atmospheres and consists of 96.3% hydrogen and 3.25% helium. It is located at 9.5 times Earth's distance from the Sun, has an orbital period of over 29 years and a rotation period of around 10 hours. The escape speed is not as large as Jupiter but is still much greater than on Earth at 35.5 km/s. Like Jupiter, Saturn may have a rocky core surrounded by liquid metallic hydrogen and molecular hydrogen layers. The presence of a smaller (than Jupiter) liquid core gives rise to a weaker magnetic field than Jupiter. The beautiful yellow appearance of Saturn is believed to be caused by the presence of ammonia in the atmosphere, the clouds of which are quite thick, uniform and cooler than on Jupiter, giving rise to little visible features. Saturn also has very high wind speeds, faster in fact than on Jupiter, and they reach speeds of up to 500 m/s near the equator.

The planet is surrounded by a complex ring structure held in place by Saturn's massive gravity, which is clearly visible from any basic telescope and is probably the reason why Saturn holds such fascination (having few surface features). The rings are around 46 m (150 ft) thick, stretch to 273,000 km and consist of 1,000 narrow, closely spaced ringlets with names such as A and F ring. The rings are composed of mostly frozen water and ice-coated rocks that range in size between 1 mm and several meters. Several features do appear in the rings due to the nearby moons, which exert their own gravitational influence on them, helping to shape the orbits of the bodies within the rings. Like Jupiter, Saturn has many moons. To date there are 63, which range in size from a few kilometers to that of small planets. One of the moons, called Enceladus, is a 500-km wide geologically active body and is one of the brightest objects in the Solar System. Ice particles and water vapor fall to the moon's surface. The volcanoes produce geyser-like plumes of ice particles and water vapor that erupt from the surface vents, renewing the surface material. Some scientists believe that Encedalus may be a suitable environment for living organisms, given it has a combination of water ice and volcanism.

The biggest moon in the Saturn system is Titan, which may contain dry riverbeds carved by flash floods of liquid methane over a cold surface of $-180°C$. Recent discoveries suggest that oceans of liquid methane tens of miles across may be present today. The Cassini-Huygens probe entered the opaque-layered hydrocarbon atmosphere of Titan in 2005 and took pictures of features that resembled liquid oceans and the clear presence of surface weathering and erosion carved from water ice and hydrocarbons. The identification of surface channels is also suggestive of falling rain, probably methane, down onto the surface. Titan is an important world for another reason – that it may have resembled the early-formed Earth billions of years ago. Titan has to rank as one of the top visitor spots for the first human spacecraft missions to the outer Solar System. Several spacecraft have visited the Saturn system, including Pioneer 11 in 1979, Voyager 1 in 1980, Voyager 2 in 1981 and Cassini in 1997.

The planet Uranus is very large, having a diameter of 51,118 km, and is over 14 times more massive than Earth. It has an atmosphere mainly made up of 82.5% hydrogen, 15.2% helium, 2.3% methane along nitrogen and hydrogen compounds. Its surface temperature is around −218°C, and it has a surface pressure of greater than 1,000 atmospheres. Uranus exhibits prolonged seasons, which can last for over 20 years, but despite this Uranus is so far from the Sun that the seasonal temperatures do not vary widely. The blue green color of the planet is largely a result of scattering in the largely cloud free cold atmosphere (some methane clouds are present). When reflected sunlight from the clouds passes back through the methane gas it absorbs the red part of the light spectrum but allows other wavelengths through, giving rise to the color that we observe. It is located around 19 times the distance of Earth from the Sun, has an orbital period of nearly 84 years and a rotation period of a little over 17 hours. It is believed that Uranus has a liquid core made up of mostly icy water, methane and ammonia and this core makes up around 80% of the total mass of the planet. The planet also has its own magnetic field, but it is nearly 50 times stronger than that of Earth although its strength varies.

The Voyager 2 spacecraft visited Uranus in 1986 and made the discovery that the planet has 11 rings, the thickest of which is around 10 km wide; they consist of particles of a coal-like substance. Unusually the rings are perpendicular to the planet's orbital path through the Solar System, largely because of the planet's unusual orientation 98° to the orbital plane. This also results in retrograde motion of the planet and the large planetary wind speeds, between 40 and 160 m/s, flow primarily in the same direction as the planets direction of rotation. It is not known how Uranus got its unusual tilt but is thought to be due to some catastrophic impact event in the distant past.

The planet Uranus has 27 moons at last count. Miranda is a small moon only 470 km in diameter, but it has huge canyons 20 km deep, which are a mixture of old and young geologically, possibly as a result of impact events during its evolution. Miranda would be a geologist's dream to explore. The moon also has considerable tectonic activity and may experience tidal heating effects due to the gravitational field of Uranus that could internally heat up the moon. This also may be responsible for the motion of icy material on the cold −187°C surface.

The discovery of the planet Neptune did not come about through a direct observation but instead from studying perturbations in the orbit of Uranus. As such it was a theoretical prediction for the existence of a planet, which was then proved in 1846. Neptune is located right at the outer limits of our Solar System, and we did not know much about this world until the arrival of spacecraft in the late 1980s. It is a large planet with a diameter of 49,528 km and is around 17 times more massive than Earth. The atmosphere is largely made up of 80% hydrogen, 19% helium and 1.5%

methane. It has a surface temperature of −218°C and a surface pressure of over 1,000 atmospheres. It is located at around 30 times the distance of Earth from the Sun, has an orbital period of around 165 years and a rotation period of just over 16 hours. The bluish color of the planet is due to the presence of methane in the atmosphere, which absorbs the red part of the visible spectrum, the same as Uranus. Neptune and Uranus are very similar in size, but Neptune is around 20% more massive and the axis of rotation has a tilt of only 30° to the orbital plane compared to Uranus's 98°. Neptune has its own giant storm known as the 'Great Dark Spot.' White cloud features can also been seen on the surface of Neptune, where winds have carried methane gas high up into the cool upper part of the atmosphere. The gases then condense into crystals of methane ice.

An interesting fact about Neptune is that it is believed to be still slowly contracting gravitationally today. This is due to gravitational energy being converted from deep in the planet's core into thermal energy that continues to heat it, slowing down the collapse. Like all of the other gas giants, Neptune also contains rings, five of them, which are less than 50 km to up to 4,000 km in width and are located at various distances from the planet, between 42,000 km and 63,000 km. Neptune has 13 moons. One of its moons is called Triton. This is an icy world with a young surface and tenuous atmosphere. It is believed that Triton did not form originally within the orbit of Neptune but instead was a visitor from the Kuiper Belt, which was then gravitationally captured. Triton even has its own thin atmosphere, and evidence suggests that the atmosphere is warming up. Triton is an interesting moon with surface cracks around 80 km wide spewing up nitrogen through geysers 8 km high. Triton has a reddish appearance that could be due to irradiation of methane on the surface of the moon, which is in gaseous, or ice, form. It is a very cold world with surface temperatures of −235°C. Most peculiarly for a moon, Triton orbits Neptune in the opposite direction to the planet's rotation. The Voyager 2 spacecraft visited Neptune in 1989.

6.5 Mining He-3 from the Gas Giants

During the Project Daedalus study (see Chap. 11) designers settled on the atmosphere of Jupiter for the choice of helium-3 acquisition. It has also been suggested that helium-3 could be mined from the Moon, approximately one million tons deposited by the solar wind, although it may require far more energy to extract the helium-3 than is gained from its actual use. Alternatively, there is direct mining of the solar wind itself, but there are not large quantities to be gained from this. Then there are the asteroids, although due to their small surface area and low gravity, very low levels are expected to be present on their surface. The U.S. scientist John Lewis has considered the question of gas giant mining for helium-3, specifically for Jupiter, Saturn, Neptune and Uranus [8]. The composition for all these planets is expected to be about 45 parts per million (similar to the Sun). There are two basic problems with gas giant mining. The first is overcoming the huge gravity wells. The escape velocities for the various gas giants are 21.3 km/s   (Uranus), 23.5 km/s (Neptune), 35.5 km/s (Saturn) and 59.5 km/s (Jupiter).

The second issue is extraction of helium-3 from the background of helium. Launching a spacecraft to orbit Uranus, which would arrive in around 7 years, and then dropping a probe down into the atmosphere could solve the problem of fuel extraction. An inflatable gasbag would then be used to move through the vertical atmosphere until a few atmospheres of pressure had been reached. Neutral buoyancy would then be maintained. Small fission reactors would be used to power pumps and any refrigeration for the liquefaction and extraction of the helium-3 by separating it from the helium-4. Any other excess gases would be either dumped or used to cool the systems on board. The equipment would then be jettisoned and the fuel tank essentially launched back into orbit to rendezvous with an awaiting vehicle. Alternatively, the material could be launched on a highly elliptical orbit and returned to either Earth or Mars after several years. Any atmospheric probes would also have to avoid the hazards of lightning strikes if Jupiter is involved.

Historically the distance to the gas giants would also have been perceived to be an issue, but with today's technology practically all of the Solar System is reachable. From Earth the gas giants are located at 5.2 AU (Jupiter), 9.5 AU (Saturn), 19.2 AU (Uranus) and 30.1 AU (Neptune). However, it is worth noting that the distances from Mars are slightly closer, namely 3.7 AU (Jupiter), 8 AU (Saturn), 17.7 AU (Uranus) and 28.6 AU (Neptune). Overall, all of the gas giants have great potential, although Uranus is approximately half the distance of Neptune and has a much lower gravity well. Despite being further, it would seem that Neptune and Uranus are acceptable to mine, because the escape velocities are reasonable (approximately twice that of Earth).

If helium-3 mining turns out to be as important as it appears to be presently, this could open up a whole new economic frontier. Any colony established on Mars would be in a prime position to exploit this opportunity and so would likely dominate mining activities on the asteroids or gas giants. By this time, private entrepreneurs should have developed sufficient space transport capabilities for orbital and probably lunar exploration. In fact, it is most likely that private initiatives would have overtaken government programs in the majority of space operations. A competitive market in space exploration will emerge based around the areas of commercial tourism and economic opportunities such as the acquisition of rare materials. Seeking new opportunities to make business returns, they will want a part of the helium-3 mining industry of the future.

6.6 The Outer Solar System

Pluto is unfairly known as the mythological god of the underworld. It is unfortunate that it has now been given the status of a dwarf planet. Pluto is a small world of only 2,320 km in diameter and has a mass of order 0.002 Earth masses. It is thought to have an atmosphere made up of methane and nitrogen. It has a surface temperature of $-223\,^{\circ}\mathrm{C}$ and a surface pressure of only -3 micro-atmospheres. It is located around 40 times the distance of Earth to the Sun. It has an orbital period of around 248 Earth years and a rotation period of over 6 Earth days. The escape velocity is even smaller than Earth's Moon, a tiny 1.2 km/s.

The planet's orbit is unusual in that it is more elliptical and more steeply inclined to the ecliptic plane than any of the other planetary bodies in the Solar System, being at an angle of more than 17°. The American Clyde Tombaugh discovered Pluto in 1930, and at the time there were rumors of a 'Planet X' as far back as the 1800 s due to perturbations in the orbit of Neptune that could not be accounted for. The appearance of Pluto is not well known, but images obtained from the Hubble Space Telescope in 1994 clearly show what appear to be bright polar caps and a dark equatorial belt. The same images are suggestive of the presence of surface basins and impact craters. It is believed that a methane, nitrogen and carbon monoxide atmosphere, giving rise to the high surface reflectivity and allowing us to see it from within the inner Solar System, covers the surface of Pluto. The origin of Pluto is unknown, but it seems likely that it originates from the Kuiper Belt, the outer boundary of our Solar System.

Pluto is now known to have several moons. The largest is Charon, which is around 1,200 km in diameter (half the size of Pluto) and orbits Pluto's equator in the same period as Pluto rotates on its axis, so that Charon is always hovering over the same spot on Pluto. Charon is in very close proximity to Pluto, with a separation distance of just under 20,000 km. The other moons are known as Hydra and Nix, and a fourth moon has recently been discovered. No spacecraft have visited Pluto to date, but the U.S. probe New Horizons (discussed in Chap. 8) is on its way there and due to rendezvous with it in 2015.

In recent years Pluto underwent a controversial re-categorization to a dwarf planet, which some see as a metaphorical demotion. This goes to the heart of what is a planet. But wider discussions in the astronomical community began to ask if this sort of definition was sufficient. Those charged with considering a more appropriate definition raised several issues. A planet must have a mass below that of a brown dwarf so that the deuterium-burning limit is not attained. So this places a maximum mass on a planet of around 8% of the Sun. The minimum mass of a planet is a bit trickier to define, but it should be massive enough so that its own gravity maintains a spherical-like shape. A planet must not be a satellite of another world and may have satellites of its own. It must of course orbit a central star (although there are objects called 'free floating planets' that inhabit space but do not orbit any star). A planet must also be big enough to dominate its own orbit, and unfortunately this is where Pluto apparently fails the test, where part of its orbit is partly captured by the planet Neptune.

Other dwarf planets have been discovered. This includes Ceres, located around 413 million km from the Sun. It has a rotation period of around 9 hours and an orbital period of 4.6 Earth years. With a diameter of around 965 km it is the largest object in the Asteroid Belt and was discovered as long ago as 1801. Many of the dwarf planets reside within the Kuiper Belt. This odd-sounding name derives from Gerard Kuiper who first proposed its existence in 1951 as the origin of many objects from the outer reaches of the Solar System. It is also the source of short-period comets, which take less than 200 years to orbit to the Sun, and the comets travel along the ecliptic plane where most of the planets are located. Also in the Kuiper Belt is Eris, which was only discovered in 2005 and was one of the objects that started the whole Pluto debate. Amusingly Eris is named after the Greek goddess of discord. Eris is the largest object in the Kuiper Belt so far discovered, with a diameter of around 2,413 km (larger than Pluto), and is located 10 billion km from the Sun. It has an orbital period of around 560 Earth years. It is an interesting world, thought to be made of mainly rock and water ice and methane ice that makes it visible through a powerful telescope, being brighter than Pluto. Evidence suggests that Eris even has its own moon.

Another Kuiper Belt object is Sedna, named after the Inuit goddess of the sea. It is around 1,500 km in diameter, or three quarter the size of Pluto, and is located around 130 billion km from the Sun, or 900 times Earth's distance from the Sun. It is also three times further from the Sun than Pluto, which gives it the distinction of being the most distant object yet discovered within the gravitational reach of the Sun. It is thought to have a rotation period of between 25 and 30 days and may have a moon of its own. It has a highly elliptical orbit, which extends between 76 and 900 AU, and its orbit is tilted at around 12° to the ecliptic plane. Amazingly, it takes around 10,000 years to orbit the Sun.

There are many other objects outside of Pluto, including Quaoar, an object approximately 2,500 km in diameter, located at 42 AU in a nearly circular orbit, which is tilted at 7.9° to the ecliptic plane. Finally, it is worth mentioning XR190 or Buffy, an object also located three times further from the Sun than Pluto, with a diameter of less than 1,000 km. It is located at around 58 AU in a near circular orbit, which is tilted at 47° to the ecliptic plane. Other authors discuss a short but useful review of planetary objects particularly beyond the orbit of Neptune [9].

Named after the Dutch astronomer Jan Hendrik Oort (after the revival of his idea in 1950), the Oort Cloud is an immense spherical cloud surrounding the Solar System. It is located outside the Kuiper Belt and extends from around 2,000–50,000 AU from the Sun. The outer boundary of the Oort Cloud therefore marks the true edge of the Solar System, where the gravitational influence of the Sun becomes so weak that objects can easily escape out into deep space. The Oort Cloud is largely made up of icy objects that are dispersed at tens of millions of km from each other. When another star comes near, its gravitational influence can affect the orbits of the Oort Cloud objects, sending some of them towards the inner part of our Solar System; this is what we call long-period comets. It is not known for sure how many objects reside in the Oort Cloud, but the total mass is predicted to be around 40 times the mass of Earth. Outside of the Oort Cloud and the bounds of our Solar System lays the diffuse interstellar medium. A collection of particles of mainly hydrogen and helium spread throughout space. When a spacecraft gets to here then it will surely be an interstellar voyager.

Before the first interstellar probe can be launched, demonstrator missions to the Kuiper Belt and Oort Cloud will firstly have to be attempted. Proposals for this have been made [10]. One is called Icarus Pathfinder, which is a mission to 1,000 AU within 10–20 years and could use a VASIMR engine. This demonstrator mission would test out various technologies required for longer-range missions. Another is called Icarus Starfinder and is a mission to between 10,000 and 50,000 AU and would have a full up fusion-based engine that would be used for the actual interstellar flight. These concepts are illustrated in Plates 18 and 19.

6.7 Practice Exercises

6.1. Construct a map showing the orbits of the different planets as concentric spheres. The orbits should be drawn to scale. For each planet identify all the spacecraft missions that have visited such worlds historically. Plot these on the map and include the year of spacecraft arrival in the planetary system. From this information identify the diffusion rate (number of probes per year per AU) from Earth of robotic expansion out to the Solar System and beyond and on the basis of this information extrapolate the arrival of the first interstellar probe at the Alpha Centauri system, assuming technology was to scale linearly.

6.2. In this chapter we have discussed the possibility of mining lunar ice from the Moon, helium-3 from the solar wind and gas giants, iron ore from Mercury. What other precious materials, minerals or gases can you identify as potential commercial incentives for driving Solar System exploration and thereby being an enabler for interstellar progress?

6.3. Imagine a future 200–300 years from now where humankind has fully colonized the Solar System. We have small bases or colonies on most of the planets and moons and economic trade exists between those colonies. Describe the inter-colony communications, legal system, and trade activities. If the Solar System is fully colonized and humans exist in the outer reaches, is this infrastructure a sufficient platform from which to embark upon manned interstellar missions? If so how and if not why not?

References

1. Von Braun, W (2006) Project Mars a Technical Tale, Apogee.
2. Hoffman, SJ, Ed et al., (1997) Human Exploration of Mars: The Reference Mission of the NASA Mars Exploration Study Team.
3. Zubrin, R (1996) The Case for Mars, the Plan to Settle the Red Planet and Why We Must, Simon & Schuster.
4. Cockell, C et al., (2005) Project Boreas A Station for the Martian Geographic North Pole, BIS publication.
5. Halyard, RJ (2009) Growing a Colony From An Outpost: The Colonization of Mars in the 22nd Century, JBIS, 62, 9.

6. Chang-Diaz, FR, (1995) Rapid Mars Transits with Exhaust-Modulated Plasma Propulsion, NASA TP3539, March 1995.
7. Fogg, M, (1995) Terraforming: Engineering Planetary Environments, SAE International, Warrendale, PA.
8. Lewis, JS, (1997) Mining the Sky, Untold Riches from the Asteroids, Comets and Planets, Basic Books.
9. Williams, I.P (2007) Exploring The Solar System Beyond Neptune, JBIS, 60, 10, pp387–388.
10. Long, K.F (2011) Starships of the Future, Spaceflight, 52, 4.

Chapter 7
Exploring Other Star Systems

*There are countless planets, like many island Earths...man
occupies one of them. But why could he not avail himself of
others, and of the might of numberless Suns?... when the Sun
has exhausted its energy, it would be logical to leave it and look
for another, newly kindled, star still in its prime.*

Konstantin Tsiolkovsky

7.1 Introduction

In this chapter we learn some fundamental astrophysics and consider the types of
stars that exist in the universe. This is important, as the determination of a suitable
destination is a fundamental prerequisite to launching any interstellar probe. Such a
venture will require an enormous amount of effort, time and money; and mission
planners need to be sure that they are sending the probe to the right place. When it
finally reaches its destination, perhaps a century later, it will enter the orbit of
another star and even of an extrasolar planet with a suitable biosphere for life. This
could become a future home for pioneering humans, representing our first true
expansion out into the cosmos. Alternatively, we may find life is there already and
shed light on a question that has haunted our dreams since we had first looked to the
heavens – are we alone in the universe?

7.2 The Stars and the Worlds Beyond

To begin, it is worthwhile reviewing what we know about our own star. The Sun is
quite an ordinary star, one of possibly 400 billion in our galaxy, which we call the
Milky Way, and taking around 220 million years at a speed of 220 km/s to orbit
the galactic center located around 2.5×10^{17} km away. The Sun contains 99.8% of

K.F. Long, *Deep Space Propulsion: A Roadmap to Interstellar Flight*,
DOI 10.1007/978-1-4614-0607-5_7, © Springer Science+Business Media, LLC 2012

the total mass in our Solar System, and is on the edge of a galactic spiral arm. It has a beautiful solar corona that can be seen during an eclipse. It is believed that perhaps something like 50% or (even higher) of stars in the galaxy come in pairs, so-called binary systems, but our Sun is alone. The Sun is much bigger than any of the planets in our Solar System, with a diameter of 1,390,000 km and a mass of 2×10^{30} kg, which is nearly 335,000 times the mass of Earth. It is at a distance of 1.5×10^{8} km from Earth, which is 1 AU. It has a surface temperature of 6,000°C and a core temperature of around 15 million C where the pressure and density becomes greater. The Sun has a surface gravity 27.9 times stronger than Earth. The escape velocity from the Sun is an astonishing 618 km/s, although this is reduced to 42 km/s in the reference frame of Earth. Heat is transported in stars through the processes of convection (hot buoyant mass is carried outwards and cool mass is carried inwards), conduction (small amount of collisions between particles) and radiation (electromagnetic energy generated from nuclear reactions at the core that gradually diffuse from the center towards the surface in a time frame of order 50,000 years) [1].

Most people now understand that stars are essentially atom factories and that every atom that makes up our bodies has originated at some point from the depths of a star – the stars that preceded our existing Sun. We are literally children of the stars. The Sun has an enormous atmosphere that stretches right out into the Solar System and surrounds all of the planets. It blows off a huge charged particle plasma stream known as the solar wind that travels at speeds of between 300 and 1,000 km/s and interacts with the planetary magnetic fields, causing the aurora phenomena. The solar wind is a clear demonstration that although the Sun exhibits an overall state of equilibrium (a necessary condition for a life-habiting planet such as Earth), it is also a dynamic object. A view through a specially designed solar telescope reveals the presence of short-lived cool temperature 4000°C regions on the surface which are known as sunspots and which range up to 100,000 km in size around the location of magnetic bipolar fields. The number of sunspots appear in periodic frequency in an 11-year cycle. Flares are generated in regions of high magnetic field intensity around the sunspots, which are then revealed as massive 'tongues of fire' emanating from the solar surface.

The mystery of how a star stays stable for long periods of time was finally solved by German-born physicist Hans Bethe in the 1930s. He correctly described the fusion reaction sequence that maintains a balance over billions of years. The Sun is around 5 billion years old with a temperature and pressure at the core sufficient to cause the generation of a total solar energy output rate of 4×10^{26} W mainly by the proton-proton and carbon–nitrogen–oxygen chain of nuclear reactions. Throughout its history, the Sun has been on the brink of collapsing gravitationally due to its enormous mass, but if it does so even a tiny bit, the core heats up slightly, increasing the fusion reaction rate and therefore radiation pressure on the collapsing atmosphere. This pushes back out and prevents the collapse. Similarly, if for some reason the Sun was to increase some of its core-produced energy, it would begin to expand outwards, but then the core reaction rate would slow down, reducing the pressure and allowing the atmosphere to collapse back to its original size. The Sun is in effect a giant self-sustaining pressure cooker with an automatic sensor on the stove switch,

Table 7.1 Stellar fusion burn cycle

Fuel (ignition temperature)	Typical reaction products
Hydrogen burning (10^7 K)	Helium
Helium burning (10^8 K)	Carbon, oxygen
Carbon burning (5×10^8 K)	Oxygen, neon, sodium, magnesium
Neon burning (10^9 K)	Oxygen, magnesium
Oxygen burning (2×10^9 K)	Magnesium, sulfur
Silicon burning (3×10^9 K)	Iron

and all stars work on the same principle. This is known as a state of hydrostatic equilibrium. How a star gets to this state of equilibrium is a complex process of cloud collapse and fragmentation until fusion conditions are satisfied. A star will evolve throughout its life, burning hydrogen, helium and heavier elements until sufficient energy is produced to maintain the hydrostatic equilibrium state. The typical burning cycles are illustrated in Table 7.1.

A star begins as a giant molecular gas cloud at a low temperature of around 10 K, which is cold enough for hydrogen atoms to form molecules. These are spread throughout space and typically contain a total mass in the range 10^5–10^6 solar masses and have a density of 10^9 molecules/m^3 or greater. They are around 10–40 pc in radius (1 parsec = 3.26 light years), and it is the combination of low temperature and high density that makes them the perfect place for the seeds of star formation. This process is described by something in astrophysics known as the Virial theorem, which states that for a molecular cloud in space if twice the total internal energy of the cloud exceeds the absolute value of the gravitational potential energy, the force due to the gas pressure will dominate over the force of gravity, and the cloud will expand. But if the internal kinetic energy is too low the cloud will collapse until it adjusts thermally to its new state as a main sequence star in a process that takes around 10 million years. This leads to the formation of pre-nuclear burning objects called protostars. Such objects, such as the Eagle Nebula, have been seen in images taken by the Hubble Space Telescope.

Protostars with masses of less than 8% than that of a solar mass evolve into objects where gravity is countered by the pressure of degenerate electrons – brown dwarfs. Light is emitted primarily in the near infrared, and these are faint objects. To sustain hydrogen burning through to helium burning, the core must be greater than 3×10^6 K and the star must have a minimum mass of around 75 times that of Jupiter or 7% that of the Sun. Brown dwarfs only need another 1% of a solar mass to effectively become a star. These are less massive than stars but more massive than planets. If a collapsed cloud has a final mass of more than ~8% of the Sun, fusion reactions can be initiated with the thermonuclear ignition of hydrogen to helium. It is not known for sure how massive a star can be, perhaps between 100 and 150 solar masses, although recent reports of astronomical observations indicate that a star may go as high as 300 solar masses. The limit to the mass of a star is due to the presence of significant radiation pressure (pressure due to photon gas) generated in stars much larger than our Sun. Small changes in total energy are accompanied by large changes in the internal and gravitational energies. In effect, large stars are easily disrupted and therefore are rare.

A planet is generally much smaller than a star and lies in orbit about a star at distances of between 0.1 and 100 AU. It has its own self-gravity and is illuminated by a star and any nearby moons. No nuclear fusion takes place, and the planetary core is formed through a process of sedimentation and coagulation of dust grains to large sizes. It is supported by the chemistry of its materials, which usually gives it a solid surface. In contrast, a star is generally much bigger than a planet and is the center of orbital attraction for the system being positioned at the foci of an ellipse. The star also has its own self-gravity, which dominates the system, and the star is the main source of illumination for that system.

People often wonder about Jupiter's potential for becoming a star, not just because of its enormous size but also because its atmosphere is mostly hydrogen and helium, the same as stars. However, the main difference is that the heavy elements are about three times more abundant in Jupiter than in a star, which is characteristic of a body formed by the planetesimal process. It turns out that in order for Jupiter to be a star, it would have to be around 75 times more massive than it currently is. This could only occur if another planetary body (equivalent to that mass addition) were to come into near contact with Jupiter and be gradually accreted, thereby causing the planet to gravitationally collapse so that fusion ignition could occur. The total mass of the combined body would have to be around 8% of the mass of the Sun, which is the minimum mass required to achieve the ignition of hydrogen and helium elements. More than 75 times its own mass sounds like a lot and is an unlikely scenario. However, a more likely (but still speculative) possibility is that another planet of around 13 times the mass of Jupiter might be accreted. This would then allow the combined mass to be around 7% of the Sun mass, which is the requirement for reaching the deuterium burning limit and thereby produce a brown dwarf. The whole question of Jupiter becoming a star was dramatically demonstrated in the film *2010: The Year We Make Contact*, directed by Peter Hyams and based upon the novel by Arthur C. Clarke.

The radius of a star is easily determined from the luminosity relation

$$L = 4\pi\sigma R^2 T_s^4 \tag{7.1}$$

where σ is a physics constant called the Steffan-Boltzmann constant equal to a value of 5.67×10^{-8} Wm^{-2} K^{-4}. For the Sun the radius R is 6.95×10^5 km and the surface temperature T_s is 5,780 K. This allows for example one to determine that if two stars have the same temperature but one of them is 100 times more luminous, then its radius must be 10 times greater. For the Sun the luminosity is around 4×10^{26} W, which is how much energy, it radiates per second. There is also a limit to a star's luminosity, known as the Eddington luminosity limit, named after the British astronomer Arthur Eddington. This is when the luminosity is so great that the star will begin to lose mass, and increased radiation pressure will lead to an intensely driven solar wind. For massive stars the outer envelopes are only loosely bound so they are prone to lose material, which results in a variability in their luminosities. Many stars, of course, will end their life in a catastrophic collapse of the core, which results in a supernova explosion visible throughout the universe.

This is accompanied by a rapid rise in the solar luminosity for a short period before decreasing. The amount of energy output from a supernova can be as much as 10^{44} J, which is about as much as the Sun produces throughout its whole life.

What we are really interested in is whether other stars in the universe are similar to our own and if not, how they are different. We know that our star has a habitable planet with life on it within its orbit. So if we find a similar star to ours, which also has a planet in a similar orbit, the chance for life would seem good. But many of the stars in our galaxy are not like the Sun, and so we have to understand them better. The other problem is that it is difficult to determine whether or not they have a planet, let alone a habitable one, because they are so far away. However, this has all changed in recent years with the discovery of many extrasolar planets around other stars.

At this time, however, let us concentrate on stars within our near neighborhood, which are potentially reachable with an unmanned probe in the next century or so. This will help to inform us about where we may like to go on our first missions to the stars.

A credible distance for any interstellar probe to travel in under a century is probably out to around 15 light years or less. In this distance there are 31 stars of varying spectral type, which includes G, M, K, D and F class stars. The spectral types are a form of stellar classification going from the hottest stars (>33,000 K surface temperature) to the coolest stars (<3,700 K surface temperature) in the order O, B, A, F, G, K and M, respectively. These are further designated with numbers from 0 to 9 to indicate subcategories in stellar class. The 20 nearest stars are shown in Table 7.2 and many others are illustrated in Fig. 7.1 with exact distance scaling. It is worth discussing just a few of these as potentially interesting targets for a future space mission.

Alpha Centauri is located 4.3 light years away, or 1.34 pc, and is the nearest star to our Sun and the brightest star in the southern constellation of Centaurus. It is actually a binary star system, with its companion Alpha Centauri B located around 11 AU distance at the closest approach and 36 AU distance at the farthest. Centauri B is not resolvable with the naked eye. Centauri A and B are believed to be between 5 and 6 billion years old and so predate the Sun. Alpha Centauri A is slightly bigger than the Sun with 10% more mass and over 20% larger in size. It is a main sequence star of spectral type G2. Alpha Centauri B is slightly smaller than the Sun at 0.9 solar masses with a radius of 13% less and is of spectral type K6. Centauri B is also interesting because the observed light curve varies over a short time scale and gives rise to flare events, one of which has actually been observed.

The binary system also has a stellar companion in orbit around it called Proxima Centauri, which is located at 0.21 light years (around 13,000 AU) from the pair and an angular separation of 2.2° (around four times the angular diameter of the Moon in the sky), which is larger than the separation distance between the binary pair thought to be gravitationally associated with Proxima Centauri. This combination means that the Centauri AB-C is a triple star system, although some still argue that Proxima is not gravitationally bound to Centauri A and B. Proxima Centauri is a small red dwarf star of 0.12 solar masses and is of spectral type M5. It is thought by astronomers to be

Table 7.2 Stellar data for nearby stars

Star	Distance (light years)	Spectral type	Relative mass (radius)
The Sun	0	G2	1.0 (1.0)
1. Proxima Centauri	4.3	M5	0.1 (0.14)
2. Alpha Centauri A and B	4.4	G2/K6	1.10 and 0.89 (1.23 and 0.87)
3. Barnard's Star	5.9	M5	0.15 (0.12)
4. Wolf 359	7.6	M8	0.2 (0.04)
5. Lalande 21885	8.1	M2	0.35 (0.35)
6. Sirius 48915 A and B	8.7	A1/DA (double)	2.32 and 0.98 (1.8 and 0.022)
7. Luyten 726-8	8.9	M6	0.12 and 0.1 (0.05 and 0.04)
8. Ross 154	9.5	M5	0.31 (0.12)
9. Ross 248	10.3	M6	0.25 (0.07)
10. Epsilon Eridani	10.7	K2	0.85 (0.72)
11. Luyten 789-6	10.8	M6	0.25 (0.08)
12. Ross 128	10.8	M5	0.31 (0.1)
13. 61 Cygni A and B	11.2	K5/K7 (double)	0.59 and 0.5 (0.7 and 0.8)
14. Epsilon Indi	11.2	K5	0.71 (1.0)
15. Procyon 61421 A and B	11.4	F5/DA (double)	1.77 and 0.63 (1.7 and 0.01)
16. +59° 1915 A and B	11.5	M4/M5 (double)	0.4 and 0.4 (0.28 and 0.2)
17. Groombridge 34 A and B	11.6	M2/M4 (double)	0.38 and 0.16 (0.38 and 0.11)
18. Lacaille 9352	11.7	M2	0.47 (0.57)
19. Tau Ceti	11.9	G8	0.78 (0.79)
20. Luyten BD + 5° 1668	12.2	M4	0.38 (0.16)

a flare star and can vary in magnitude suddenly. It is slightly closer to us than Centauri A and B and located at a distance of around 4.22 light years.

Because Centauri A and B are very similar to the Sun, including in metallicity, it is thought that they may be host to several planets. Stable orbits have been found in the inner systems for both stars but within a few AU of the center. So far, however, no gas giants or even brown dwarfs have been located. It may be that the astronomical technology is not yet sufficiently advanced to resolve any terrestrial planets within the system, although with the rapidly improving techniques we should be optimistic that positive detection surveys will be completed in the near future. For any such planets with liquid water present they would have to be at a distance from each star of 1.2 AU (Centauri A) and 0.7 AU (Centauri B), in the habitable zone. There is optimism in the astronomical community that some terrestrial planets may yet be found in the Centauri system, making the case for the first mission target for any interstellar probe. In particular, having three different stars of varying spectral type to visit (even in flyby mode) will maximize the science return of any such mission.

Barnard's Star, located 5.9 light years away, or 1.8 pc, is the fourth nearest star to our Sun. It was the mission target for the BIS Project Daedalus [2] (discussed in Chap. 11). The main reason for choosing this target was because at the time it was believed that astronometric evidence indicated the presence of one or two large Jovian gas giant planets by observing perturbations in the proper motion.

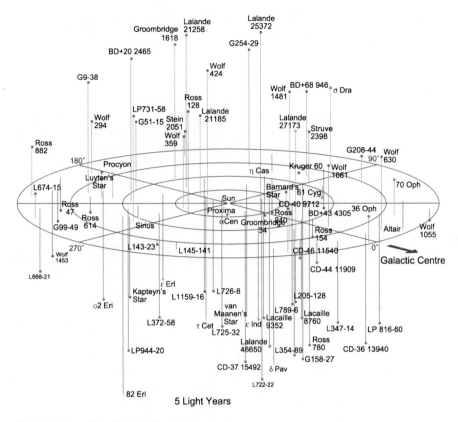

Fig. 7.1 Map of local interstellar neighborhood

The planets were thought to have a mass 1–2 times that of Jupiter and located in an eccentric orbit within a few AU of the parent star. But later this claim was found not to be true [3].

Barnard's Star is an M4 class red dwarf population II star located in the constellation of Ophiuchus; although untypical of other population II stars it has a high metallicity, with an abundance of chemical elements higher than helium. It is around 0.15 solar masses in size and around 0.12 solar radii, much smaller than the Sun. It is beyond the range of human visibility and requires a telescope to view it. It is thought to be much older than our own Sun, perhaps by several billion years. Some observations indicate that it may be flaring today, which would certainly make it an interesting stellar object of scientific study if you can get a probe out to it. By the year A.D. 11,700 Barnard's Star will have moved so close to us that it will only be 3.8 light years away, closer than the current nearest neighbor, Alpha Centauri. Perhaps we could wait until then before launching a mission there, but that seems a long time to wait for new discoveries. The situation today however is that unless smaller planets are found in orbit around the parent star Barnard's star is unlikely to be a priority target for the first interstellar mission.

Tau Ceti, located at 11.9 light years, or 3.65 pc, and is the 25th nearest star to our Sun or 19th system. This is a single G-class star with a mass just slightly less than the Sun at 0.78 solar masses and a slightly smaller size of 0.79 solar radii. As of today, no giant exoplanets have yet been discovered in this system. However, infrared surveys have detected a prominent debris disc that extends from around 30 AU out to 55 AU and would contain objects at least as large as comets in an abundance ten times more than in our own Solar System and perhaps much greater in the form of rocky planets [4]. It is possible that terrestrial-sized planets exist in this system, but they are yet to be detected, and the presence of so many cometary objects would make the stability of any world very questionable. Life would have to evolve in a changing environment due to the frequency of impact events. Despite this, if any Jupiter-sized planets do exist and deflect a sufficient amount of these impacts then any smaller world may have a chance. But to date radial velocity surveys have not detected any planets. The star is in the constellation Cetus and is visible with the naked eye. It has a similar mass and spectral type to our own Sun. The metallicity of Tau Ceti is only about one third of the Sun despite being twice as old, so the formation of terrestrial planets is made less likely. Because of the similarities with our Sun it has frequently been the target of SETI-type searches, with no positive results yet obtained.

One of the more interesting potential targets is Epsilon Eridani; located 10.7 light years away or 3.28 pc it is the star 12th nearest star to our own Sun. Despite its great distance from us, Epsilon Eridani is likely to be a system of intense scientific interest in the coming decades. This is because radial velocity observations have detected the existence of a giant gas planet Epsilon Eridani b at around 1.5 Jupiter masses with a period of around 7 years [5]. It is in an eccentric orbit with a semi-major axis of around 3.4 AU from the parent star and represents the nearest exoplanet discovery found to date. Another much smaller planet at around 30 Earth masses and with a period of 280 years may also be present in the system with a semi-major axis or around 40 AU, approximately the distance of Pluto from our Sun.

Because the star is so young, around 600–800 million years old, it is likely to be the site of early Solar System formation, and this is evidenced by the detection of dust rings or debris belts mainly composed of asteroids in the system with a structure that suggests collisional evolution of the system, similar to how our own Kuiper Belt was formed [6, 7]. The belts were detected from infrared emissions using NASA's orbiting telescope IRAS. There are two asteroid belts located at 3 AU and at 20 AU; the third dust disk is much further out in the system, between 0.35 AU and 100 AU. The structure of these belts may even imply the presence of other planets undergoing early formation, perhaps in the habitable zone, which is within 0.48–0.93 AU from the star. The star is of spectral type K2 and located in the constellation of Eridanus. Due to its young age the stellar winds from the star are expected to be tens of times higher than for our own Sun. The star has a mass of 0.85 solar masses and a radius of 0.72 of the Sun. The metallicity is quite low.

There is no doubt that scientists have many reasons for why we would want to send a space probe to this system. There is so much happening and by definition so much to find, although the age of the star and early forming Solar System probably makes this a weak candidate for the discovery of any life. Indeed, a radio wave search (and later a microwave search) conducted in the 1960s for this system (and Tau Ceti) detected no intelligent signals, but this shouldn't be a surprise. In terms of an interstellar probe and mission planning, the problem will always be the amount of propellant required and the vehicle performance in order to reach such a great distance within a reasonable amount of time. Epsilon Eridani is drifting towards the star Luyten 726-8, and they will come within less than a light year's distance of each other in around 31,000 years, perturbing any outer Oort Cloud-like objects.

In recent years the field of astronomy has seen the detection of several hundred planets around other stars, the number now at over 500. Figure 7.2 shows the masses of the exosolar planets discovered as of December 2009. For the objects plotted where the mass (lower bound) is known there are 394 plotted out of a total of 432 identified at the time this data was collated. Of this total 59 (14.9%) have a mass <100 Earth masses, 38 (9.6%) have a mass <30 Earth masses, 17 (4.3%) have a mass <10 Earth masses and 9 (2.3%) have a mass <6 Earth masses. If we assume that there are 100 billion stars in the galaxy, and we further assume (conservatively) that each one has at least one planet then there are around 100 billion planets (but probably a lot more). If the 2.3% estimated above represented an accurate sample (in all probability it does not, but this is due to the limitations on current technology and observational techniques) this would equate to 2.3 billion planets below 6 Earth masses.

Most of the planets discovered so far by astronomers are gas giants, as these are easier to detect using the radial velocity method. Of the M class stars surveyed around 1% are believed to possess a giant planet. This number increases to 7% for stars of a spectral class F, G and K. One of the main observation platforms making these discoveries using the planetary transit technique is the French mission called CoRoT, which was launched in 2007. The transit technique looks for changes in the brightness of a star due to a planet moving across the line of sight to the observer. It even discovered the first so called 'superEarth' planet. Another platform using the planetary transit technique is Kepler, which uses a camera to cover 105° of the sky in the constellation Cygnus so as to examine 100,000 stars at distances of between 150 and 2,500 light years. The main advantage of instruments such as CoRoT and Kepler is they are space-based platforms not limited to optical aberrations due to Earth's atmosphere, a problem that has limited Earth-based telescopes for centuries. Other more exciting projects are on the discussion table, including NASA's Terrestrial Planet Finder mission and the ESA's Darwin mission. We live in exciting times, where the full power of technology is being utilized for astronomy. The detection of habitable worlds and possible life-bearing worlds is likely to be made within decades if not sooner, and then we will know for sure to what locations we should send our first space probes.

When considering what target stars to send a spacecraft probe to, one must consider several critical factors, such as the mission duration and the potential

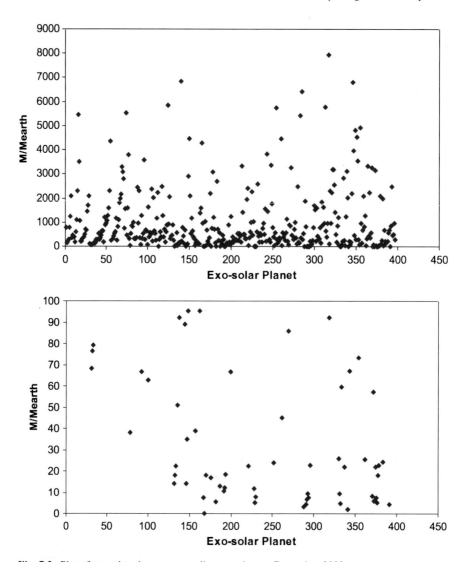

Fig. 7.2 Plot of exosolar planet masses discovered up to December 2009

engine performance for a stated mission (e.g., maximum cruise velocity possible). One can examine this question by simply considering the linear distance in light years to a set destination at different velocities in appropriate units, which is given by the product of the velocity (fraction light speed) and the time (years). Table 7.3 shows the results of some mission speed-distance optimization possibilities. One simply reads across the mission duration and fractional light speed to arrive at a contour, which is then traced to derive the mission range. This can be compared to Table 7.2 shown earlier.

Table 7.3 Typical linear mission range to different star systems as a function of maximum cruise velocity and mission duration

Mission duration (years)	Max cruise velocity 10%c	Max cruise velocity 20%c	Max cruise velocity 30%c	Max cruise velocity 40%c	Max cruise velocity 50%c
50	5 ly, 2 systems	10 ly, 8 systems	15 ly	20 ly, 68 systems	25 ly
60	6 ly, 3 systems	12 ly, 19 systems	18 ly, 48 systems	24 ly	30 ly
70	7 ly, 3 systems	14 ly, 27 systems	21 ly, 74 systems	28 ly	35 ly
80	8 ly, 4 systems	16 ly, 40 systems	24 ly	32 ly	40 ly
90	9 ly, 7 systems	18 ly, 48 systems	27 ly	36 ly	45 ly
100	10 ly, 8 systems	20 ly, 68 systems	30 ly	40 ly	50 ly

7.3 The Discovery and Evolution of Life

The prospect for discovering planets around other stars is a very exciting one. The prospect of discovering planets with life is even more exciting. Should we ever encounter intelligent life on such worlds, this will inevitably change the course of our future and have enormous implications for some of the world's most cherished belief systems. So far astronomers have discovered over 500 planets orbiting other stars in our galaxy. These discoveries do seem to suggest that planets of some form accompany many Sun-like stars in our galaxy. Any Earth-like worlds are thought to inhabit what is known as the habitable zone, a region within a Solar System thermally compatible with life. The thermally life-friendly environment may be provided by radiative heating (sunlight) or tidal heating (due to planetary resonance). Several of the large Jupiter-like extrasolar planets have been found within this region around other stars. The chances that some of those planets or their moons existing within a tidally heated habitable zone (like Jupiter's Moon Europa) appear to be good. Typically, extrasolar planets are located in an orbit around the primary star at distances from 0.03 to 40 AU.

One of the first exosolar planets to be discovered was a companion to the star 51 Pegasi, a Sun-like star beyond our own Solar System. Another is a companion to the 1-million-year-old 6Q Lupi, where images show a faint companion 100 AU distant. The mass of this companion is uncertain due to the crudeness of the planetary evolution models. Another image captured was of the brown dwarf 2M 1207, which showed the existence of a faint redder companion 2M 1207b gravitationally bound to its host star. It is located at 55 AU from its host star and has a mass of around 5 times that of Jupiter. Extrasolar planets have also been suggested in the form of free-floating planets [8]. These are not in orbit around any star but perhaps have been ejected out into deep space. They are thought to have masses of between $0.1M_{Sun}$ and $1M_{Jupiter}$. Alternatively they may have originated in interstellar space and formed in a similar manner to stars with mass ranges between 1 and 13 $M_{Jupiter}$. The number of unbound planets could be as common as brown dwarf stars or even exceed the number of stars, but this is speculation. It is thought that if they are many in number, most will consist of low mass rock or icy planetary embryos ejected from the host star system. This is a field of active research. It is entirely possible that the first star systems our robotic probes will visit is a brown dwarf system, not visible to us today but being at a distance much closer than Alpha Centauri.

The detection of life on another world depends heavily on the accuracy of the scientific instruments making the measurements. In particular, a high-speed flyby probe has a small window of opportunity upon which to make any observations (perhaps hours to days), hence deceleration at the target destination is preferred. Detection techniques were demonstrated in the vicinity of Earth at the instigation of the American astronomer Carl Sagan helping to kick-start the field of astrobiology. When the Galileo spacecraft passed by Earth in December 1990 during its first flyby, it looked at the planet to see if it could detect signs of Earth. It detected strong absorption of light in the red part of the visible spectrum, and this was particularly strong over continents. It is believed that this is caused by absorption of chlorophyll in plants by photosynthesis. It also detected absorption bands of molecular oxygen, a result of the activity of life on Earth. It detected infrared absorption bands due to the presence of methane. Methane is produced by either volcanic eruptions or life. Finally, it identified modulated narrowband radio transmissions unlike any natural source, a signal perhaps characteristic of the existence of intelligence with technology. These four observations are now known collectively as the Sagan life criteria. Observations of distant worlds should focus on these sorts of characteristic, for evidence of any life-dwelling occupants.

On planet Earth there is a variety of different terrains including land, ocean, atmosphere, ice, deserts, forests, rocks and an abundance and variety of life forms. It is not known for sure how life first started, but it may have been deep down in the ocean vents living off the volcanic heat. With such a wonderful world, why would we ever want to leave it? Indeed, as far as we know today Earth remains the only world that harbors life in the entire universe. Not just intelligent life, but any form of life. It may be that our own Solar System is in fact teaming with different types of life forms, although the intelligent variety is less likely. Historically the definition of life was clear. Fundamentally it was based upon several important parameters, such as reproduction, growth, nutriment, respiration, excretion, senses, environmental stimuli and locomotion. Reproduction is generally the process by which a new organism is produced from a pair of parent organisms, although asexual reproduction also occurs. Growth refers to the increase in size and complexity of an organism during its development from an embryo to a mature state, which includes cell division, enlargement and differentiation. Nutriment is the ability of an organism to ingest nourishment, which for animals is food and water. Respiration refers to the reactions by which an organism releases the chemical energy from the food it has taken. Excretion is the elimination of waste products of the metabolism by the living organisms. In theory, any structure that metabolizes and self perpetuates can be considered alive, and these are then the important signs of life for biological detection on any planet. Any organisms will be dependent upon the ability to move from one place to another (locomotion) as well as any senses that allow it to respond to any changes in its environment such as heat, sound, light.

All of life as we know it is carbon-based, that is, the main elements (compounds) that make up the structure are based largely upon the element carbon, although it is quite possible that in the vast universe we may find alternative forms of life based upon other elements such as silicon. But carbon has the most versatile chemistry compared to other elements. Carbon atoms can form chemical bonds to create especially long and complex molecules. Among these are 'organized molecules' upon which living organisms are made. Also, the main constituents of organic molecules (carbon, hydrogen, nitrogen, oxygen, sulfur, phosphorus) are among the most abundant elements in the universe, leading us to believe that life elsewhere will also be based upon organic (mainly carbon) chemistry. Organic molecules are scattered throughout the galaxy and are found in giant molecular clouds of interstellar space.

Radio astronomers have detected many dozens of carbon-based chemicals from the observations of spectral emission lines of light. These include ethyl alcohol such as formaldehyde. Falling meteorites called carbonaceous chondrites are also often found to contain a variety of organic substances. So from the earliest days, Earth has been bombarded with organic compounds. Ultimately, what we need is the light spectrum of an extrasolar planet to pick out molecules that indicate the presence of chemicals such as water. There is a proposal (currently not funded) for a space infrared telescope called the Terrestrial Planet Finder. The hope would be to produce a spectrograph of an extrasolar planet, which clearly shows evidence for water vapors, ozone, and carbon dioxide. Although non-biological processes can create all of these molecules, the presence of life will change the relative amounts of each molecule in the planet's atmosphere. Thus, the infrared spectrum of such planets will make it possible to identify worlds on which life may have evolved.

In 2010 astronomers successfully measured the atmosphere of the exoplanet GJ 1214b, an object thought to be over six times more massive than Earth and nearly three times larger. Using the Very Large Telescope in Chile astronomers detected a thick featureless spectrum, indicating a thick hydrogen atmosphere blanketed by a cloud layer like either Venus or Titan. Alternatively it may be the result of an icy world sublimating due to the proximity to the star forming an atmosphere of steam. The point is that we can already detect the atmospheres of alien worlds.

When one examines the background noise level in the sky at various radio and microwave frequencies, there is a range of frequencies of between around 1–10 GHz that has very little cosmic noise. There is also the region between 1.42 and 1.66 GHz. This range is known as the 'water hole,' as it is thought to contain hydrogen and hydroxide, which combined together make up H_2O, or water. At even lower frequencies there is substantial emission from interstellar gas making it noisy. Similarly, at higher frequencies the radio waves tend to be absorbed by Earth's atmosphere also making lots of noise. Hence the range of frequencies that comprise the 'water hole' would appear to be ideal for picking up interstellar messages by radio waves. More specifically, the ideal transmission is expected to be at a wavelength of 21 cm, which corresponds to a frequency of 1,400 MHz. Time will show if this fundamental assumption is correct, but so far no definitive signals have been detected.

In 1952 two American scientists demonstrated that simple chemicals could combine to form pre-biological compounds under conditions thought to be similar to the primitive Earth's environment. This was known as the Miller-Urey experiment. In a closed container they prepared a mixture of hydrogen, ammonia, methane, and water vapor and exposed this mixture of gases to an electric arc (simulate lightening bolts) for a week. At the end of this period, the inside of the container had become coated with a reddish brown substance rich in amino acids and other compounds essential to life. The results of this experiment suggest that life could have originated as a result of chemical processes. It also suggests that because it forms from common compounds found in the universe, life may have originated on other planets, too.

This exciting prospect was bought home to scientists in 1996 with the discovery of the Mars rock known as Alan Hills 84001 that had fallen to Earth 13,000 years ago, when the rock had been ejected from the Martian surface during a collision with a large body. Tiny fossils of bacteria about five times smaller than the diameter of a human hair were found on the rock. When scientists cracked this rock open they discovered the fossils were made of calcium carbonate ($CaCO_3$), which is the same material that is found in seashells.

The evidence for the fossils being life was threefold. Firstly, lifelike shapes were found in the rocks. Secondly, the chemical composition was consistent with life. Thirdly, the existence of the tiny magnetite crystals was similar to crystals produced by some bacteria on Earth. When comparing the grains of magnetite crystals to those produced on Earth, it was initially concluded that they were identical. Approximately 25% of magnetite crystals have a shape identical to that produced by the terrestrial aquatic bacterium called MV-1. The crystals are classified as being truncated hexa-octahedral in shape. But the main discoverers claim that terrestrial bacteria haven't penetrated the Mars rock since its fall to Earth some 13,000 years ago. The grains could have formed by a blast of heat, perhaps when the rock was chipped off Mars. However, the fact that the crystals were so similar to those found on Earth raised the suspicions of some, and even today there is large disagreement among planetary scientists on the origin of the fossilized bacteria. Some believe that the grains could even have been formed by the process of blasting the rock off the surface of Mars under extreme heat. The best way to solve this debate is to go to Mars and find more such rocks, proving the question definitively.

When we examine planets around other stars we are compelled to ask if it is likely that they harbor any form of life. This may be largely determined from the properties of the atmosphere, using our own Earth biosphere as a baseline. If we are led to the belief that life does exist on such worlds, then we are also driven to the possibility of establishing communications, assuming they are intelligent. To address this, the International Astronomical Union has set up a special commission to concentrate upon the Search for Extraterrestrial Intelligence (SETI). At the general assembly in 1991 it even published a declaration giving instructions as to the procedure to be followed in the event of an alien contact. Even if we received a signal, it would take a finite time to reply, and then a finite time for the signal to reach them, and a further finite time to acknowledge our reply. So there is a time delay factor, due to the size of the universe and the vast distances between worlds. So even if we pick up a signal, it is unlikely that we can establish a regular dialogue with them.

One approach to the existence of intelligent life in the galaxy was proposed by Frank Drake in 1961 and is a way of estimating how many intelligent life forms there are. This takes the form of what is known today as the Drake equation:

$$N = R_* \times f_p \times n_e \times f_l \times f_i \times f_c \times L \tag{7.2}$$

where N = the number of technologically advanced civilizations in the galaxy whose messages we might be able to detect; R_* = the rate at which solar type stars form in the galaxy (~1/year based on observations and statistical studies); f_p = the fraction of stars that have planets (most stars are thought to have planets ~1); n_e = the number of planets per solar system that are Earth-like, e.g., suitable for life (we suppose one in ten stars have Earth-like worlds so ~0.1); f_l = the fraction of those Earth-like planets on which life actually arises (~1 if we suppose that life evolves purely as a function of chemistry and conditions); f_i = the fraction of those life forms that evolve into intelligent species (~0.5 if we suppose that around half will lead to intelligent life); f_c = the fraction of those species that develop adequate technology and then choose to send messages out into space (~0.5 if we suppose that around half then develop the technology to communicate); L = the lifetime of that technologically advanced civilization (~1,000 may be typical from history of Earth). Putting these numbers into the equation we find

$$N = \frac{1}{year} \times 1 \times 0.1 \times 1 \times 0.5 \times 0.5 \times 1000 \; years \approx 25 \tag{7.3}$$

So with these numbers we estimate there are ~25 technically advanced civilizations in our galaxy alone. If we include all the galaxies in the known universe (around 100 billion), then the number runs into the billions and many working in the field of SETI would consider this estimate of 25 to be ultra conservative. The original numbers assumed by Frank Drake in 1960 led to an estimate of around ten civilizations and others have computed numbers close to 10,000. The computation is really dependent upon the assumptions made and how well one knows each of the parameters, underscoring the value of observational astronomy in particular.

In 1950 the Italian physicist Enrico Fermi posted an important question. If there has been sufficient time for galactic colonization to occur, where is everybody? This has since been termed the Fermi Paradox, and many potential answers have been proposed. If there had ever been a single advanced civilization in the cosmological history of our galaxy, dedicated to expansion, it would have had plenty of time to colonize the entire galaxy via exponential growth. No convincing evidence of present or past alien visits to Earth are known to us, leading to the standard conclusion that no advanced expanding civilization has ever existed in the Milky Way.

To examine the Fermi Paradox further, members of the British Interplanetary Society set up Project Daedalus in the 1970s. This was a design study of an engineered interstellar vehicle and is discussed in Chap. 11. The motivation behind the project was to demonstrate if a credible design for interstellar flight could be produced. It feasibility could be demonstrated just at the outset of the Space Age, then this implied that in the coming centuries the design would be greatly refined and interstellar travel would become possible. This directly implied that interstellar travel must therefore be possible, and so the absence of other intelligent life in our galaxy must be explained by other more sophisticated reasons. One of the leaders of the Daedalus project was Alan Bond, who the design study later led to consider the evolution of biological life from simple single-celled organisms all the way through to complex life such as our own. Bond constructed a biological model and concluded that although life was probably frequent in the universe, complex life was probably very rare, something like one in every ten galaxies would have an intelligent civilization like our own [10].

Another explanation for the absence of intelligent life is known as the 'zoo hypotheses,' where other intelligent civilizations simply choose to avoid our Solar System and have no desire to interact with us. This could be due to a fear of our perceived war-like mentality or due to a fear that interaction will bias our cultural growth, as has happened several times with ancient cultures on Earth coming into contact with a perceived more advanced culture (e.g., the Spanish colonization of the Americas and the fall of the Aztec Empire). It is also possible that it is the nature of intelligent life to destroy itself or others that it meets. This could be a fundamental result of our clan-like mentality, which may be ubiquitous among civilizations.

Others have suggested that technological civilizations tend to reach a point of exponential advancement, also known as a technological singularity. Science fiction writers such as Vernor Vinge have discussed ideas along these lines for some time [9]. This technological growth leads to a marriage between biological and artificial intelligence that allows even more intelligent beings to be created. Such beings may then become disinterested with lower intelligence or advance to an existence that is beyond the technology of others to comprehend or detect with current instruments.

Whatever the answer to the Fermi Paradox, it is a thoroughly interesting topic with perhaps a combination of answers. We can only hope to address it by theoretical methods, advanced astronomical techniques, the exploration of space or by gaining a deeper meaning of the question 'What is life?'

7.4 Practice Exercises

7.1. Using the data from Table 7.2, Eq. (7.1) and other information given in this chapter, compute the luminosity of the following stars assuming these surface temperatures: 3,042 K (Proxima Centauri); 5,790 K (Centauri A); 5,260 K (Centauri B); 3,134 K (Barnard's star); 5,300 K (Tau Ceti); 5,084 K

(Epsilon Eridani). What challenges do these different solar irradiance environments present to a spacecraft passing close to the star trying to make observations of it?

7.2. Conduct some research on the next generation of space telescopes coming online or proposed within the next 20 years. What is the projected detail (resolution) of a nearby star system that these telescopes will be able to provide? How does this compare to the level of detail and amount of information obtained by a flyby (non-decelerated) probe that travels to these star systems and passes through at a cruise velocity of 10% of light speed? Will the projected advances in astronomical techniques make the need for an interstellar probe obsolete? What in situ information could a visiting flyby probe obtain that an advanced astronomical telescope could not?

7.3. Examine each term in the Drake equation. Do some basic research into the background of each term. On the basis of this research perform an optimistic and pessimistic calculation to determine the likely upper and lower bounds for planets in the galaxy that may harbor life. What key advances need to be made in astronomy and other fields in order to tighten up the estimates of each term in the Drake equation?

References

1. Phillips, AC (1999) The Physics of Stars, 2nd Edition, Wiley.
2. Bond, A et al., (1978) Project Daedalus – The Final Report on the BIS Starship Study, JBIS.
3. Gatewood, GD (1995) A Study of the Astrometric Motion of Barnard's Star, Astrophys. Space Sci, 223, 1–2, 91–98.
4. Greaves, JS, et al,. (2004) The Debris Disc around Tau Ceti: a Massive Analogue to the Kuiper Belt, MNRAS, 351, L54–L58.
5. Hatzes, AP et al., (2000) Evidence for a Long Period Planet Orbiting Epsilon Eridani, Astrophys.J.,L, 544, 145–L148.
6. Quillen, AC & SD.Thorndike (2002) Structure in the Epsilon Eridani Dusty Disk Caused by Mean Motion Resonances with a 0.3 Eccentricity Planet at Periastron, Astrophys. J., L, 597, 149–L152.
7. Backman, D (2009) Epsilon Eridani's Planetary Debris Disk: Structure and Dynamics Based on Spitzer and CSO Observations, Astrophys. J., 690, 1522–1538.
8. Fogg, M (2002) Free-floating Planets: Their Origin and Distribution, Master's Thesis, University of London.
9. Vinge, V (1993) The Coming Technological Singularity, Vision 21 Interdisciplinary Science & Engineering in the Era of Cyber Space, Proceedings of a Symposium held at NASA Lewis Research Center, CP-10129.
10. Bond, A (1982) On the Improbability of Intelligent Extraterrestrials, JBIS, 35, 5, 195–207.

Chapter 8
Solar System Explorers: Historical Spacecraft

Interstellar flight demands a gut-check for our entire civilization,
a journey so preposterously long and difficult that it dwarfs all
prior engineering. We'll do it – in decades or, more likely, in
centuries – because we have no choice. We're a driven species
and every bit of our recorded history says our most insatiable
compulsion is to explore.

Paul Gilster

8.1   Introduction

In this chapter we discuss some actual spacecraft designs that have been launched into the outer parts of the Solar System. These spacecraft represent the first ambassadors from Earth and move outwards with ever-greater speed and to an ever-greater distance. Although these missions are only within the Solar System or just beyond it, they are technology demonstrators for eventual interstellar missions. This is what in the industry is termed 'interstellar precursor' missions. Some of these visit the outer planets but others go beyond to the edge of our Solar System, through the interstellar heliopause and into the Kuiper Belt. In the history of space exploration there have been many successful missions, and it is not possible to cover all of them, an entire book in itself. Instead, we briefly discuss only a select few to give the reader an overview of the achievements of these explorers.

8.2   Precursor Mission Probes

Before we begin to explore some of the design concepts for reaching the nearest stars, it is worth reviewing some of the historical space exploration missions that have traveled across the enormous distances through space to far off destinations.

K.F. Long, *Deep Space Propulsion: A Roadmap to Interstellar Flight*,
DOI 10.1007/978-1-4614-0607-5_8, © Springer Science+Business Media, LLC 2012

Table 8.1 Masses (kg) of science instruments on typical spacecraft designs [1]

Spacecraft → Instrument ↓	IIE	Pioneer	Voyager	New horizons	Ulysses	IBEX
Vector helium magnetometer	8.8	2.7	5.6		2.3	
Fluxgate magnetometer		0.3			2.4	
Plasma wave sensor	10.0		9.1		7.4	
Plasma	2.00	5.5	9.9	3.3	6.7	
Plasma composition					5.6	
Energetic particle spectrometer	1.50	3.3	7.5	1.5	5.8	
Cosmic-ray spectrometer: anomalous and galactic cosmic rays	3.50	3.2	7.5		14.6	
Cosmic-ray spectrometer: electrons/positrons, protons, helium	2.30	1.7				
Geiger tube telescope		1.6				
Meteoroid detector		3.2				
Cosmic dust detector	1.75	1.6		1.6	3.8	
Solar x-rays and gamma-ray bursts					2.0	
Neutral atom detector	2.50					12.1
Energetic neutral atom detector	2.50				4.3	7.7
Lyman-alpha detector/UV measurements	0.30	0.7	4.5	4.4		
Infrared measurements		2.0	19.5			
Imaging photopolarimeter		4.3	2.6	8.6		
Imaging system			38.2	10.5		
Common electronics, harness, boom, etc.						5.4
Totals	35.2	30.1	104.4	29.9	54.9	25.2

The probes that represent these missions are the results of teams of engineers, physicists and mission planners often widely distributed throughout the world, all working towards the common success of the robotic ambassador reaching its destination and achieving its scientific goals. These probes have helped us to gain a better understanding for our place in the universe by the value of the science data that they transmit back to us. The construction of these probes and the contribution of the information they provide to the progression and technological development of our civilization cannot be emphasized enough.

The instruments they carry vary remarkably in the physical properties they intend to measure or even the technique by which those measurements are achieved. Table 8.1 shows the various instruments used on several spacecraft missions. Often a spacecraft will have several of the same instrument or two different spacecraft may measure a similar property; allowing for duplication in the measurement ensures confidence in its value or parameter range. For typical spacecraft probes the total instrument mass will vary between around 20 and 100 kg. A team of specialists has carefully designed each instrument. Careers will literally be built upon the success of the instrument's data return, feeding into technical publications for the academic literature and advancing our overall knowledge of the properties of space.

The total mass of an unmanned spacecraft can be divided into three classes known as Reconnaissance, Exploration and Laboratory class [2]. A Reconnaissance class probe such as the Voyager spacecraft would have a mass of between 0.2 and 0.8 tons and be intended for a flyby encounter with limited science return. An Exploration class probe such as Galileo would have a mass of between 0.75 and 1.5 tons and would usually include an atmospheric re-entry probe with the main orbiter. Laboratory class probes such as the Mars Viking spacecraft would have a mass typically greater than 1.5 tons and include and orbiter and robotic lander. An interstellar probe such as the fusion powered Project Daedalus discussed in Chap. 11 would ideally encompass all three of these classes.

When the former administrator Daniel Goldin was in charge of NASA he initiated the Discovery program, which is a series of low cost but highly focused missions, modeled on the motto "faster, better, cheaper." Under this program, NASA undertook the NEAR Shoemaker mission to study the asteroid 433 Eros, the Mars Pathfinder mission to deploy a rover, the Deep Impact mission that sent an impactor into a comet, the Lunar Prospector mission to study the mineralogy of the Moon and the Stardust mission to collect samples from the comet 81P/Wild. The Discovery program was highly successful. After the success of this program, NASA also formed the New Frontiers program that entailed a series of medium cost (<$700 million) highly focused missions to explore the Solar System. Some of the proposals launched or under consideration includes the New Horizons mission to Pluto, the Juno mission to Jupiter, as well as missions to study the Moon, the planet Venus and a comet. There is also the Flagship program, which is for much more expensive missions ($2–3 billion).

In consideration of highly ambitious missions for interstellar travel, we first must be aware of the missions that have ventured to the furthest reaches of our Solar System and to consider what separates such missions from the more ambitious ones. For the sake of comparison we shall categorize the different spacecraft mission types as:

- Interplanetary probe
- Near distance precursor probe
- Far distance precursor probe.
- Long range precursor probe
- Interstellar probe

An interplanetary probe is a spacecraft that is designed to visit any of the planets or other objects within our Solar System – from the planet Mercury to the dwarf planet Pluto and any moons, asteroids or comets that happen to be located in between. An interstellar precursor probe is a spacecraft that pushes the technological boundaries of our capability to the furthest distances that can be reached at the highest speeds. Such a probe may go to the edge of our Solar System and the Kuiper Belt (out to 150 AU) or even as far as the Oort Cloud (>10,000 AU). An interstellar probe is a spacecraft that is design specifically to approach near the orbit of another star system, such as to Alpha Centauri.

Table 8.2 describes the typical characteristics of such missions from an optimistic point of view. Duration is the time period for which the mission is expected to last, from launch to end of mission as determined by successful completion of the mission goals. The specific impulse and exhaust velocity have already been discussed in Chap. 5, and the numbers used are those required to complete the mission within the specified duration. Risk in the context used here refers to the probability of a negative performance of a specific system or sub-system such as by failing to function within the allocated design margins and impacting the success of the overall mission. Risk is often quantified as the probability of a specific event occurring multiplied by the impact of that event; thus it is possible to usefully quantify and minimize it. In general, any spacecraft mission should be designed to minimize risk, and we characterize the current level of risk management used in modern day spacecraft as the minimal baseline for this exercise.

Although a mission may be expensive and with a lot of hard effort gone into making it a success, if it is lost within the Solar System, there is always the possibility of building another one and to see it through to successful completion within the lifetime of the designer. However, for longer duration-longer distance missions with more ambitious goals risk must be managed more tightly. It would be a sad result indeed if after traveling for a century a probe was to have a catastrophic system's failure that in hindsight could have been prevented. A concept closely related to risk is system redundancy. This is the need to ensure completion of overall system functionality, even if a part of that system fails. An example could be in a specific science instrument such as an optical imager, where two to three may be included in case one fails.

To increase the probability of achieving the mission goals, a high degree of redundancy is desirable and may be expected for longer distance-longer duration missions. The only problem, however, is that the doubling or even tripling of any instruments will increase the mass of the spacecraft payload. So a clever choice of redundancy is required for critical systems but not necessarily for all systems. Alternatively, one can have a design with a very high degree of redundancy but with fewer instruments so that a trade-off with the overall payload mass is achieved. Overall, however, for an interstellar mission, one can expect a requirement for long duration functionality, environmental longevity/survivability, and extreme risk management with maximum engineering system redundancy. For any mission launched towards the nearest stars, engineering failure modes and external environment conditions are the biggest threat to mission success and must be mitigated using 'clever' technology and sensible design decisions in the mission planning. Such technology would be particularly focused on the need for self-repair and possibly even self-replication, to a degree much larger than any conventional space mission previously launched.

Table 8.2 Description of mission types for spacecraft probes and the projected capability of our civilization in the decades hence

	Interplanetary	Near distance precursor	Far distance precursor	Long range precursor	Interstellar
Distance (AU)	<100	100–1,000	1,000–10,000	10,000–100,000	>100,000
Target	Planets, moons, short period comets, Kuiper Belt	Heliopause, ISM, long period comets	Oort Cloud, ISM	Oort Cloud, ISM	Other Solar Systems
Mission duration (years)	<10	<20–50	50–500	500–1,000	>1,000
I_{sp} (s)	<1,000	1,000–3,000	3,000–10,000	10,000–100,000	100,000 – 1,000,000
V_e (km/s)	<100	100–500	500–1,000	1,000–5,000	5,000–10,000
Risk	Low	Medium	High	very high	Maximum
Redundancy	Minimal	Medium	High	very high	Maximum
Technological maturity (years from present)	Current	Current	30–50	50–100	>100

All duration estimates are based upon near-linear extrapolation of current technology such as the Voyager spacecraft, except for the mission duration, specific impulse and exhaust velocity, which are engine dependent and assumes more advanced propulsion is utilized for more difficult missions

8.3   Pioneer Probes

Pioneer 10 was launched in March 1972 and the NASA Pioneer program (which included Pioneer 6–11) is one of the most successful in space history. An artist's illustration of a Pioneer probe is shown in Plate 5. In particular the Pioneer 10 and 11 probes are the first robotic explorers to visit the outer planets and to travel beyond the orbit of the dwarf planet Pluto. Pioneer 10 is effectively the first interstellar spacecraft because it was the first to leave the Solar System. It passed Neptune in June 1983 and eventually left the Solar System 11 years later. As of 2009 it was at a distance of around 100 AU from the Sun. Currently traveling at a speed of around 2.6 AU/year it will reach the nearest stars in around 2 million years. The main purpose of the Pioneer 10 mission was to study the interplanetary and magnetic fields as well as the solar wind interaction with the solar heliosphere. Because the spacecraft was performing a flyby of Jupiter it also studied the Jovian atmosphere and satellite system.

The spacecraft had a mass of 258 kg of which around 29 kg comprised the science instruments. As with the Voyager spacecraft, the Pioneer probe was powered by a Radioisotope Thermoelectric Generator (RTG), providing around 155 W of power during the launch and 140 W by the time of the Jupiter flyby encounter. The spacecraft carried 11 instruments, including magnetometers, charged particle detectors, photometers, Geiger counters, plasma analyzers and optical imaging cameras. The spacecraft also had both a medium and high gain antenna. Six hydrazine thrusters were included to provide velocity, attitude and spin-rate control. The Pioneers 10 and 11 spacecraft also carried a message to any would-be space travelers that might encounter the probes. This was in the form of a 120 g, 349 cm^2 pictorial plaque constructed from gold and anodized aluminum and included the naked picture of a male and female human silhouetted by the spacecraft and other symbols depicting our knowledge of the atom and our position within the galaxy relative to 14 distant pulsars. Contact with the Pioneer 10 probe was lost in 2003 due to power reduction and long-range communications reliability.

Pioneer 11 was nearly identical to the Pioneer 10 spacecraft except for an additional science instrument known as a flux gate magnetometer, which measures the strength and direction of magnetic fields in the vicinity of the instrument, particularly around a planet's magnetosphere. It was launched in April 1973 and left the Solar System 17 years later in February 1990. Unfortunately communications were lost in November 1995, 22 years after launch. Pioneer 11 also used a Jupiter gravity assist to pick up velocity and obtained dramatic images of the Great Red Spot as well as accurately determining the mass of the Jovian moon Callisto. Pioneer 11 had similar mission objectives to the Pioneer 10 probe, although the mission also included a close flyby of the planet Saturn, and it obtained spectacular images of the ring system, discovering a new ring as well as a new moon.

One cannot mention the Pioneer probes without mentioning the Pioneer anomaly. This is the result of radio tracking Doppler shift observations of the Pioneer 10 and 11 spacecraft that consistently indicate a constant acceleration of around 8.74×10^{-10} ms^{-2} in the direction of the Sun being applied to the spacecraft.

It is possible that this is a result of fuel leakage or some anomalous drag. Alternatively it may derive from a more exotic origin and may indicate that our understanding of gravity, particularly at high velocity, requires modification. One person has speculated that the acceleration can be attributed to a modification of the spacecraft's mass due to an interaction with the zero-point field, or so called Unruh radiation [3]. Others have speculated that the acceleration can be attributed to the local expansion of the universe by equating it to the Hubble expansion, where the acceleration is given by –cH, where c is the speed of light in a vacuum and H is the Hubble constant [4]. If the Pioneer anomaly turns out to be a real reflection of some unknown physics occurring it will potentially revolutionize our understanding of the universe and be yet one more reason why missions such as the Pioneer probes should be launched every few years where possible. Both the Pioneer 10 and 11 probes will take around 2 million years to reach the nearest stars along their direction of travel at their current speeds.

8.4   Voyager Probes

Launched in September 1977 Voyager 1 (just 2 weeks after Voyager 2) remains the most distant manmade object ever sent into space. An artist illustration of the Voyager probes is shown in Plate 6. Currently traveling at a speed of 17.1 km/s or 3.6 AU/year it is also one of the fastest manmade objects. Its primary mission was to reach and explore the Jupiter and Saturn systems. Voyager 1 got very close to Jupiter at a distance of 349,000 km from the center of the planet that it reached in March 1972, two years after launch. It also had an extended mission to locate and study the outer boundaries of the Solar System and enter the Kuiper Belt. Despite being launched in the 1970s, none of the current space probes (even New Horizons) will overtake Voyager 1, due to the benefit of several gravity assists from the outer gas giants. Both the Voyager probes were 722 kg in mass. They had a 3.7 m diameter high gain antenna and 16 hydrazine thrusters, all run from an RTG supplying 420 W. The spacecraft were covered with thermal blankets to protect them. Voyager 1 had many instruments on board, including an ultraviolet spectrometer, infrared spectrometer and radiometer, photopolarimeter, cosmic ray detector and magnetometers.

In December 2004 it finally crossed the termination shock of our Solar System, where the heliosheath meets the interstellar medium and the solar wind compresses up against interstellar space. As of 2005 Voyager 1 was in the heliosheath and would have reached the heliopause by 2015, by which time it would technically become the first manmade object to have left the Solar System. As of November 2008 it was at a distance of 108 AU from the Sun with radio signals taking nearly 15 hours to reach Earth. Although Voyager 1 has not been aimed at any particular star system, it is moving in the general direction of a star in the Ophiuchus constellation, which it will reach in tens of thousands of years.

Voyager 1 made an interesting observation just prior to crossing the termination shock, when the intensity of the low energy particles being observed suddenly increased. This was believed to be due to the effect of crossing the magnetic field lines where charged particles were beamed along the field directions. The spacecraft also measured increased levels of turbulence, adding to the evidence that Voyager 1 had indeed passed through the termination shock, where the solar wind bulk velocity was also believed to have dropped.

Voyager 2 was launched in August 1977 and is moving out of the Solar System at a speed of 3.28 AU/year or 15.6 km/s. It completed flybys of all the main gas giant planets Jupiter, Saturn, Uranus and Neptune. It completed its closest approach to Jupiter in July 1970 coming within 570,000 km of the planet's outer atmosphere. It made many discoveries, including active volcanism on the Moon Io and the existence of Jupiter's Great Red Spot as being a storm. It reached Saturn in August 1981, Uranus in 1986. As of September 2008 Voyager 2 was at around 87 AU from the Sun. For both Voyager 1 and Voyager 2 the nuclear power was expected to run out 30 years from launch, which was around 2007, due to the atomic decay of the radioactive elements. The probes should reach the heliopause in 2014 and in theory the Oort Cloud in 24,000 years.

None of these Voyager probes exceeds around 17 km/s. At this speed any probe would take 18,000 years to reach 1 light year and 76,000 years to reach 4.3 light years (i.e., Alpha Centauri). Although Voyager 2 is not aimed at any specific star system, it is heading in the direction of the star Sirius around 8.5 light years away and will reach its vicinity in about 150,000 years. It is also of fundamental importance to the subject matter of this book to realize that both of the Voyager probes should remain in radio contact until about 2025, when their electrical power is too weak to transmit. By this time, the probes would have been on mission durations of around 48 years.

Both the NASA Voyager probes also contain a golden record, which is a recording of sounds and images of life on Earth, for any extraterrestrial that may find it. This was a suggestion by the astronomer Carl Sagan. A message from the former U. S. President Jimmy Carter reads:

> This is a present from a small, distant world, a token of our sounds, our science, our images, our music, our thoughts and our feelings. We are attempting to survive our time so we may live into yours.

The Voyager space probes are likely to become important pieces of history hundreds of years from now. It would be a sad state of affairs if when we next encounter them with future space missions; we still have no idea as to whether we are alone in the universe. The recording will be an echo from a time at the beginning of the Space Age, a time when humankind wasn't very far from self-destruction.

8.5   Galileo

Named after the Italian Renaissance man, Galileo was launched in October 1989 and was headed for the Jupiter system. This was largely a NASA project, although Germany supplied the engine. To arrive at the Jupiter system, Galileo conducted two gravity assists around the orbits of Venus and Earth. The spacecraft had a total mass of around 2,380 kg. The propulsion on Galileo was a 400 N thrust engine along with twelve 10 N thrusters using 925 kg of monomethyl hydrazine and nitrogen tetroxide, although these were used for path corrections as the velocity to reach Jupiter was imparted from the initial orbital insertion launch and enhanced by the gravity assists. Galileo used two RTGs supplying 570 W of power, each of which was mounted on the end of a 5-m boom. The spacecraft carried many instruments on board, including a plasma wave antenna, magnetometer sensors, energetic particle detectors, ultraviolet spectrometer, a star scanner, and heavy ion counter and dust detector. All of the instruments combined made up 118 kg of mass. The magnetometer was mounted on the end of an 11-m boom to avoid spacecraft interference.

During the mission the high gain antenna on the Galileo spacecraft failed to fully deploy during its first flyby of Earth, possibly due to storage issues. Several attempts were made to initiate the deployment, such as maximum spinning of the spacecraft, thermal cycling the antenna, turning off the deployment motors continuously – all to no avail. If such a thing happened to an interstellar probe this would be a disaster, and underlines the need for system redundancy and ideally a truly autonomous repair system such as the warden probes proposed in the Project Daedalus study (discussed in Chap. 11). The low gain antenna was protected by the presence of a Sun shield. The spacecraft also carried a small 339 kg 1.3 m-wide probe to be launched into the Jovian atmosphere at a speed of 47.8 km/s, halting transmission after about an hour when it reached 150 km into the atmosphere where the pressure was around 23 atmospheres.

The Galileo mission was highly successful, making the first observations of ammonia clouds within the atmosphere, confirming volcanic activity on the moon Io and further suggesting the presence of a liquid ocean under the moon Europa. It even found that the moon Ganymede had its own magnetic field – no other satellite in the Solar System has a magnetic field. Remarkably, the mission also discovered the presence of a thin ring system around Jupiter, made up mainly of dust. Some of the most remarkable achievements of Galileo, which were unplanned, were its observations of Comet Shoemaker-Levy 9, which crashed into the atmosphere of Jupiter, providing real close up images of this catastrophic event. Designers were concerned that when the probe reached the end of its life, it would de-orbit onto the surface of the moon Europa, which is thought to harbor a liquid ocean and may contain life. If Galileo entered this ocean any Earth-based bacteria present on the spacecraft could contaminate any life forms already present on the moon. To prevent this scenario, Galileo was plunged into the Jupiter atmosphere at a speed of 50 km/s, where it was eventually crushed by the massive atmospheric pressures. What would Galileo have thought had he known that a spacecraft bearing his name would enter the atmosphere of that world that he first observed through a telescope?

8.6   Ulysses

This was a joint initiative between NASA and ESA. Its original launch was delayed by the space shuttle Challenger accident, but it was eventually launched in October 1990 aboard the space shuttle Discovery. The primary mission was to study the Sun at all latitudes, and this was achieved by using an encounter with Jupiter so as to change the plane of the spacecraft to an inclination of over 80°. This is known as an out of the ecliptic plane mission. Previously, all observations of the Sun had been performed from low latitudes and so Ulysses was to characterize the heliosphere as a function of solar latitude, particularly above 70° at both the Sun's south and north poles. Among its many discoveries includes the fact that the outward solar magnetic flux in the solar wind did not vary greatly with latitude, demonstrating the importance of magnetic pressure forces near the Sun's surface.

Because the spacecraft went out to Jupiter it used an RTG for power generation. The spacecraft was also covered with blanketing material and electrical heating systems to help to maintain the temperatures from the relatively warm inner Solar System to the cold temperatures among the gas giants. The spacecraft mass at launch was 370 kg including 33 kg of hydrazine monopropellant for the eight thrusters to use in attitude control course corrections around Jupiter. The spacecraft had a 1.6-m high gain antenna dish. This could be used to search for gravitational waves by the Doppler shift effect. It had several instruments on board including an experiment boom that contained an x-ray instrument, a gamma ray burst experiment, a vector helium magnetometer and a fluxgate magnetometer. The boom also included a two-axis magnetic search coil antenna for measuring magnetic fields. The spacecraft was used to assist in the triangulation of gamma ray burst measurements from telescopes back on Earth. During its flight the spacecraft also passed through the tail of Comet Hyakutake and Comet McNaught. The mission ended in June 2009 when it was no longer possible to keep the attitude control fuel warm and thereby in a liquid state. Ulysses is another good example of international cooperation in space exploration with the spacecraft being built in Germany and the instruments provided by the United States and other European countries.

8.7   Cassini-Huygens

Launched in October 1997 by a Titan IVB booster rocket from Cape Canaveral in Florida, the Cassini orbiter spacecraft was sent on its long journey to Saturn and its surrounding satellites. After well over a decade the spacecraft is still operating and uncovering the secrets of Saturn and its ring system, one of the primary mission objectives. The spacecraft has a mass of 2,500 kg of which 350 kg is the Huygens probe. When the propellant is added in the total launch mass is around 5,700 kg, making it one of the heaviest spacecraft ever launched into space. The orbiter is quite large with a total height of nearly 7 m. The spacecraft carries an array of 12 instruments designed for no less than 27 diverse science investigations using technology such as spectrometers, cosmic dust analyzers, magnetometers as well as radar and imaging technology. It also carries three RTGs to convert radiatively generated heat into electrical power, similar to other spacecraft such as Galileo, Ulysses and New Horizons also discussed in this chapter. Communications of its discoveries back to Earth are enabled by the use of a high-gain antenna and two low-gain antennas. The spacecraft uses a cluster of 16 small rocket thrusters powered by hydrazine propellant to control the spacecraft's orientation, although the main engine provides the thrust for trajectory changes using monomethyl-hydrazine fuel combined with nitrogen tetroxide as the oxidizer.

The most interesting element of this mission was the inclusion of the 1.3-m Huygens probe, which reached the moon Titan in January 2005 nearly seven years after the mission launch. It descended into the nitrogen-methane covered atmosphere for over 2 hours and landed on its surface. To reach the surface safely it first performed atmospheric re-entry braking using its 2.7-m diameter heat shield followed by the deployment of a parachute. The probe successfully transmitted its data back to the main orbiter. Since then the main orbiter has detected lakes of liquid hydrocarbon in the form of methane and ethane in the northern parts of Titan; an exciting discovery. As well as the main mission objectives, the spacecraft also discovered several new moons around Saturn and performed flybys of several other objects such as the planet Venus plus Jupiter and its moons. During its outward trajectory it also performed a unique radio wave test of Einstein's theory of gravity, general relativity. Again, Cassini-Huygens represents a real success in international cooperation in space mission design because NASA designed the main spacecraft whereas the Huygens probe was designed by the ESA. Plate 7 shows an image of the Cassini-Huygens probe during pre-launch assembly.

8.8   Deep Space 1

This unique and advanced spacecraft was launched from Cape Canaveral in Florida on board a Delta II rocket and was produced from conception to launch in just over three years. It was designed with no less than 12 high-risk revolutionary technologies and was part of the NASA New Millennium Program. Launched in October 1998 the spacecraft went on to perform a flyby of an asteroid and later on a comet as the mission was extended. The main propulsion for Deep Space 1 was the NASA Solar Electric Propulsion Technology Application Readiness (NSTAR) developed by the then Hughes Electron Dynamics. It is an electrostatic ion engine using xenon fuel and it was the primary mission objective to demonstrate long duration use of this propulsion technology. It made history in this regard. With continuous thrusting for 678 days at 0.092 Newton's its performance enabled a specific impulse of between 1,000 and 3,000 s, an order of magnitude improvement over chemical systems. The exhaust velocity from the engine was around 35 km/s. It also employed hydrazine-based attitude control thrusters although eventually the hydrazine was used up so the mission had to resort to using the main ion engine with the surplus fuel remaining. The spacecraft had a mass of around 374 kg of which only 74 kg constituted the xenon propellant for the ion engine.

The spacecraft used solar arrays to provide the required power, using linear Fresnel lenses to concentrate the sunlight. The arrays would provide 2.5 kW at 1 AU, 84% of which was used to power the ion engines. Combining this technology with solar cells and advanced microprocessors allowed an improvement over historical solar array technology with the system used on Deep Space 1, costing less and containing less mass than conventional solar arrays. The Autonav navigational tracking system was designed as an improvement over the already burdened Deep Space Network transmitters. It took images of the asteroids as it was passing at a sufficiently high speed, and from these photographs and the motion of the asteroids it was able to determine its position accurately. Deep Space 1 was also unique in that it used an artificial intelligence based control system called 'Remote Agent' to diagnose and fault correct component malfunction, simulated throughout the flight. Such technology is essential for an interstellar mission, too. Because many of the experimental systems on board were tested rigorously, this technology can be applied to other spacecraft with confidence in future missions. The mission officially ended in December 2001, and despite some technical challenges that the team had to overcome it was a very successful flight and an amazing achievement considering the experimental nature of much of the on board technology.

8.9 Cosmos I

Cosmos 1 was unique in that it was a project organized by the space advocacy group The Planetary Society on a shoestring budget of only $4 million. Although the mission did not proceed due to a failure of the rocket to reach the required deployment altitude, having been launched in June 2005 from a submarine using a Volna rocket based in the Barents Sea, it is still worth mentioning this creative spacecraft concept that would have been a key demonstrator mission for the concept of solar sails. The sail itself was comprised of eight separate 15-m length triangular sails, it had a total mass of 100 kg. The sail would unfurl once in orbit at an altitude of around 800 km and then structural tubes would inflate the sails so that photons from the Sun would provide the necessary pressure to propel the spacecraft to a higher orbit, potentially up to 900 km over one month. With a typical acceleration of around 0.0005 m/s^2 the sail would achieve a velocity relative to Earth of 45 m/s or if it remained in position for three months 4.5 km/s. Arguably, high speeds are possible enabling missions to the outer planets with this technology in less than a decade. An exciting and possible secondary mission objective for Cosmos-1 would have been the demonstration of microwave beam propulsion. This would be achieved by using the 70 m NASA Goldstone dish by beaming 450 kW microwaves from the ground to the sail.

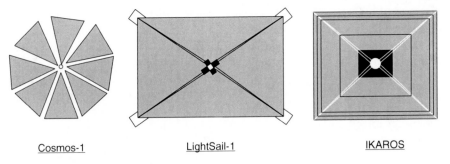

| Cosmos-1 | LightSail-1 | IKAROS |

Fig. 8.1 Various sail demonstrators (not to scale)

Initially, The Planetary Society planned to launch a second attempt known as Cosmos-2; however, changes in the availability of technology led to a new proposal known as LightSail-1, which was announced in 2009 and is expected to cost around $2 million. This is also a solar sail demonstrator mission, and the planned launch is sometime in 2011. Its main objective is to test the idea of using sails for space propulsion, and it uses a Mylar material for the sail to propel the spacecraft once it has been deployed into its 800 km orbit. It will use CubeSat technology as part of the mission architecture. Follow up missions known as LightSail-2 and LightSail-3 plan to demonstrate solar sail control and a mission to the Lagrange-1 point. This is an orbital position where the situation is such that a small spacecraft will be stationary relative to two larger objects such as a moon or planet. The L1 point lies on either the Sun-Earth line (ideal for solar observations) or the Moon-Earth line (ideal for accessing lunar or Earth orbit). The Japanese Space agency JAXA has also recently launched a 20-m diameter solar sail mission known as IKAROS for Interplanetary Kite-craft Accelerated by Radiation of the Sun, which became the world's first solar-powered sail employing photon propulsion. JAXA plans to launch a 50-m diameter sail. These achievements and the current LightSail attempts are stupendous for the field of solar sail research. These concepts are illustrated in Fig. 8.1 along with the IKAROS spacecraft discussed more in Chap. 10.

8.10 New Horizons

The latest robotic probe to be sent out into the Solar System is the New Horizons mission as part of the NASA New Frontiers program. It was launched in January 2006 using an Atlas V rocket from Cape Canaveral, Florida. It is a follow on to the originally planned (but canceled) mission called the Pluto Kuiper Express. The trajectory for the probe will take it past the dwarf planet Pluto and out into the Kuiper Belt to a distance of 55 AU. As the spacecraft is moving on a trajectory towards Pluto it is capturing images from afar of the dwarf planet, demonstrating a capability to track distance targets while also in motion, a critical requirement for an interstellar probe. When the spacecraft finally arrives at the Pluto system it will fly within 10,000 km of Pluto at a relative velocity of 13.8 km/s during the closest approach. The main science objectives for the mission include a flyby of Pluto and its three moons Charon, Nix and Hydra. It will arrive at the Pluto system in July 2015. The primary goal is to characterize the global geology of Pluto and its moons as well as map the composition of the surfaces. Another area of interest is the possible presence of an atmosphere on Pluto and the rate at which it escapes.

New Horizons reached Jupiter in February 2007 and Saturn in June 2008 and attained a velocity on passing of 21 km/s. As of March 2008 the probe is located 9.37 AU from the Sun and is traveling at 16.3 km/s or 3.4 AU/year, although it will eventually slow down to around 2.5 AU/year so will never catch up with either of the two Voyager probes. The probe has a mass of around 478 kg, and it uses an RTG system for power generation in the range 200–240 W. The propulsion system is comprised of 16 large and small hydrazine thrusters proving a capability for up to 0.29 km/s velocity increment from a thrust range of 0.9–4.4 N. These are used for trajectory changes and attitude control. The spacecraft uses a 2.1-m diameter high-gain antenna for communications as well as several medium- to low-gain antennas. There is a wide assortment of instruments on board the spacecraft including imagers, plasma and particle spectrometers, radio science instruments and a dust detector. New Horizons also carries some of the ashes of the American discoverer of Pluto, Clive Tombaugh, as well as over 400,000 names of people stored on board a disc and other artifacts such as an American flag.

Pluto is of course the subject of recent controversy, having been demoted from a planet to a dwarf planet. It is interesting that some of the science team leading the New Horizons mission have openly stated that they regard Pluto as a planet and not a dwarf planet, on the basis that other planets in the Solar System such as Jupiter and Neptune have not completely cleared their orbits either.

The discoveries that New Horizons will make of the Pluto system are likely to be very exciting indeed. Without doubt, New Horizons is the most ambitious space-craft mission yet launched to the outer part of the Solar System.

8.11   Dawn

The NASA Dawn spacecraft is a mission to the asteroids Vesta and Ceres that are located between Mars and Jupiter. It was launched in September 2007, although it won't arrive at Vesta until 2011 and Ceres until 2015. It was launched from Cape Canaveral in Florida using a Delta II rocket. The mission goal is to study these two objects as remnants of the Solar System and characterize the conditions and processes in the early formation, with the contrasting structures of icy Ceres and rocky Vesta. Because Ceres has now been categorized (since 2006) as a dwarf planet, the Dawn spacecraft will become the first mission to visit such a planetary object.

The spacecraft has several instruments, including cameras, optical and infrared spectrometer, gamma ray and neutron spectrometers. It will determine the surface elemental composition of each object, including the abundance of elements such as oxygen, magnesium, silicon and iron. Radio tracking of the spacecraft will enable mission designers to determine the mass of the two objects, their gravity field, information about their rotation axis and moments of inertia. Power for the spacecraft is provided by a 10 kW photovoltaic solar array and it uses a high-gain antenna. The total spacecraft mass is around 1,210 kg. A unique element to the Dawn mission is that it will actually go into orbit around the first asteroid before powering out to the next one; it is not just a flyby mission. The mission follows an outward spiral that includes a flyby gravity assist at Mars coming within 550 km to the surface. The total mission will last around eight years.

The propulsion system for Dawn is three electrostatic ion thrusters, the same engine used for Deep Space 1. There are three because they fire one at a time for each phase of the mission. Each has a specific impulse of 2,100 s and a thrust of 0.009 N. The propellant is a total of 425 kg of xenon, of which 90% is allocated to reaching Vesta and Ceres, the rest being available for a possible mission extension. If Deep Space 1 is the first spacecraft to employ ion engines, then Dawn is the first spacecraft to use them in an exploratory mission. Any remaining propellant would be used for secondary operations, which could include visiting other asteroids.

8.12   Interstellar Boundary Explorer

The NASA Interstellar Boundary Explorer (IBEX) mission was launched in October 2008 by use of a Pegasus XL launch from Marshall Island to an altitude of 200 km. A solid rocket motor then burns to raise this apogee to around 37 times the radius of Earth. It is by no means comparable in distance to the Pioneer, Voyager or New Horizons missions. However, its mission and associated technology are worth mentioning in the context of robotic explorers. IBEX has a mission to study the interstellar boundary, the region between our Solar System and interstellar space and in particular how the solar wind interacts with the interstellar medium. This will allow an improved understanding of how the large atmosphere of our Sun interacts with the interstellar wind passing through the galaxy.

It was mentioned earlier how the Voyager 1 probe had passed through the termination shock, determined by increased levels of low energy particles. However, the data from Voyager 1 was for a single point in space and at a given time. IBEX will fill in the picture by providing information on the global nature of the heliosheath-termination shock interaction. It plans to make observations by using energetic neutral atom cameras to image the interaction between the heliosheath and the interstellar medium. The two cameras on board IBEX will measure energetic neutral atoms with energies in the range 10 eV up to 6 keV. The heliosheath-interstellar medium interactions are studied by sending the spacecraft on highly elliptical orbits up to 322,000 km from Earth (or 80% of the way to the Moon) that allows it to go beyond Earth's magnetosphere away from Earth's radiation environment. IBEX will build up an image of the termination shock and the surrounding interstellar space. Another area of investigation for IBEX will be to study the galactic cosmic ray particles emanating from beyond our Solar System. This is a critical issue in determining whether humans could ever undergo a journey between the stars.

The IBEX probe itself was a low cost, rapidly developed spacecraft. The spacecraft has a dry mass of 80 kg with an additional 107 kg fuel mass and 26 kg payload mass. At a size of 58 × 71 cm, IBEX is a small but neat spacecraft. It uses 66 W of power during normal operations with an additional 16 W being used by the payload. The on board solar arrays provide 116 W. For orbital maneuvering it has a hydrazine propulsion engine and can communicate using two hemispherical antennas.

8.13   Summary Discussion on Probe Design

It is worth conducting a brief examination of some historical spacecraft missions. None of the above spacecraft will be heading out to the distant stars. However, even without any power the Voyager probes should reach the heliopause in 2014 and the Oort Cloud in 24,000 years. None of these spacecraft will travel at a speed much exceeding 17 km/s. At this speed it would take 18,000 years to reach 1 light year and 76,000 years to reach 4.3 light years, the location of the Alpha Centauri system, some 271,932 AU distance. Similarly, if it took Pioneer 10 around 18 years to get to 50 AU, at that speed of around 2.8 AU/year it would take around 98,000 years to reach the Alpha Centauri system. And this is assuming that there are no obstacles in between to degrade the spacecraft trajectory (e.g., magnetic fields, particles, dust, asteroids, comets).

Table 8.3 shows a matrix of required linear mission velocities in units of AU/year as a function of trip duration and distance to destination. Current and near future maximum mission velocities are of order 10 AU/year or less. Any faster and this will either increase the distance traveled in a fixed time or reduce the trip time for a fixed distance. Now we have to remember that the nearest star is 272,000 AU away, so at a

Table 8.3 Linear determination of cruise velocity (AU/year) as a function of distance traveled (100–1,000 AU) and mission duration (10–100 years), ignoring the acceleration phase

| | Years | | | | | | | | | |
AU	10	20	30	40	50	60	70	80	90	100
50	5	2.5	1.7	1.3	1	0.8	0.7	0.6	0.6	0.5
100	10	5	3.3	2.5	2	1.7	1.4	1.3	1.1	1
200	20	10	6.7	5	4	3.3	2.9	2.5	2.2	2
300	30	15	10	7.5	6	5	4.3	3.8	3.3	3
400	40	20	13.3	10	8	6.7	5.7	5	4.4	4
500	50	25	16.7	12.5	10	8.3	7.1	6.3	5.6	5
600	60	30	20	15	12	10	8.6	7.5	6.7	6
700	70	35	23.3	17.5	14	11.7	10	8.8	7.8	7
800	80	40	26.7	20	16	13.3	11.4	10	8.9	8
900	90	45	30	22.5	18	15	12.9	11.3	10	9
1000	100	50	33.3	25	20	16.7	14.3	12.5	11.1	10

modest speed of 10 AU/year it would still take over 27,000 years to get there. Clearly, to allow significant advancements in spacecraft speed and therefore distance attained, different forms of propulsion technology will be required to those being used today or even being proposed for the next two to three decades. In the next chapters we will begin to review some of these proposals that have the engineering aim of delivering high performance for minimum mission duration.

Now that we have considered many of the spacecraft that have been launched to the outer parts of the Solar System, it is worthwhile discussing what this tells us about the future. We can address this by considering some pivotal milestones in both robotic and manned exploration of the Solar System and beyond and compare the respective history. Eight key milestones are defined:

1. Probe or human reaching the orbit of Earth.
2. Probe or human reaching the surface of the Moon.
3. Probe or human reaching the surface of a planet within the inner Solar System.
4. Probe or human reaching the surface of a planet or moon within the outer Solar System.
5. Probe or human reaching the edge of the Solar System defined to be around 100 AU.
6. Probe or human reaching the undisturbed interstellar medium defined to be around 200 AU.
7. Probe or human to outer boundary of gravitational solar lens, point defined to be 1,000 AU.
8. Probe or human to inner Oort Cloud defined to be 2,500 AU.

The phrase 'planetary object' only refers to planets, dwarf planets or moons and does not include asteroids or comets. The boundary between the inner and outer Solar System is assumed to be the main asteroid belt between Mars and Jupiter. To reach distances of a lot greater than 1,000 AU, it is likely that a combination of technologies will be required. This includes a chemical rocket for achieving Earth escape velocity. Technology such as solar sails would enable the attainment of a high velocity. Gravity assists would also be required from either one or more of the gas giants to add to the required Solar System escape velocity. The on board systems would be powered by a radioisotope generator that could also be used to power an electric engine to augment the sail and gravity assists.

We can examine the developments in the speed of planetary spacecraft and the mission distances they attain after the journey is complete. Examining human exploration we note that the first human carrying spacecraft into orbit was the Russian Vostok 1 in 1961. The first manned landing on the Moon followed this in 1969, the American Apollo 11. Human past recorded expansion stops here but we can make a fairly reasonable assumption (optimistically) that the first manned landing on the planet Mars at 1.5 AU distance might occur by 2040. This then allows one to form an extrapolated trend fit, which based upon the first lunar landing and an assumed Mars landing turns out to be approximated by a linear or low-growth exponential form and projecting a human expansion speed of ~0.01 AU/year. Extrapolating such a line forward suggests that human missions to the outer gas giants will not occur until the twenty-fourth century – a projection which is hard to believe and perhaps illustrates that these types of projections may not be appropriate.

Now let us briefly look at robotic probes. The first probe into orbit was the Russian Sputnik 1 in 1961. The first robotic landing on the Moon was Luna 9 in 1966. The first probe onto the surface of another planet was Venera 7 onto the surface of Venus in 1970 and the Viking probes onto the surface of Mars in 1976 (other missions also occurred during this period such as the Mariner and Russian Mars spacecraft launched in the 1960s and 1970s). The first probe to a planetary body out as far as Saturn was the Cassini-Huygens probe onto the surface of Titan in 2005. The first probe to reach 100 AU was the Pioneer 10 mission in 2009. When this data is examined it is shown that unlike human exploration, robotic exploration appears to follow a trend that approximates a high-growth exponential speed improvement. Robotic missions to 1,000–2,000 AU for example, should be possible by the end of the twenty-first century.

From the above very simple analysis we can summarize some important conclusions:

1. Human expansion appears to follow a linear or low-growth exponential trend, the gradient of which will only increase if 'game changing' propulsion technology is designed, tested and applied to human space missions.
2. Robotic expansion appears to follow a medium to high-growth exponential trend and progresses ahead of human expansion.
3. The first ambassadors to the nearest stars sent from Earth are likely to be robotic in type.
4. Any human missions to the nearest stars will take centuries to thousands of years and so will therefore require large World Ship type vessels if colonization is to be successful.

What will change the human expansion possibilities is investment in propulsion technology today? The required cruise speeds of 100–500 km/s are achievable in the near future using medium-term 'game changing' propulsion technology such as ion drives, nuclear thermal, nuclear electric or plasma drive propulsion technology. Other longer-term 'game changing' propulsion technology is also on the horizon such as solar sails, laser beaming, fusion propulsion and antimatter – all discussed in the next chapters. Table 8.4 shows this author's own personal projections for what is possible in the coming decades just assuming medium 'game changing' technology schemes. The opportunities for humans to follow our robotic explorers out into the solar system and beyond are clearly there should we choose to take this path.

Any space probes expanding at an ever increasing rate out into the galaxy will be a direct consequence of a trend in artificial intelligence, producing more intelligent and more efficient probes. There will come a point where such probes will avoid the mass ratio issue for large interstellar issues by simply re-fueling en route and even manufacturing spare parts en-route. It is then not a large jump to expect that such probes would be capable of completely rebuilding themselves – so called self-reproducing and self-repairing probes. Unless faster than light speed propulsion technology comes to fruition, human expansion into the galaxy is likely to be slow and require World Ships (see Chap. 1) to be successful.

Table 8.4 Actual and projected (optimistic) space mission arrival times with estimated launch date and peak cruise velocity

Progress level	Definition	Robotic mission arrival date	Human mission arrival date	Primary motivation for human mission
1	Probe/human in to earth orbit	Sputnik 1, 1957	Vostok 1, 1961	Reach earth orbit
2	Probe/human on to lunar surface	Luna 9 (moon), 1966	Apollo 11, 1969	Reach the moon
3	Probe/human on to inner solar system planetary object	Venera 7 (Venus), 1970	Assumed Mars, 2040	Reach Mars, first 'flag' landing, establish conditions for follow up first colony.
4	Probe/human on to outer solar system planetary object	Huygens (Titan), 2005	Assumed gas giant moon, 2060 (launched 2055, ~21 AU/year ~100 km/s)	Exploration of gas giant moons, search for life, establish colony.
5	Probe/human on to exo-solar system defined to be 100 AU	Pioneer 10, 2009	Assumed 2080 (launched 2070, ~42 AU/year ~200 km/s)	Exploration of dwarf planets and solar heliosphere.
6	Probe/human to 200 AU	Assumed 2040 (launched 2020, ~10 AU/year ~47 km/s)	Assumed 2100 (launched 2090, ~84 AU/year ~400 km/s)	Exploration of dwarf planets, comets, interstellar medium.
7	Probe/human to 1,000 AU	Assumed 2070 (launched 2040, 33 AU/year ~158 km/s)	Assumed 2130 (launched 2115, ~100 AU/year ~470 km/s)	Exploration of interstellar medium and testing of deep space long duration missions.
8	Probe/human on to Oort Cloud defined to be 2,500 AU	Assumed 2100 (launched 2050, ~50 AU/year ~237 km/s)	Assumed 2160 (launched 2140, ~150 AU/year ~710 km/s)	Exploration of Oort Cloud objects.

For both the robotic and human missions the final cruise velocities are likely to be achieved by reliance on gravity assists and boosted assists to supplement the main engine thrust

Robotic probes however will not be constrained by the same limitations and instead will colonize the entire galaxy in only hundreds to thousand of years. And even if the robotic expansion was also quite slow, the probes have the luxury of tolerating large journeys. In contrast humans may get bored with the journey or even the attempt and this could give rise to a loss of purpose over multiple generations, which could halt the overall exploration attempt of our civilization or force a premature colonization selection point. The potential for full robotic expansion of the galaxy leads to another interesting question that arises from this analysis – if other civilizations do exist in the galaxy and they also embark on similar robotic exploration attempts, how long would it take them to arrive in our own Solar System? Sir Arthur C. Clarke provided a tantalizing answer in his 1948 story "The Sentinel": [5]

> "I can never look now at the Milky Way without wondering from which of those banked clouds of stars the emissaries are coming. . .I do not think we will have to wait for long."

8.14   Practice Exercises

8.1. For some of the spacecraft discussed in this chapter do some research into the history of how the proposal for the mission and spacecraft design was first made, noting the date. Then determine when the proposal became an actual mission and assess the development time through the design stage and into manufacture and launch. How long does it typically take to go from initial concept development through to mission flight and completion for each spacecraft examined? On the basis of this assessment how long would it take to develop a new propulsion technology in particular that had not yet been tested in the environment of space?

8.2. Three spacecraft set off in a race at the start of the year 2020 to be the first to reach 50 AU (1) the first spacecraft *Explorer* accelerates for 1 year to a cruise velocity of 3 AU/year, which it maintains for 2 years. It then fires up its engine to give an instantaneous boost of 0.3 AU/year (2) the second spacecraft *Diver* accelerates for 1.5 years to attain a cruise velocity of 4 AU/year, which it maintains for 1 year. It then does a flyby of Jupiter and picks up an additional velocity boost of 1 AU/year, assumed to be instantaneous (3) the third spacecraft *Ascender* accelerates for 0.9 years to a cruise velocity of 2 AU/year, which it maintains for 3 years. It then fires up its engine to receive an instantaneous velocity boost of 0.5 AU/year. Using the linear equations of motion described in Chap. 3, namely $x = vt$ and $x = at^2/2$, calculate each stage of the race for each spacecraft, ensuring you only work in the units of AU (distance), years (duration), AU/year (velocity) and AU/year^2 (acceleration). Sketch the mission profile for each spacecraft. Determine the year each spacecraft crosses the finish line at 50 AU? Which spacecraft wins the race and which one comes second?

8.3. For interstellar mission proposals the most convenient units to use are years (time), light years (distance) and fractions of the speed of a light wave in a vacuum (speed). Because velocity equals distance over time, this relation can be extended for an interstellar mission as light speed fraction equals light years over years. Gaining an intuition for these units is crucial. Assess the distance attained by an interstellar spacecraft moving at speed of 0.01c, 0.1c and 0.5c assuming the travel time was 1 year and 10 years and 50 years for each case. The acceleration phase and relativistic effects are to be neglected in this analysis. A spacecraft accelerates for 5 years and then attains a cruise velocity of 0.008c. It travels a distance of 1.5 light years and then relays a radio signal back to Earth to inform mission control of its scientific discovery at the destination. How long did it take the spacecraft to reach its destination and how long will it be before mission control receives the images of the discovery?

References

1. McNutt, RL, Jr (2010) Interstellar Probe, White paper for US Heliophysics Decadal survey.
2. Garrison, PW (1982) Advanced Propulsion for Future Planetary Spacecraft, J. Spacecraft, Vol.19, No.16, pp.534–538, November-December.
3. McCulloch, ME (2008) Can the Flyby Anomalies be Explained by a Modification of Inertia, JBIS, 61, 9.
4. Lenard, R (2010) A General Relativistic Explanation of the Pioneer Anomaly And the Utility of A Milli-c NEP Mission on Validating the Observations, Presented 61st IAC Prague, IAC-1-. C4.2.5.
5. Clarke, AC (2000), The Sentinel, Voyager.

Chapter 9
Electric and Nuclear-Based Propulsion

We can take it for granted that eventually nuclear power, in some form or other, will be harnessed for the purposes of space flight. ...The short-lived Uranium age will see the dawn of space flight; the succeeding era of fusion power will witness its fulfillment.

Sir Arthur C. Clarke

9.1 Introduction

The science that goes behind nuclear physics is very mathematical and based upon the laws of quantum mechanics. Over many centuries and with determined effort humans have gradually learned what exactly an atom is and how to tame it. It is true that nuclear energy carries negative side effects through which history records the consequence. However, it is also true that the energy released from atoms is at least a million fold greater than can be obtained from any chemical reaction. To consider the control of the atom for the peaceful purposes of electrical power generation will also lead naturally to the consideration of alternative applications. One of those is space propulsion. In this chapter, we shall firstly given an overview of electric propulsion and then discuss what is considered the likely technology for future Mars missions, nuclear electric and nuclear thermal propulsion schemes derived from the process of fission.

9.2 Electric Propulsion

One of the common forms of space propulsion used today is electric propulsion, which falls into three types: electrothermal, electrostatic and electromagnetic. In essence all will heat up a fuel electrically and then use electric and/or magnetic fields to accelerate charged particles to provide thrust. But they all have differences in how they achieve

K.F. Long, *Deep Space Propulsion: A Roadmap to Interstellar Flight*,
DOI 10.1007/978-1-4614-0607-5_9, © Springer Science+Business Media, LLC 2012

this, which are distinguished principally by the type of force that is used to accelerate the charged particles. Different electric engines will also vary in their burn periods, where some will burn for extended periods (continuous) and others will burn in set impulses (unsteady). Because some electric engines do not contain an exit nozzle to direct the exhaust velocities, they differ from a conventional rocket engine, although both still exploit the action-reaction principle to obtain forward motion.

One of the first people to suggest electric propulsion technology was the Russian Konstantin Tsiolkovsky, who wrote a paper in 1911 where he first suggested that electricity could be used to produce a large velocity from the particles ejected from a rocket. This was particularly the case for electrons ejected from a cathode ray tube that would attain high velocities. The fact that such large velocities could potentially be used for rocket propulsion was clear when it was believed that an order of magnitude more energy release can be obtained compared to chemically energized fuels. Much later, Tsiolkovsky and others realized that positively charged ions could also be used for the same purpose. In 1906 the American physicist Robert Goddard also considered the electric acceleration of electrons for rocket propulsion in his notebooks. His later notes clearly demonstrate that he also appreciated that the acceleration of ions would be more advantageous than that of electrons. The name ion rocket was not coined until the end of the World War II, when a young American student discussed the applications of such propulsion [1].

One of the advantages of electric engines is that they offer the potential for a high velocity increment with a higher specific impulse or exhaust velocity than can be obtained using a chemically based rocket engine. Whereas a chemical engine will rely upon the thermodynamic expansion of the gases to attain high velocity, an electric engine instead obtains high velocity by directly applying electric forces to the charged particles to accelerate them. This offers the advantage that high temperature particles are not in physical contact with the spacecraft and thus are not limited by the melting temperatures of the surrounding materials. Also, chemical fuels will tend to store their energy in the propellants, but with an electric propulsion system the energy is usually generated by the use of solar panels. This is known as solar electric propulsion – where electricity is generated from solar cell arrays and linked to an engine. Such propulsion may be capable of attaining an escape velocity of 10–15 km/s. Solar electric designs would use typical power levels of 100 kW, and because this is quite low they would need to be supplemented by chemical propulsion, for example, if intended for interplanetary missions [2]. So a solar electric scheme is a Sun-dependent one.

As part of the NASA New Frontiers program the JUNO mission, named after the goddess sister-wife of the Roman god Jupiter, was due to be launched in 2011 on a 5-year mission to Jupiter, arriving in 2016, to study the planet's composition, gravity field, magnetic field and to determine if it has a rocky core. This is a solar-powered spacecraft but also utilizing an Earth flyby gravity assist to pick up velocity. There are three solar arrays on the spacecraft, stowed on launch but unfurled upon reaching space.

Whereas a chemically based engine is energy limited (reactants have a fixed amount of energy per mass) electrically based engines are not. This is because the charged particles are their own source of kinetic energy and so the reaction rate is independent of the mass. This means that high power levels and high specific impulse can be reached by using an electric propulsion-based drive. However, an electric engine is power limited, due to the limited mass of the power system. This limits the thrust of the engine and hence is the reason why electric engines only produce low thrust levels and so moderate accelerations. In essence this means that an electric engine will produce a long specific impulse which results in a large exhaust velocity. However, a chemical engine will produce a high thrust but with a small specific impulse resulting in a moderate exhaust velocity, only sufficient for Earth escape speeds.

In the electrostatic approach positively charged particles such as ions (or colloids) are accelerated in a static electric field by the Coulomb force in the direction of the acceleration, producing thrust. Typical fuels for electrostatic engines include inert gases such as argon and xenon. Historically, mercury was also considered but this has been largely done away with due to fears over corrosive and environmental impact issues. Examples of electrostatic thrusters include (1) electrostatic gridded ion thrusters, (2) Field Effect Electrostatic Propulsion (FEEP), and (3) colloid thrusters.

A Hall Effect Thruster (HET) is really a combination of an electrostatic and electromagnetic engine because the typically xenon ions are accelerated by the electrostatic field but yet the electrostatic field itself is generated by electrons interacting with a magnetic field. Xenon is used due to the efficiency by which it can be ionized and then electrically pushed around for eventual thrust generation. A HET was used on the SMART-1 spacecraft. Another example of where electrostatic thrusters have been used includes the Intelsat VII spacecraft, which had 10 xenon 25 mN thrusters. This was ESA technology for the SAT-2 mission. Typically, an electrostatic thruster can produce up to >1 N.

In an electrothermal engine electromagnetic fields are used to generate a plasma state, which then thermally heats up the propellant and is accelerated through either a solid or magnetic nozzle, which increases the particle's kinetic energy. The typical fuels include low molecular weight materials such as hydrogen, helium and ammonia. There are different types of electrothermal engines, which include (1) DC arcjet, (2) the microwave arcjet, (3) the pulsed plasma thruster and (4) the Helicon Double Layer Thruster.

In electrodynamic propulsion (also known as magnetoplasmadynamic) is an approach where positively charged ions are accelerated either by the Lorentz force or by an electromagnetic field that may not necessarily be in the direction of the acceleration but results in the particles being accelerated out of the plane of the crossed fields. There are different types of electromagnetic engines, including the (1) Electrodeless Plasma Thruster, (2) Hall Effect Thruster, (3) MPD thruster, (4) Pulsed Plasma Thruster and (5) VASIMR.

An ion engine is a gridded electrostatic accelerator, which accelerates ions that have been ionized such as by radio frequency methods. The ions exit the ionization chamber and must pass through a double grid that has sub-millimeter holes, across which an electric potential is applied. Ions are then extracted from the grid to provide acceleration and exhaust velocities of around 30 km/s and specific impulse

3,000 s. An ion thruster was used on the ESA EURECA spacecraft in 1992 and in many other geostationary satellites. An excellent review of the early history of electric propulsion developments from 1906 has been published elsewhere [3]. The required specific power p to accelerate a spacecraft with an ion engine is given by:

$$p = \frac{aV_e}{2\eta} \tag{9.1}$$

where V_e is the exhaust velocity, η is the thrust efficiency. For a spacecraft accelerating at 100th of gravity to achieve an exhaust velocity of 100 km/s and assuming 100% efficiency the power required is 5 kW/kg. This suggests that such propulsion schemes are impractical for outer interplanetary missions. However, more modest accelerations of order 10,000th of one gravity and a more modest velocity requirement of order 30 km/s results in a power requirement of around 2 kW/kg.

One of the issues with attaining better performance from electrically powered engines is the need for an internal nuclear power source that gives a power output of 5 kW or ideally more for manned missions to the planets. Until then, electrical propulsion systems will likely be only used for satellite control and orbit raising and perhaps missions to Mars. As of today, over 200 solar powered satellites have benefited from the use of electric propulsion. Electric propulsion was used on the NASA Deep Space One mission discussed in Chap. 8. The ESA SMART-1 spacecraft also used an electric engine.

The specific impulse of an ion engine is given by consideration of the charge q and ion mass m as well as the electric potential difference V_a through which the ions are accelerated. These are related by:

$$I_{sp} = \frac{V_{ex}}{g_o} = \frac{1}{g_o}\sqrt{\frac{2qV_a}{m_i}} \tag{9.2}$$

For current ion thruster designs V_a would typically be around 2.5 kV. If we assume a xenon propellant, which has an atomic mass of 131.293 atomic mass units (were 1 amu $= 1.5505 \times 10^{-27}$ kg), then the specific impulse and exhaust velocity for this configuration is 6,179 s and 60.6 km/s. The power of the rocket exhaust per unit time will give an estimate of the energy expenditure of the engine, and this is termed the 'jet power.' This can be related to the exhaust velocity and mass flow rate $\dot{m}$ by the following equation:

$$P_j = \frac{\mu}{2}\dot{m}V_{ex}^2 \tag{9.3}$$

The term μ is an efficiency conversion term, which we can assume to be unity for simplicity. Assuming a mass flow rate of 4×10^{-6} kg/s and the exhaust velocity calculated above this would result in quite a high jet power of 7.34 kW. We can easily calculate the thrust for this system as follows:

$$T = \dot{m}V_{ex} \tag{9.4}$$

This computes to a thrust of 0.24 N.

The British propulsion engineer David Fearn has considered the application of ion engines to an interstellar precursor mission that could be completed in 25–30 years [4]. He discusses an advanced form of a 4-gridded ion thruster that can provide a specific impulse as high as 150,000 s with thrust levels in the Newton range and a velocity increment exceeding 37 km/s. Fearn identified the requirement for a nuclear power source with a mass to power ratio of 15–35 kg/kW and an output power of several tens of kW. Such a cruise velocity would allow a mission out to 200 AU with a performance competitive with alternative propulsion schemes such as solar sails (discussed in the next chapter).

Unlike chemical propulsion systems, electric propulsion systems are not energy limited. In theory, a large amount of energy can be delivered to a given mass of propellant so that both the exhaust velocity and specific impulse are very high. However, electric propulsion systems are power limited. This is because the mass of the power system itself places a limit as to how much energy can be delivered to a propellant. Ultimately, this places a constraint on the maximum thrust levels attainable (and therefore acceleration), and so the thrust to weight ratio is low. In theory conventional electric propulsion will attain an exhaust velocity of up to 30 km/s (0.001c) corresponding to a specific impulse of thousands of seconds, allowing missions to Alpha Centauri in around 42,000 years.

An extension of this technology is to employ a fully ionized gas that has been heated to a much higher temperature so that the gas becomes a plasma state. Such plasma propulsion designs have been proposed as ideal for Earth-Moon and Earth-Mars missions in future decades. One of the exciting technology developments in recent years is the Variable Specific Impulse Magnetoplasma Rocket (VASIMR), or what is known as a Plasma Rocket [5, 6]. The former astronaut Franklin Chang-Diaz and his team at the Ad Astra Rocket Company in Texas in conjunction with NASA developed this. The engine is unique in that the specific impulse can be varied depending upon the mission requirement. It bridges the gap between high thrust-low specific impulse technology (e.g., like the space shuttle) and low thrust-high specific impulse technology (e.g., like electric engines) and can function in either mode. The company who designs the VASIMR engine talks about possible 600 ton manned mission to Mars powered by a multi-MW nuclear electric generated VASIMR engine, reaching Mars in less than 2 months.

The VASIMR drive could in theory attain an exhaust velocity of up to 500 km/s (0.002c) corresponding to a specific impulse of around 50,000 s and could reach Alpha Centauri in 2,200 years. Although this technology is impressive, it still won't get us to the stars in a short duration. This type of engine is also quite convenient because it is considered to be a scaled down fusion development engine. Several elements of the design, namely the ionizing and energizing of a gas using an helicon RF antenna, the use of electromagnets to create a magnetic nozzle and the storage of low mass hydrogen isotopes (including helium and deuterium) are similar technology that would be used in a fusion-based engine. Developing this technology will also assist in understanding how to control plasmas for a space propulsion engine, although large improvements are required in the power, field control and shielding.

So far we have discussed electric propulsion technology, because it is relevant to interplanetary vehicles within the inner Solar System. However, it has little application in the outer parts of the Solar System or beyond. Instead, the technology must be enhanced in some way, and this is where the nuclear-electric rocket comes into its own.

9.3 Nuclear Thermal and Nuclear Electric Propulsion

To some the idea of using nuclear power to propel a spacecraft represents a future of hazardous radiation risks. But to others it represents an opportunity to free up the Solar System to not only robotic but full human interplanetary travel. This is because the energy release available from nuclear reactions is of the order a million times greater than that from a typical chemical reaction. This is by the process of nuclear fission where a fissionable isotope is induced to undergo a process of fission (atom splitting) by the absorption of neutrons. A nuclear reactor will contain this fission process in a controlled way so that only a small neutron population persists in the system to maintain a barely critical state.

As long ago as the 1950s the science and science fiction writer Arthur C Clarke was discussing the idea of a nuclear powered spacecraft in his book "*Interplanetary Flight*". The unusual configuration that he proposed bore the resemblance of a 'dumb-bell' shape with the nuclear reactor power system being separated by distance from any crew or payload section to ensure they are not exposed to any radiation. Figure 9.1 depicts this design layout, along with the cylindrical spacecraft

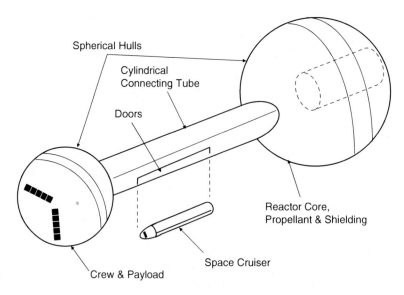

Fig. 9.1 The 'dumbbell'-shaped spacecraft as proposed by Arthur C. Clarke

to represent the landing craft to be deployed on arrival at the mission target. It is interesting to note that the 'dumb-bell' shaped design appears to be the basis for the Discovery spacecraft depicted in the movie *2001: A Space Odyssey* and shown in Fig. 1.2 of Chap. 1. Let us briefly consider how a nuclear reactor may be used in a spacecraft by both thermal and electric means and then we shall review some examples of actual spacecraft proposals.

The basic idea of a nuclear thermal rocket (NTR) is to use a nuclear reactor to produce energy that is then moved by a heated working fluid (gas or liquid) such as hydrogen, which can then be expanded for thrust generation through a rocket nozzle. Because the more efficient energy release is higher, this means that the exhaust velocity is also much higher than a chemical-based system, allowing for reduced mass overall (propellant and reactor). Thus, a spacecraft cannot only have a higher performance but can also carry a more massive payload.

Fundamentally there are three types of nuclear thermal propulsion systems, known as solid, liquid and gas core reactors. A solid core engine will use a solid reactor material to heat the working fluid, and these are considered to be simple designs. High temperatures up to around 3,000 K will limit the performance to around 1,000 s of specific impulse, however, due to the melting point of the materials being reached. A liquid core engine will involve mixing the nuclear reactor material with the working fluid so that the reactions take place within the fluid. In theory this approach can lead to higher temperatures and so a more efficient design with a specific impulse up to 1,500 s. A gas core engine could in theory achieve the highest performance – up to 6,000 s specific impulse – because the fluid is a gas that can circulate more rapidly while containing the reactor material inside a central gas pocket and preventing it from interacting with the surrounding material structure.

Theoretical research into nuclear thermal propulsion was initially made by the German scientist Ernst Stuhlinger working in the United States after the war and members of the British Interplanetary Society such as A. V. Cleaver and L. R. Shepherd who discussed atomic rockets where nuclear fission energy is used to heat an on board fuel [7]. They discussed exhaust velocities of 10–100 km/s as obtainable from ion rockets, although they said that such propulsion schemes would have massive power requirements, which rendered the technology impractical at the time.

Nuclear thermal propulsion began its ground based experimental history in 1953 with Project Rover, which was a joint project of the U.S. Atomic Energy Commission and the U.S. Air Force [8]. Then it was in the 1960s that Rover came to fruition in a series of Nuclear Engine for Rocket Vehicle Applications (NERVA) tests in the Nevada test site in the United States [9]. The liquid hydrogen based NERVA engine rocket was nearly 44 m in length and had a dry mass of 34 tons capable of generating 334 kN of thrust in a vacuum with a specific impulse of around 850 s and a burn time of over 1 hour. The specific impulse was approximately double what can be obtained from any chemical system.

Because a nuclear thermal rocket will use fission as the energy source, this has a maximum conversion efficiency of 0.0007, that is, the fraction of mass converted into energy per reaction. As described in Chap. 5 the exhaust velocity is given by the square root of 2ε, where ε is the fraction energy release. This leads to a maximum theoretical exhaust velocity of 0.0374c, or around 11,225 km/s. If we assume a typical thrust maximum associated with the NERVA program of around 1,000 kN then the power required for the engine in the units of Watts is given by:

$$P(W) = \frac{\mu T V_{ex}}{2} \qquad (9.5)$$

where μ is the approximate conversion efficiency, assumed here to be around 0.1%. This then computes to a power required of 5.6 GW. Note that for the NERVA program power levels of around 5 GW were being considered so the program would had to have had a highly optimized design to perform this well.

Instead of a thermally heated fluid to provide the propulsion, the other suggestion is to directly use the energy release from a nuclear reactor to produce electrical energy in nuclear electric rockets. These have the capability to deliver payloads to the outer planets with an order of magnitude greater mass than for solar electric systems. An electric thruster would typically be based on an Argon fuel using ion engines and operating in the 5,000–6,400 s specific impulse range. These are typically low acceleration systems with potential power level requirements in the 200 kW to 1 MW range and capable of achieving heliocentric escape velocity of 50–60 km/s [10]. As a rule of thumb designers will aim for a dry nuclear electric propulsion mass 30–35% of the mass of the spacecraft.

One study looked at using nuclear electric propulsion for a 1,300-day Jupiter mission [11]. The authors described how the spacecraft would be launched into orbit using a space shuttle. It would consist of a 9,400 kg vehicle and a payload of 12,400 kg, which would encompass a 9,000 kg orbiting laboratory and 500 kg of landers for use in sample return operations. The design was said to require a specific mass of no less than 25 kg/kW. The engines would probably use mercury-based ion thrusters or argon-based magnetoplasma dynamic thrusters and would have a specific impulse of around 9,000 s and an effective jet power of between 270 and 840 kW. Other mission possibilities for a nuclear electric- based rocket would be to Saturn and Uranus, where a 200 kW power system could deliver a payload exceeding 3–8 tons in around a decade [12]. Going as far out as Neptune and Pluto is also possible but would require improvements in power conversion system efficiency, longer mission duration and the potential for a reduced payload mass. However, as will be shown in the next few sections, more ambitious missions have been proposed using advanced nuclear electric propulsion technology.

9.4   The Interstellar Precursor Probe

One of the first teams to properly consider an interstellar precursor mission was CalTech's Jet Propulsion Laboratory. In the late 1970s it conducted a study called the Interstellar Precursor Probe (IPP), a mission designed to explore the solar heliopause perhaps as far out as 1,000 AU [13]. It was to use nuclear electric engines and would attain a heliocentric hyperbolic escape velocity of around

50–100 km/s in order to achieve the mission in the stated timeframe of 20–50 years. The power and propulsion system dry mass at 100 km/s exhaust velocity would be round 17 kg/kW. The spacecraft initial mass would be around 32 tons, including the engine, where around 1.2 tons would be allocated to the science instruments, computation, data processing and communications systems. As well as instruments such as spectrometers, magnetometers, and various plasma detectors requiring around 100 W of power, the potential for using a 200–300 kg telescope was also considered, as were nuclear electric engines, solar sails and laser sail technology.

It was to have been launched around the year 2000. The mission was for an encounter with Pluto (including a 500 kg orbiter similar to the Galileo spacecraft) and its moon Charon and then the extended mission would continue on to the Kuiper Belt and beyond to a distance of 370 AU around 20 years after launch. After 50 years it would reach a distance of over 1,030 AU. The key mission goal was to measure the interaction between the solar wind and the interstellar medium so as to better characterize the space surrounding our Solar System. Primary objectives also included a study of stellar distances through parallax measurements, low energy cosmic rays, the interplanetary gas distribution and the mass of the Solar System. The study did discuss the potential for an interstellar mission, but concluded that the achievable escape velocity of 100 km/s for, say, a 4.3 light year Alpha Centauri mission would imply mission duration of order 10,000 years, so it did not seem worth seriously considering.

Much of the work from this project went into the design considerations for other projects such as the Thousand Astronomical Unit (TAU) mission, which was also a nuclear electric propulsion system. Many studies since, including the Innovative Interstellar Explorer, have been based upon some of the initial calculations and configuration layout considerations for the Interstellar Precursor Probe. In particular this study really established for the first time the idea of an unmanned interstellar precursor mission, laying out the science objectives, mission concepts and technology requirements. It was seen as a way of addressing the interstellar problem without actually launching an interstellar mission, while also delivering valuable science data.

9.5   The Thousand Astronomical Unit Mission

During the 1970s and 1980s scientists from the Jet Propulsion Laboratory had discussed the concept of an interstellar precursor mission to explore the solar heliopause that would last between 20 and 50 years. In its original incarnation it was known simply as the Interstellar Precursor Mission (ISP). After further consideration the concept eventually became known as the Thousand Astronomical Unit mission or TAU [14]. The idea was to design an interstellar precursor mission that was restricted to using only technology that had actually been tested. A good distance to aim for which would also stretch the technology to its limit was deemed to be 1,000 AU, hence the name of the mission. Recall, that an astronomical unit is

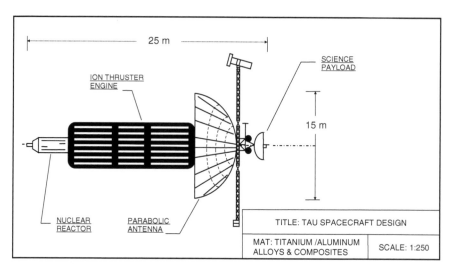

Fig. 9.2 Illustration of the TAU proposal

the mean distance from the Earth to the Sun. The mission was for a planned launch date around the year 2000.

Although solar sails were considered on option for propulsion if unfurled sufficiently close to the Sun, the final concept used Nuclear Electric Propulsion (NEP), where a nuclear reactor electrically ionizes the gas of a low thrust-high specific impulse ion engine. This technology has since been clearly demonstrated in the Deep Space 1 spacecraft discussed in Chap. 8. This would allow the vehicle to reach a cruise velocity of 106 km/s (0.00035c) after a 10 year burn, sufficient to escape the Solar System. After jettisoning the main engine, the vehicle would then cruise to its destination in under 50 years from initial launch. An alternative option considered in the initial discussions was to launch the spacecraft into low Earth orbit and then use gravity assists to reach an escape velocity of around 50 km/s and utilizing RTG to power the spacecraft systems throughout the flight. This option was rejected in the final choice, but it is worth noting that there are several ways of performing a mission of this type.

The spacecraft would employ 12 xenon ion electric thrusters, each with a specific impulse of 12,500 s and would burn for around 2 years with a specific mass of 4 kg/kW. The vehicle systems would be powered by 1 MW nuclear reactor with a specific mass of 12.5 kg/kW. The ion thrusters and reactor would be located towards the rear of the vehicle. The total mass of the spacecraft would be about 60 tons, where 5 tons of this would be the actual spacecraft and instruments and 15 tons would be for the engine. The science payload and instrumentation would be around 1.2 tons, which would include spectrometers, magnetometers and optical cameras. Most of the mass would be attributed to 40 tons of liquid xenon. The vehicle was about 25 m long and contained a 15-m diameter main antenna for communications, located near the front of the vehicle. It is shown in Fig. 9.2.

A nice addition to the spacecraft was a Pluto orbiter, which could be dropped into Pluto orbit as the spacecraft went past, as well as making observations of the moon Charon. The orbiter would be located in front of the main antenna as well as a massive instrument boom with a telescope and auxiliary antenna located at either end. The front and rear of the spacecraft would be separated by a large cylindrical column, which would include a radiator on its surface to remove any waste heat generated.

A spacecraft reaching 1,000 AU would provide very valuable insights into the outer Solar System and the solar heliopause. Specifically, finding any further Kuiper Belt objects that may be lurking in the darkness would be of tremendous scientific interest. One of the primary objectives of the mission was to perform long baseline parallax measurements and allow the distances to nearby stars to be measured with greater accuracy out to 50 kpc distance, which extends to the Magellenic Clouds. Another interesting science objective of TAU was to test aspects of Einstein's general relativity theory by using a laser transponder so that Earth and the spacecraft can act as end masses of a gravitational wave detector, with potentially much more accurate observations compared to those limited to the location around Earth. Scientists back home studying the spacecraft's trajectory as communicated by radio signals would achieve this. All scientific observations would be transmitted back to Earth using an advanced 1 m optical telescope combined with a 10 W laser transponder, allowing a transmission rate of 10–20 kB/s to a 10 m telescope located in Earth orbit.

One of the advantages of a probe like TAU is that it would pass through the Sun's gravitational focal point located between 500 and 1,000 AU, where distant light rays are bent around the Sun and converge to an amplified image. This would provide for a magnification of future interstellar targets. However, we must remember that the nearest star system, Alpha Centauri, is nearly 272,000 AU away, much further than TAU was designed to reach. At the specified maximum cruise speed for TAU, it would take the vehicle around 12,000 years to reach Alpha Centauri. This suggests that although a nuclear-electric engine may be useful for achieving missions just outside of our Solar System, it is not adequate for much further distances. But precursor missions such as TAU are clearly required before attempts to launch probes to the nearest stars can be properly considered. It is a shame that space agency managers didn't realize the enormous potential in such a credible spacecraft mission that TAU represents, especially since it could have been built and launched by the year 2000 as proposed. The spacecraft would have been well into the Kuiper Belt by now.

9.6 The Innovative Interstellar Explorer

The Innovative Interstellar Explorer (IIE) is a NASA proposal to launch a 34 kg payload to around 200 AU. It largely grew out of an earlier study for a Realistic Interstellar Explorer (RIE) proposed by Ralph NcNutt, who has been evolving the concept of an interstellar precursor mission in collaboration with the Jet Propulsion

Laboratory for some years and is the principal investigator behind the IIE [15–18]. He has argued the case for an interstellar precursor mission as a high priority for science. For such long duration space missions the biggest problem has always been the propulsion necessary for the required high performance. To send a space probe to a distance of 200 AU in a time scale of 15, 20 or 30 years requires an average velocity of approximately 13, 10 and 7 AU/year respectively. This is equivalent to 63, 47 and 32 km/s, respectively, and compares to the moderate 17 km/s or around 3.3–3.6 AU/year experienced by the Voyager probes. Because these velocities are so large, conventional propulsion systems based upon chemical engines cannot meet the performance requirements. Hence alternative technology is required.

The initial plan for the mission was for it to be launched in 2014 and arrive at its destination in 2044, around 30 years after launch. By the year 2020 the Voyager 1 and 2 spacecraft would have ceased transmitting and be at around 150 and 125 AU from the Sun, respectively. So although they may have passed the heliopause it is unlikely that they will have reached the undisturbed interstellar medium, which is thought to be beyond 200 AU. Hence the spacecraft should be able to reach it and transmit data using its high gain antenna. By the year 2057 the spacecraft would be expected to reach 300 AU, some 43 years after launch. By the year 2147, the spacecraft would reach 1,000 AU from the Sun, and be the furthest and fastest manmade object ever launched into space. This will be a century after the planned launch data. One century for 1,000 AU sounds like a long time, and if we multiply this by the distance to Alpha Centauri at 272,000 AU, such a probe would take many tens of thousands of years.

One of the primary science goals of the mission is to reach and measure the properties of the interstellar medium as well as the exact location of the termination shock. It would also measure the properties of the solar magnetic field and incoming cosmic rays, important measurements for future spacecraft that leave our Solar System. These measurements also have tremendous implications for understanding the path of our supersonic Solar System through the cosmic ray intense Milky Way. Our Solar System is protected from these rays by the solar heliopause, the boundary between the solar wind generated by our own Sun and interstellar space. Innovative Interstellar Explorer measurements may also shed light upon the role of cosmic rays on biological mutation rates of life on Earth or even the cosmic ray link to global climate change. Another science goal could be the measurements of the Big Bang nucleosynthesis in terms of the abundance of helium-3, deuterium and lithium-7. Exotic physics issues such as the detection of gravitational waves or further tests of our theories of gravity would also be desirable objectives.

The Innovative Interstellar Explorer is shown in Fig. 9.3. It uses a combination of a heavy launcher such as a Delta 4, to place it into orbit and a kW power ion engine that burns xenon propellant and is powered by an RTG, similar to the Voyager probes. This is necessary for long duration missions and to achieve the required vehicle speed of around 10 AU/year at 100 AU. Any power source that reduces the overall vehicle mass is also a bonus. In earlier versions of the mission plan the trajectory would obtain velocity by dropping deep into the Sun's gravitational potential and then executing a velocity burn at perihelion. But this was

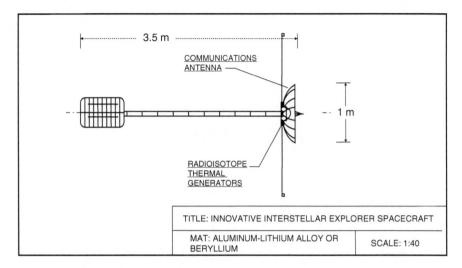

Fig. 9.3   Illustration of the innovative interstellar explorer

abandoned in favor of a Jupiter flyby where the gravity assist will boost the spacecraft velocity by around 25 km/s or 5.2 AU/year. It would obtain a maximum cruise velocity of 37.5 km/s or 7.8 AU/year. With the heliopause boundary speculated to be located between 100 and 150 AU, the Innovative Interstellar Explorer would be the first probe to enter interstellar space – a milestone achievement and within our technological grasp. It could credibly be built and launched today. As stated by mission planners in a 2006 paper [19]: "The time is right for the first step into interstellar space."

9.7   Project Prometheus

Project Prometheus was a NASA 2003 attempt to develop a deep space vehicle for long duration outer Solar System robotic exploration mission that would combine nuclear reactor technology with electric propulsion [20]. It was formerly named the Jupiter Icy Moons Orbiter (JIMO) Project and was part of the NASA Nuclear Systems Initiative. The motivation for such a spacecraft is to launch missions that allow spacecraft to go out to large distances in the Solar System while conserving vital energy resources until the destination is reached. This would particularly enable missions to the outer planets, where current spacecraft are very limited. This is due to the fact that solar energy is not available in sufficiently intense levels at large distance from the Sun and so cannot power any on board systems, including the propulsion system. Historically, some of these probes used RTGs as a power source. Project Prometheus was a step up in power generation in that it proposed to use a nuclear reactor as the power source. This would enable longer duration missions as well as continuous operation of the propulsion system.

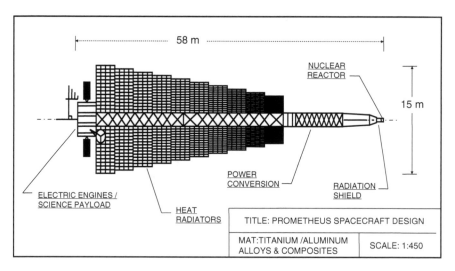

Fig. 9.4 Illustration of project prometheus

Missions to Jupiter and its moons were seen as primary targets, such as Callisto, Ganymede and Europa. Europa was particularly of interest given the potential for finding ocean life below the icy crust. Prometheus would also investigate how the satellites interact with Jupiter. The spacecraft would escape from Earth using a chemical propulsion system. The original JIMO proposal had a trajectory that did not involve any planetary gravity assists. For Prometheus, however, options were considered to allow an increased payload mass to the destination and decrease the flight time. The final Prometheus trajectory would involve a flyby gravity assist at Jupiter in 2021 and then a gravity assist at Callisto prior to being captured by Callisto for a period of around 120 days. The spacecraft would then move on to Ganymede for 120 days followed by Europa for 60 days. The mission requirements were for a payload mass of not less than 1,500 kg. This would include science instruments such as for topographic mapping, subsurface mapping, optical cameras, spectrometry, particle and field analyzers, altimetry and magnetic field measurements.

The spacecraft systems would be powered by an on board high temperature nuclear fission gas-cooled reactor for massively improved power compared to non-nuclear systems. Connected to turbo-alternators for power conversion it would generate around 200 kW of electrical power, a staggering amount compared to historical space probes such as Voyager and Pioneer, and much greater than the minimum initially thought to be required from looking at electric thruster technology, which was in the range 20–30 kW. A conical radiation shield would be positioned directly behind the nuclear reactor to protect the other spacecraft systems. The reactor was to be designed by the U.S. Navy, who had considerable experience in this area. Figure 9.4 shows a schematic of Project Prometheus.

Aft of the reactor module is the spacecraft module, which is dominated by a 43-m-long main boom assembly. Large arrays of heat radiators were positioned between the spacecraft power source and the ion engine thrusters. This array of panels would come off the main central truss. The propulsion system was ion thrusters positioned at the rear of the spacecraft and would use 12 tons of xenon propellant. For Prometheus this would allow a performance in the specific impulse range of 6,000–9,000 s. A stowed spaceship would be located towards the back, which could be deployed at the destination. The total spaceship mass at launch would be around 38 tons. Unfortunately Project Prometheus funding was cut in 2006, and a mission utilizing nuclear power combined with an ion engine would have to wait for another day.

9.8   Practice Exercises

9.1. A Hall Effect Thruster uses 80 kg of xenon propellant and thrusts for a period of 5,000 hours. Assuming a potential difference of 0.2 kV, using the equations of this chapter show that the specific impulse, exhaust velocity and thrust for this engine is around 1,748 s, 17.1 km/s and 0.08 N, respectively. Note that xenon has a molecular mass of 131 atomic mass units.

9.2. Repeat the above calculations but for a monopropellant thruster using 52 kg of hydrazine fuel thrusting for 46 hours with a potential difference of 0.0006 kV. Note that hydrazine has a molar mass of 32 g/mole or molecular mass of 32 atomic mass units. Show that the specific impulse, exhaust velocity and thrust is around 194 s, 1.9 km/s and 0.6 N, respectively.

9.3. An experimental nuclear thermal engine has a specific impulse of 759 s, using up 144 tons of propellant in 25 minutes. It has a conversion efficiency of 0.15%. Calculate the exhaust velocity and thrust and then show that the power is 3.85 MW.

References

1. Radd, H (1945) A Survey of Spatial Problems, Journal of the American Rocket Society, 62, pp28–29.
2. Jaffe, LD et al., (1980) An Interstellar Precursor Mission, JBIS, Vol. 33, pp.3–26.
3. Choueiri, EY (2004) A Critical History of Electric Propulsion: The First 50 Years (1906–1956), Journal of Propulsion & Power, 20, 2.
4. Fearn, D (2006) Ion Propulsion: An Enabling Technology for Interstellar Precursor Missions, JBIS, 59, pp88–93.
5. Chang-Diaz, RR, (2002) Fast, Power-rich Space Transportation, Key to Human Space Exploration and Survival, 53rd IAC.

6. Chang-Diaz, FR, (1995) Rapid Mars Transits with Exhaust-Modulated Plasma Propulsion, NASA TP3539, March 1995.
7. Shepherd, LR & AV Cleaver (1949) The Atomic Rocket, JBIS, 8.
8. Finseth, JL (1991) Rover Nuclear Rocket Engine Program, Overview of Rover Engine Tests, Final Report. NASA CR-184270.
9. Bowles MD & RS Arrighi (2004) NASA's Nuclear Frontier, The Plum Brook Reactor Facility. Monographs in Aerospace History No.33, SP-2004-4533.
10. Jaffe, LD et al., (1980) An Interstellar Precursor Mission, JBIS, 33.
11. Pawlik, EV, et al., (1977) A Nuclear Electric Propulsion Vehicle for Planetary Exploration, J.Spacecraft, Vol.14, No.9.
12. Phillips, WM (1980) Nuclear Electric Power Systems for Solar System Exploration, J.Spacecraft, Vol.17, No.4, July–August.
13. Jaffe, LD, et al., (1977) An interstellar Precursor Mission, JPL Publication 77–70.
14. Nock, KT (1987) TAU – A Mission to a Thousand Astronomical Units, presented at 19th AIAAIDGLR/JSASS international electric propulsion conference.
15. McNutt, RL, Jr et al., (1999) Phase I Final Report; A Realistic Interstellar Explorer, NIAC CP 98–01.
16. McNutt, RL, Jr et al., (2003) Phase II Final Report; A Realistic Interstellar Explorer, NIAC 7600–039.
17. McNutt, RL, Jr et al., (2003) Low-cost Interstellar Probe, Acta Astronautica, 52.
18. McNutt, RL, Jr et al., (2004) A Realistic Interstellar Explorer, Advances in Space Research, 34.
19. Gruntman, M & RL McNutt, Jr et al., (2006) Innovative Explorer Mission to Interstellar Space, JBIS, 59, 2.
20. Taylor, R (2005) Prometheus Project Final Report, NASA 982-R120461.

Chapter 10
Sails & Beams

> Whatever the source of the beam (power supply plus antenna, the "beamer"), the ability to move energy and force through space weightlessly is key to a genuinely twenty-first century type of spacecraft. The expensive part of this utility is the beamer, which can project energy anywhere within its range and also drive one sail after another. Like the nineteenth century railroads, once the track is laid, the train itself is a small added expense. Compared with rockets, sails are very cheap once the beamer is built. Just as railroads opened up the American West, a beamer on Earth – or, at shorter range, in orbit – could send entirely new kinds of missions throughout and even beyond the solar system. Beamed sails (light or microwave) offer the promise of interstellar flight.
>
> James Benford

10.1 Introduction

This chapter will discuss various types of propulsion schemes that have resulted from a paradigm shift in thinking about interstellar travel. These schemes depend upon the exploitation of solar or laser energy for propulsion. The utilization of natural energy sources in space seems like a prudent way to conduct exploration missions, having the advantage of minimizing fuel mass for the spacecraft or even negating entirely the need to carry a reaction mass. We firstly discuss the subject of solar sailing and then move on to laser beam sails and microwave sails. We discuss some of the physics, applications and some past projects.

K.F. Long, *Deep Space Propulsion: A Roadmap to Interstellar Flight*,
DOI 10.1007/978-1-4614-0607-5_10, © Springer Science+Business Media, LLC 2012

10.2   Solar Sailing

The Sun ejects mass at a rate of 150 billion tons per day, mainly consisting of charged particles, which are spewed out radially into the surrounding solar system. It would therefore seem a perfect opportunity to exploit this phenomenon for propulsive means. The outer most region of the Sun's atmosphere is called the corona and it extends from the top of the chromosphere out to a distance of several million kilometers where it gradually becomes the solar wind, a continuous outflow of over 10^9 kg of matter per second. The corona is hot at ~10^6 K by comparison with the photosphere and chromosphere which is at a temperature of ~10^3 K.

While the mechanism by which the corona is heated is the subject of decades of inconclusive research, it is well known that the temperature of the corona plays a key role in the acceleration of the solar wind. Suffice it to say, thermal acceleration is accomplished by the difference between the relatively high pressure near the surface of a star and the negligible pressure of the interstellar medium. The main source of the solar wind comes from coronal holes. In coronal holes bright spots of x-ray emission appear and disappear in a matter of hours. Weaker x-ray emission coming from coronal holes is characteristic of the lower densities and temperatures that exist in those regions. The existence of coronal holes is linked to the Sun's magnetic field. Closed field lines form loops that go back into the Sun but in coronal holes the field lines can become open and extend out to large radial distances where any emitted charged particles are forced to spiral around the magnetic field lines whilst flowing away from the Sun. Hence the origin of the solar wind is particles escaping from coronal holes where the magnetic field configuration is diverging.

The solar wind is accelerated out from the solar corona into interplanetary space, extending far beyond the orbit of the Earth and terminating after having hit the weakly ionized interstellar medium around 100 AU. Evidence for the existence of the solar wind derives from four main observations. Firstly, it is observed that comet tails always seem to point away from the Sun, hinting that solar radiation alone cannot be responsible. Secondly, ions collide with atoms in the Earth's upper atmosphere resulting in the aurora displays in the North and South magnetic poles. The ions become trapped within the field, bouncing back and forth between the two magnetic poles and creating the Van Allen radiation belts. Thirdly, studies of cosmic rays produced in solar flare events correlate with studies of the Earth's magnetic field. Finally, actual measurements of the solar wind have been obtained from orbiting spacecraft. This includes measurements of the solar wind temperatures, velocities and densities that help us to produce mathematical models of its dynamic and transient global behavior.

Typical measurements for the solar wind properties at 1 AU shows a proton and electron number density of 7 cm^{-3}, radial velocity between 400 and 500 km/s, magnetic field strength of 7 nT, temperatures of 10^5–10^6 K, gas pressure 30 pPa, magnetic pressure 19 pPa, sound speed 60 km/s. Particle interactions will occur every 46 days for proton-proton collisions and every 3–4 days for electron-electron collisions. The solar wind will take around 4 days to reach 1 AU from the Sun. The solar wind typically contains 95% protons, 4% alpha particles (fully ionized helium) and 1% minor ions, the most abundant of which are carbon, nitrogen, oxygen, neon, magnesium, silicon and iron. However, the solar wind also sweeps up material from the regions of interplanetary space as it passes. So the solar wind also contains abundances from entrained sources such as from comets, asteroids, planetary atmospheres and satellites. Observations by the Helios, WIND and Ulysses spacecraft indicate that helium is generally the faster major ion species in the solar wind. The mean particle flux from the Sun is around 10^{12} m^{-2} s^{-1}.

It was in the 1950s that the first dynamic theory of the solar wind was determined by Parker [1], which showed how it could be accelerated to supersonic speeds by an equation which describes the radial outflow velocity v of the solar wind as a function of the local sound speed c_s, radius r and critical radius r_c where the solar wind goes supersonic. This solution that Parker found was can be written in a shortened form assuming that the speed is large compared to the local sound speed:

$$v^2 \approx 2c_s^2 \left[2\ln\left(\frac{r}{r_c}\right) + 2\left(\frac{r_c}{r} - 1\right) + \frac{1}{2} \right] \qquad (10.1)$$

We can solve this to obtain an estimate of the solar wind velocity at 1 AU. Taking r at 1 AU $= 1.5 \times 10^{11}$ m, $c_s = 10^5$ m/s, $r_c = 7 \times 10^9$ m we get an estimate of around 300 km/s. Observations at 1 AU give the speed of the quiet solar wind as between 300 and 400 km/s. This falls remarkably well within the Parker prediction. From a propulsion point of view, the solar wind appears very attractive. Indeed, it would seem highly sensible for a spacefaring civilization to come up with a method of exploiting the massive amounts of energy that are dumped into space continuously from the Sun at speeds ranging from 300 to 1,000 km/s.

One suggestion for using the solar wind as in spacecraft propulsion is to use a magnetic sail to deflect the charged solar particles and impart momentum. The spacecraft generates the magnetic sail by using a loop of superconducting wire with a current passing through it. In one configuration proposed by the American aerospace engineers Robert Zubrin and Dana Andrews [2, 3] a sail located near Earth orbit would require a field strength of around 50 nT using a loop of wire 50 km in radius to create a bubble 100 km in size. As the spacecraft moves away from the Sun the solar wind flux will fall off inversely with distance and so will any thrust generated at the spacecraft. Although as well as the solar wind, the spacecraft could be used around the magnetospheres of planets.

Another suggestion is to use an inflated plasma-injected generated magnetic field analogous to the Earth's magnetosphere so that any incoming solar wind particles exert a dynamic pressure and are deflected. Such a scheme has been proposed under the name *Mini-Magnetospheric Plasma Propulsion* (M2P2) [4, 5] as illustrated in Fig. 10.1. Using a helicon plasma generator that ionizes a gas like Argon or Helium with high power radio waves can generate the field. The idea for

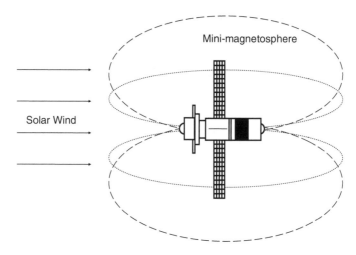

Fig. 10.1 Illustration of mini-magnetospheric plasma propulsion

plasma expanded magnetic fields has been tested in ground-based experiments and the proponents of this propulsion system are optimistic that given time they can demonstrate the plausibility of this approach. One of the advantages of M2P2 is that you can also use the magnetic field to shield any crew from cosmic rays and solar flares, provided the field is strong enough. Scientists working on the scheme say that they like to think of it as the first externally powered fusion engine where the Sun itself forms the engine – a nice thought at least.

People have been travelling across the world's oceans in sailing ships for centuries. For much of history, this has been the main (and only) method of transporting cargo between nations and allowing free economic trade. Over time, this has brought the world together and the human race has become one. It is easy to ignore this history when living in an age of aeroplanes, rockets and satellites and we tend to think of sail technology as being part of the past and now left only for sport and leisure. In fact, sails may be the future for humanity but not necessarily within the world's oceans, but in the depths of space between the planets and Moons that make up our solar system. But the unfortunate thing to note about the solar wind is that the tenuous stream of charged particles exerts very little pressure on any objects in its path. To produce real pressure for thrust generation, you really need the pressure of photons or electromagnetic energy and this pressure can be easily observed by watching the long tail of a comet pointing away from the Sun. This can be derived from the Sun directly.

The idea of a *Solar Sail* is wonderfully romantic, described elegantly in the science fiction short story "The Wind from the Sun" by Arthur C. Clarke [6], originally published as "Sunjammer" in 1964. It may have been better to have named this concept the Photon Sail or the Light Sail as this would have been more appropriate and removed some of the confusion over the energy source that often arises. If you want to invent a propulsion concept that relies on the Sun as a fuel

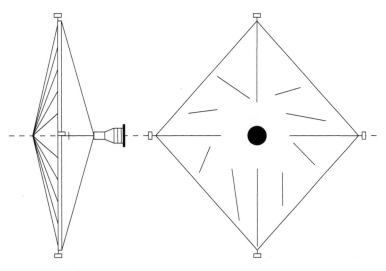

Fig. 10.2 Illustration of a solar sail pulled spacecraft

source external to the vehicle, you really have only three choices; particles, fields or photons. For a solar sail it is not the particles or fields we are interested in but photons of light. Photons are odd particles because although they do not have mass; yet they do have momentum which can be imparted to any sail. Light from the Sun will impinge on the surface of the sail. Some of this incident light will be scattered off, some will be absorbed into the material surface and some will be transmitted through it. Ideally any sail should have a high reflectivity and low absorption so that the ratio of the two incident/reflected <1. Sailing ships have several key components, which make up their design. This includes the mast, sail, hull and any rigging to hold it all together. Similarly, a solar sail for spaceflight will have a form of mast to maintain the sail rigidity, perhaps some engineering structures to store the sail and any instruments or mechanisms, as well as any cables or booms that enable the fold-ing/unfolding sequence. The sail itself would be a delicate piece of material, much thinner than sails used for oceans on Earth and having entirely different engineering requirements. A sail configuration is shown in Fig. 10.2.

The technology of a solar sail can be considered from the perspective of four key issues. The first is the density of the sail, which must be light for maximum light pressure. The second is the thickness of the sail, which must be thin for the same reason. Thirdly, there is the issue of size. A 2 m × 2 m sail will have an area of only 4 m^2, which compares to a 5 × 5 m sail that will have an area of 25 m^2. In terms of collecting the maximum number of photons of light, the larger the area the better. Finally, there is the issue of reflectivity. The more reflective the surface of the sail, the more the number of photons will be reflected (instead of absorbed) and the greater the amount of momentum imparted. The amount of light that is reflected from the surface can depend on the rigidity of the sail. This can be controlled by the centre of mass location and determines the pressure distribution over the surface.

The actual amount of solar radiation flux from the Sun varies with distance according to an inverse square law. The intensity of the solar radiation flux at the orbit of the Earth is ~1,400 W/m^2 whilst at the Sun it is around 65 MW/m^2, being a factor $1{,}400 \times (0.00465)^{-2}$ larger. The pressure of sunlight at 1 AU is around 9 N/km^2, which is sufficient to use for spacecraft propulsion. Any solar sail mission to the outer solar system and beyond that aims for maximum acceleration would ideally first perform a '*Sundiver*' maneuver into the Sun prior to heading out of the solar system. The problem however, is that solar intensity drops off with distance squared, although so does gravity, so once sufficient velocity is attained the sail should be able to maintain that motion. The solar irradiance is given by [7]:

$$S_r(\text{W/m}^2) = \frac{3.04 \times 10^{25}}{r^2} \tag{10.2}$$

The term r is the separation distance (solar radius) in meters between the centre of the Sun and the solar sail, which if positioned at 1 AU distance is 1.496×10^{11} m. This computes to a solar irradiance of 1,346 W/m^{-2}, although the figure of 1,400 W/m^{-2} is often quoted. We can then calculate the solar pressure at this distance by including a term for the sail reflectivity μ which would be equal to unity for a perfectly reflective surface and something like between 0.8 and –0.9 for a realistic sail design. The solar pressure is given by:

$$P_{rad}(\text{N/m}^2) = \frac{1+\mu}{c} S_r \tag{10.3}$$

For the 1 AU sail this computes to a solar pressure of 9.1×10^{-6} N/m^{-2}. A key parameter for comparing solar sail concepts is the sail loading which is an areal density measure given in the units of g/m^2. The mass includes all of the spacecraft systems such as the sail, science payload, communications antenna and any supporting structure. The sail loading is given by the ratio of sail mass to area:

$$\sigma(\text{g/m}^2) = \frac{m}{A} \tag{10.4}$$

We can then combine the sail loading and the solar radiation pressure to calculate the characteristic acceleration a$_c$ of the sail. This is a measure of the rate at which the sail would accelerate at 1 AU assuming it is normal to the solar radii. It is given by:

$$a_c(\text{m/s}^2) = \frac{P_{rad}}{\sigma} \tag{10.5}$$

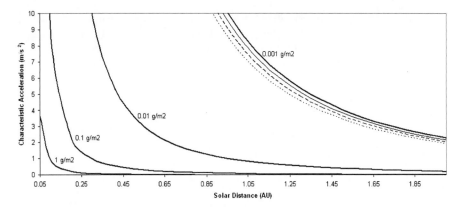

Fig. 10.3 Sail performance within the inner solar system

This can all be combined to give an equation for the characteristic acceleration as a function of the sail design (reflectivity, loading) and the distance (solar irradiance) from the Sun:

$$a_c = 3.04 \times 10^{25} \frac{1+\eta}{\sigma c r^2} \tag{10.6}$$

Figure 10.3 shows the characteristic acceleration resulting from different sail loadings at specified distances within the solar system, out as far as nearly 2 AU, which is past Mars, but not as far as Jupiter. Clearly the characteristic acceleration achievable drops with distance from the Sun although this can be increased again if the sail loading is made smaller. For the 0.001 g/m^2 sail loading curve the sensitivity to reflectivity is also shown with the upper most curve corresponding to a reflectivity of unity and then counting down to a reflectivity of 70% at the lowest curve.

The next bit of important information we need to know about solar sails is something called the Lightness number that compares the sail characteristic acceleration due to solar radiation pressure to the acceleration caused by solar gravitation at 1 AU, which is equal to 0.00593 m/s^2. The lightness number is given by:

$$\lambda = \frac{a_c}{g_{1\,\text{AU}}} \tag{10.7}$$

For a Lightness number of unity and assumed gravitational acceleration at 1 AU we can use (10.5) and re-arrange for the maximum sail loading assuming perfect reflectivity which is given by:

$$\sigma_{\max} = \frac{P_{rad}}{a_c} = \frac{9.1 \times 10^{-6}\ \text{N/m}^2}{0.00593\ \text{m/s}^2} = 0.00153\ \text{kg/m}^2 \equiv 1.53\ \text{g/m}^2$$

So this is the maximum sail loading possible within these constraints. In reality early evolution sail designs will have a Lightness number of <1 but the concepts will then evolve into more advanced designs with a Lightness number of >1.

The escape velocity for a sail is a function of the initial characteristic acceleration distance, the reflectivity and Lightness number. Assuming a parabolic sail, which is normal to the Sun, this is given by:

$$V_{esc} = \frac{42.1 \ (km/s)\lambda^{1/2}}{r^{1/2}} \tag{10.8}$$

The Lightness number can be written as a function of the sail loading:

$$\lambda = \frac{0.001574\mu}{\sigma} \tag{10.9}$$

If we assume perfect reflectivity ($\mu = 1$) and a sail loading of 0.1 g/m^2 then $\lambda = 15.74$ and if starting from a height of 0.2 AU from the Sun this corresponds to an escape velocity of around 374 km/s. At this cruise speed the spacecraft would take around 3,450 years to reach Alpha Centauri. If the sail loading were decreased to 0.01 g/m^2 then this would achieve an escape velocity of 1,181 km/s and reach its destination of Alpha Centauri in around 1,100 years. In essence, the lower the sail loading the higher the escape velocity and the shorter the mission duration to the target.

A solar sail is of much more use within a solar system in the presence of a star that it is out in the cold depths of space. It really comes into its own if it's in a solar system that has more than one star. The Alpha Centauri system has three stars, so provided you could get a sail there, such a vehicle would be of enormous benefit for moving around the solar system capturing images of the local planets; although Proxima Centauri maybe too far away (~1/5th of a light year from Alpha Centauri) to make this practical. The sail can move about by changing the orientation from the direction of motion of the local star. This causes a lateral acceleration of the sail, which can then either send the vehicle on a trajectory out of the solar system or send it back towards the inner part of the solar system. The engineering terminology of such maneuvering is referred to as orbit cranking or twisting.

Table 10.1 shows the solar irradiance, radiation pressure and characteristic acceleration as calculated from the equations in this chapter, at the distance of the different planets in our solar system, out to the dwarf planet Pluto and even as far as 100 AU. It is clear from this data that within the inner solar system the solar intensity is quite high and so utilizing sail technology, as part of a space transportation infrastructure within this domain would seem very appropriate. This would include missions to Mercury, Venus, Earth, Mars or any of the moons or Lagrange points.

In terms of an interstellar mission however, a sail would require a very low mass in order to have a sufficiently low sail loading, if it is to have any chance of reaching its distant target. The big technical issue is that in order to get a sufficiently high acceleration the sail must first fly close to the Sun in order to pick up that enormous

Table 10.1 Properties for various solar system bodies assuming perfect sail reflection

Object	Distance (AU)	S_r (W/m^2)	P_{rad} (N/m^2)	a_c (m/s^2) ($\sigma = 0.01$ g/m^2)	a_c (m/s^2) ($\sigma = 0.1$ g/m^2)
Sun	0.005	~5 × 10^7	<1	30,000	3,000
Mercury	0.387	9,070	6.0 × 10^{-5}	6	0.6
Venus	0.723	2,599	1.7 × 10^{-5}	1.7	0.17
Earth	1.000	1,358	9.1 × 10^{-6}	0.91	0.09
Mars	1.524	586	3.9 × 10^{-6}	0.39	0.04
Jupiter	5.203	50	3.3 × 10^{-7}	0.03	0.003
Saturn	9.554	15	1.0 × 10^{-7}	0.01	0.001
Uranus	19.220	3.7	2.5 × 10^{-8}	0.002	0.0002
Neptune	30.110	1.5	1.0 × 10^{-8}	0.001	0.0001
Pluto	39.540	0.87	5.8 × 10^{-9}	0.00058	5.8 × 10^{-5}
Heliosphere	100.000	0.14	9.3 × 10^{-10}	9.3 × 10^{-5}	9.3 × 10^{-6}

solar pressure and achieve sufficient escape speed for an interstellar trajectory. However, in attempting this, the sail must fly close to a very hot ionized plasma environment, the solar outer atmosphere, and risk the wrath of any coronal mass ejections as well as survive the large heat gradients. To do this would require a very special material. It is not to say that this is impossible, it's just that it requires a major research effort into the materials that are appropriate for this engineering requirement. Finally, it is worth noting that the usefulness of the sail will become more apparent as the Sun ages, because in around 5 billion years the Sun will evolve to a bloated Red Giant which is much more massive than the current solar size, allowing the sail to be accelerated from larger solar radii.

Using a Beryllium sail design as a high performing configuration the physicist Greg Matloff has estimated that for a spacecraft unfurling its sail in the region of 0.5–1 AU, interstellar cruise velocities can be obtained which is a factor 2–3 larger than for the same design unfurling close to the present Sun [8]. This is also useful for any spacecraft visiting another solar system and deploying sail sub-probes, where if a Red Giant is present the opportunities to literally 'sail' around the system are improved.

In another study with fellow physicists Les Johnson and Claudio Maccone, Matloff examined the possibility of a 20–30 year sail mission for a 500 kg spacecraft, called *Helios* [9]. It first approaches within 0.2 AU of the Sun and then after the sail unfurlment it goes into an initially elliptical orbit and then splits into two separate components. One is a scientific payload bound for the heliopause and a distance of 200 AU with the aid of a gravity assist, and the other is designed to rendezvous with a Kuiper belt object 30–40 AU from the Sun and then decelerate across the solar system using nuclear-electric propulsion based on a 3,000–5,000 s specific impulse ion engine. Such missions appear highly viable in the coming decades.

It is worth mentioning a couple of exciting solar sail projects that have occurred in recent times. During May 2010 the Japanese Space Agency JAXA launched the world's first solar sail driven spacecraft known as IKAROS or Interplanetary Kite-craft Accelerated by Radiation of the Sun. This was launched with the

Venus Climate Orbiter. The mission successfully deployed the approximately 200 m^2 area polyamide based sail, which massed over 300 kg in orbit and was able to generate solar power as well as demonstrate elements of navigation and acceleration. It did all this whilst undergoing spinning motion at 20–25 revolutions per minutes to keep it flat. The sail will eventually move on to travelling around the planet Venus whilst also investigating physics effects relating to gamma ray bursts, the solar wind and cosmic dust. The team plans to deploy another sail with an area well over 1,000 m^2, which will be integrated with ion engines, all geared towards missions to the outer planets.

The American based Planetary Society is also building a solar sail known as LightSail-1. Four triangular sails arranged in a diamond like shape built using around 32 m^2 of Mylar and having a mass of less than 5 kg. Lightsail-1 is made up of three separate cubesats. These spacecraft have an identified role. The first will serve as the central electronics component of the sail. The other two will form the basic solar sail module. The plan is to launch the spacecraft to an altitude of 800 km where it will only be subject to two forces; Earth's gravity and the intensity of solar light so as to demonstrate the principle of using photon pressure for propulsion. This is the first of three spacecraft they plan to launch from 2011 onwards using Cubesat spacecraft. The later LightSail projects, two and three aims to go further and demonstrate missions beyond Earth orbit carrying a larger scientific payload. One of the key applications of this technology is seen by the designers to be providing an early solar storm warning station.

Both IKAROS and the various LightSail projects are exciting developments, which clearly demonstrate the application of solar sail technology to interplanetary exploration and perhaps beyond. As if these projects weren't sufficient to demonstrate that the age of solar sail technology has arrived, as this book was completed, in November 2010 NASA Ames Research Center deployed a Cubesat sail to Low Earth Orbit called NanoSail-D2, which was 4 kg in mass and had a solar sail area of 10 m^2.

10.3   The Interstellar Heliopause Probe

In 2006 the European Space Agency acknowledged the growing interest in a mission that goes beyond our solar system and outside of the solar heliosphere. With this in mind they completed an initial study for the *Interstellar Heliopause Probe* (IHP) [10] to investigate the feasibility of a mission that would go to a distance of 200 AU from the Sun (see Fig. 10.4). The main motivation for this was to study potential missions that were technologically demanding but scientifically interesting and not part of the current ESA science programs. These are known as Technology Reference Studies. IHP aims to understand the nature of the interstellar medium and how it interacts with the solar system. In particular, the location of the termination shock and the heliopause are not yet known exactly.

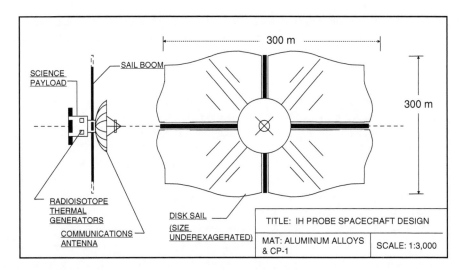

Fig. 10.4   Illustration of the interstellar heliosphere probe

IHP would be launched using a Soyuz Fregal 2B launch vehicle from Korou. The mission would take around 25 years from launch to destination and in order to keep the mission duration short it would be launched in the direction where the heliopause is closest. During the early phases of the development, designers considered several propulsion options including chemical, nuclear electric and solar sails. Chemical propulsion was dropped due to the small payload mass constraints driven by a limited specific impulse. Nuclear electric was dropped due to the large power requirements and large mass of the reactor. So a solar sail was chosen for the mission.

The plan was to use either a square sail or a spinning disk sail. The spinning solar sail was chosen for being lighter and smaller for the same acceleration requirements of around 1.5 mm/s^2. It would be around 1–2 microns thick with a sail loading of 4 g/m^2 and a sail radius of around 150 m and an area greater than 50,000 m^2. The sail would be constructed of a material called CP-1 and have a spacecraft mass of around 310 kg. One of the technical challenges of the mission is the construction, manufacturing and deployment of this fragile sail. It would be coated to protect it and ensure it has sufficient reflectivity throughout the mission. Then in order to obtain the solar system escape velocity it would make use of two gravity assists from the planet Saturn, passing within a quarter of an Astronomical Unit of its surface, the distance being a balance between achieving high velocity and maintaining a cool sail temperature. Once the sail has performed its main function after 5 years of operation, it would be jettisoned.

The spacecraft would have a total launch mass of around 624 kg, of which 210 kg would make up the spacecraft dry mass and an extra 100 kg system margin allocation. The sail would be made structurally rigid by the use of lightweight booms, which are 120 m in length. Current boom designs have a specific mass of around 100 g/m, which implies a large mass penalty for the mission. Attitude control could be enabled by the use of a gimbaled boom between the sail and the spacecraft. Either the use of an RTG or a thermal electric generator would power the spacecraft. It would carry a suite of instruments which are expected to function after the 25 year flight. This includes a plasma analyzer, magnetometer, charged particle detector, an ultraviolet photometer for measuring the hydrogen density and an energetic particle detector for measuring the cosmic ray flux. The results from the measurements would be trans- mitted back to Earth at a downlink rate of around 200 bits per second (bps) using either radio frequency or optical communication systems. The power requirements for these systems would be around 15 W.

Both the ESA *Interstellar Heliosphere Probe* (and others like the NASA *Inno- vative Interstellar Explorer* discussed in Chap. 9) is the sort of missions that the government space agencies should be undertaking. Local planetary and lunar exploration should be left to a commercialized private space industry in line with the aspirations of the US President Obama's vision for space exploration.

10.4   Interstellar Probe

In the late 1990s the NASA Jet Propulsion Laboratory (JPL) organized a project for a space probe that was designed to exceed 200 AU distance as a minimum goal in 15 years. The Interstellar Probe Science & Technology Definition Team led the study and the project was called the *Interstellar Probe* (ISP) [11] and would embark on a journey outside the solar heliosphere and study the connection between the Sun, Earth and the interstellar heliosphere. The probe would have a mass of 150 kg and its propulsion would be via a 200 m radius solar sail with a required areal density of ~1 g/m^2, designed to achieve a velocity of around 70 km/s or 14 AU/year. This is around five times the speed of the Voyager 1 and 2 spacecraft.

The key reason for selecting a sail for the mission was the mission requirement of getting to 200 AU within only 15 years. This meant that a propulsion system was required which gave rise to cruise velocities, which were several multiples of those of the Voyager and Pioneer probes. Hence after some discussion the team decided that sails were the only technology available which offered both the performance and near term technological maturity, with some initial investment. The mass of the probe and sail combined would be around 246 kg and is illustrated in Fig. 10.5.

The initial trajectory would see the spacecraft head into the inner solar system and fly within a quarter of an Astronomical Unit of the Sun in order to take advantage of the enormous radiation pressure at this distance. This would give it sufficient acceleration to boost out of the solar system, and the sail would be ejected at around 5 AU on the way out when the acceleration becomes inconsequential and to ensure there is no interference with the on board instruments. The spacecraft antenna dish would serve as the main structure and three struts around the 11 m central hole in the sail would support this. Control of the spacecraft is achieved by moving the central mass with respect to the sail centre of mass by changing the length of the three struts and it would be spin stabilized when in sail mode.

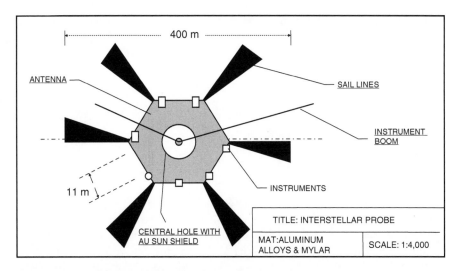

Fig. 10.5 Illustration of the interstellar probe mission

The main objective of the mission would have been to study the nature of the interstellar medium and its influence on the solar system and evolution. Another main objective was to explore the outer solar system to look for clues to its origin and understand the nature of other planetary systems should they exist. A significant penetration into the interstellar medium was seen as a key mission goal. It would carry on board a large suite of instruments designed to measure the properties of energetic particles, cosmic rays, magnetic fields and the dust at the boundary between the heliosphere and the nearby interstellar medium. These instruments were known as the 'strawman payload' and made up around 25 kg of the total spacecraft mass and required around 20 W allocation. Data would be transmitted back to Earth at 220 W power using the 2.7 m antenna via the Earth Deep Space Network.

Additional instruments considered for the mission included the use of a telescope to search for Kuiper belt objects or an instrument to identify organic molecules in space. More exotic possibilities for the mission included the measurement of low energy antiprotons emitted from primordial black holes or a search for so called Weakly Interacting Particles (WIMPs), which could make up the missing dark matter in cosmological models. Although the probe was designed to go to 200 AU or more, it was capable of going out as far as 400 AU in 30 years and indeed this was one of the goals of the designers. This really would have been an ambitious yet achievable project if it had been supported past the planning stage.

10.5 Beamed Propulsion

Because the solar intensity reduces the further out into space you go, it has been suggested that giant lasers could be built in orbit around the Sun and then pumped with sunlight, which could send a highly collimated, narrow beam continuously

Table 10.2 Laser-beam driven sail performance as determined by angular resolution or spot size (km)

Distance (s)	$D = 10$ km	$D = 100$ km	$D = 1,000$ km	$D = 10,000$ km
1 AU	0.036	0.0036	0.00036	0.000036
10 AU	0.36	0.036	0.0036	0.00036
100 AU	3.6	0.36	0.036	0.0036
1,000 AU	36	3.6	0.36	0.036
10,000 AU	359	36	3.6	0.36
100,000 AU	3,590	359	36	3.6
272,000 AU	9,766	977	98	9.8

towards the spacecraft. Just like the solar sail, the beauty of a laser driven sail is that it does not need to carry along its own propellant; the engines are left behind. Then you have the added benefit that in theory the laser beam intensity will not drop off inversely with distance, unlike pure sunlight, provided you can continue to sustain the beam power. The physicist James Benford gives insight into the potential benefits:

> The fundamental attraction of high power beams for space is simple: microwaves and lasers can carry energy and momentum (both linear and angular) over great distances with little loss. Photons lose a negligible energy when radiated out of a potential well such as Earth's. LEO chemical fuels payload cost is ~$167/MJ. Electricity to drive a microwave source costs ~$ 0.01/MJ. Microwave energy in space is cheap. In the long term, economics rules.

The point where the sail no longer receives the full output of the laser is termed the thrust run point. After this the intensity of the beam received by the sail will also drop off inversely with distance. For high acceleration and thrust, the laser would be in a position orbiting the Sun at some optimized distance, a balance between guaranteeing sufficient solar collection intensity and not actually destroying the collector array. One major advantage of keeping the energy source back within the solar system is that any components that fail (say due to over heating) can be replaced. Indeed, as the laser technology advances over the years of the flight, the laser system can be completely upgraded and eventually replaced. This gives the benefit of a potential increase in efficiency whilst the spacecraft is moving further away, or at least to mitigate any reduction in the efficiency. Any generated laser beam is fired off into deep space in the direction of the spacecraft. If the beam is made up of electromagnetic energy (light) then it would simply reflect off of some rear surface area to provide thrust in the same way as a conventional solar sail does. This is illustrated in Plate 8.

The key bit of physics to know pertaining to a laser beaming system relates to angular resolution and is defined by Rayleigh's law which refers to the balance between optical aberration (e.g. blurring due to the transmission of light over distance and a lack of convergence) and diffraction (the spreading out or divergence of light waves over distance) of light over distance. Rayleigh's law enables an assessment to balance these two effects and allows an accurate estimate of the angular resolution of an optical beam system. This is obtained by specifying the diameter of the lens aperture D and the wavelength of the light λ being passed through the lens as well as the distance s from the lens to the spacecraft. These are related to the beam spot size d as follows:

$$d(m) = 2.4 \frac{s\lambda}{D} \tag{10.10}$$

Table 10.2 shows the beam spot size results from calculating the effects of a 1 μm wavelength laser system using different lens sizes and over different distances.

As the laser beam hits the sail it will obviously heat it, implying a maximum thermal loading and thereby acceleration. This is given by combining some of the sail parameters discussed in the previous section with the law for radiation emission as follows:

$$a(m/s^2) = \frac{\varepsilon\mu}{\alpha\sigma}\left(\frac{4\sigma_{SB}T^4}{c}\right) \tag{10.11}$$

In (10.11) ε is the emissivity, μ is the sail reflectivity, α is the absorption coefficient, σ is the sail loading, σ_{SB} is the Stefan-Boltzmann radiation constant $(5.67 \times 10^{-8}$ W/m^{-2}/K$^{-4})$, T is the sail material temperature and c is the speed of light in a vacuum. The power required to push a given sail design is then given as a function of m which is the total mass of the spacecraft (structure, sail and payload):

$$P_s(W) = \frac{mca}{2\mu} \tag{10.12}$$

As an example let us imagine we have a mission called *SailBlazer*. It has a total spacecraft mass of 500 kg with a sail loading of 0.2 g/m^2.

The emissivity is 0.05 and the reflectivity is 0.9, with an absorption coefficient of 0.15. Assuming the sail is exposed to a temperature of 500 K then the acceleration will be 0.07 m/s^2 and the power required to push this sail will be around 5.83 GW. The sail diameter required for this design will be given by:

$$d_s = 2\left(\frac{A_s}{\pi}\right)^{1/2} = 2\left(\frac{m}{\pi\sigma}\right)^{1/2} \tag{10.13}$$

For this case study this computes to a circular sail diameter of 1,784 m. Let's now assume the sail is accelerated for 5 years at the value given of 0.07 m/s^2 then by linear equations (discussed in Chap. 3) the sail will achieve a cruise velocity after this time of 11,038 km/s, which is the equivalent of 0.037c. After this acceleration period the sail would have traversed a total distance of 8.7×10^{14} m or 5,817 AU. This is well into the Oort cloud and such a mission should be perfectly feasible with only a moderate sail size, spacecraft mass typical of those launched into orbit today.

Table 10.3 Laser driven sail missions for an interstellar roadmap

Mission	Kuiper belt	Oort cloud	Interstellar flyby
Total distance (AU)	100	10,000	4.2
Total duration (years)	5.3	17.6	42.2
Cruise velocity (km/s)	100 (0.0003c)	3,000 (0.01c)	30,000 (0.1c)
Sail diameter (km)	1	1	1
Lens diameter (km)	1	100	200
Spacecraft mass (kg)	200	200	100
Payload mass (kg)	66	66	33
Acceleration (m/s^2)	0.003	0.027	2.7
Power (GW)	0.1	1	25
Thrust run (AU)	5.5	550	1,100

The only difficult technical part is the deployment of a laser beam capable of providing multiple GW of power. Approximately 75% of the laser light will also miss the solar sail and so the spot sized needs to be designed so as to match the sail diameter to the maximum range the probe will go to during the mission.

In 1998 a group of physicist came together to consider the potential for beamed lightsail power systems as part of a roadmap for interstellar exploration. This included physicists such as Geoffrey Landis and Robert Forward, two of the experts in this field. The results of their deliberations were eventually published in the Journal of the British Interplanetary Society [12]. The team came up with what they called three 'strawman' missions that progressively built towards a full interstellar mission. These missions are shown in Table 10.3 but with some minor revisions to the mission duration by this author, along with the lasers and lens parameters required for each mission.

The team identified interstellar dust as a possible mechanism for performance degradation and measurements of the interstellar dust size distribution would be an important science driver in any precursor missions. In order to bring about such an interstellar precursor roadmap a simple sail demonstration was seen as the first crucial step and as discussed in the last section this has now been achieved. Future tests would include the deployment of laser lasers in space and the use of laser launched rockets. For any interstellar mission decelerating at the target star system was seen as highly desirable. Laser driven sails clearly do provide for good potential for future interplanetary, precursor and interstellar missions, provided such a program was implemented and the key technological steps demonstrated. Especially since the concept does not require the spacecraft to carry on board its own reaction mass, a disadvantage of other propulsion schemes. The specific impulse may in theory be unlimited as the lasers can continuously be replaced. The idea of using a laser beam driven sail for spacecraft propulsion was used in the science fiction story *"The Mote in God's Eye"* [13] by the author's Larry Niven and Jerry Pournelle.

It has also been suggested that a ramjet could be powered by the use of laser beams in orbit around the Sun. This is known as a *Laser Powered Ramjet* concept

[14, 15]. The laser accelerates the charged particles of the solar wind towards the spacecraft for both acceleration and deceleration. It is claimed that the performance of such a system will be more effective than the interstellar ramjet (discussed in Chap. 13), when the spacecraft is moving towards the laser but less effective than an interstellar ramjet when the spacecraft is moving away from the laser beam. The power required would be proportional to things like the spacecraft mass, acceleration, speed, size of the ramjet intake area and the density of space that the vehicle was moving through.

This technology could be used in conjunction with a 10–250 km sized sail depending on the proximity to the Sun. The sail would have the same size as the laser lens to offset beam diversion over the distance. The sail would be very thin, low density, and have the mass of several thousands of tons. To maneuver within a solar system the spacecraft would deploy high voltage wires to interact with the solar magnetic field and allow it to essentially 'tack' during its trajectory (although it would possibly take years to affect the maneuver due to the large mass of the spacecraft and the weakness of the electrostatic field). The laser would produce around 10^{13} Watts of power emitted from an aperture that was tens of kilometers in size. Accelerating at 0.35 g a spacecraft propelled by this method could reach heliocentric escape velocity and achieve 60–200 km/s [16].

It has also been suggested that a laser could be used in conjunction with a fusion engine (discussed in Chap. 11), a so-called *fusion enhanced auxiliary laser thruster*. The gas laser would be positioned on board the spacecraft and would be doped to absorb any x-rays produced from an internal fusion engine. As the x-rays leave the main body of the vehicle they enter the gas laser and pump it to a state of population inversion. The interesting thing about this concept is that the exit nozzle for the laser could be positioned at either ends, allowing it to be used in both acceleration and deceleration modes.

It is also worth noting that instead of laser energy collection from the Sun, a device could be used which directly beams matter particles at the spacecraft, such as electrons, ions or protons. This would be achieved using an electromagnetic gun located either in Earth orbit or on the Moon. Although for the charged beam one would have to deal with the inevitable dispersion of the charge density field as the beam interacts with the solar and intergalactic magnetic fields as well as the tendency of the beam to become uncollimated over distance. Finally, another alternative is not to launch particles with the intention of imparting momentum but to be used directly as fuel. The particles could be fired in advance of the spacecraft, so that it picks them up on route and uses them in some propulsive process (e.g. fusion) or the particles can be beamed towards the spacecraft at a faster speed, arriving at some rear engine collection intake to be then use in a thrust generating process. However, it is an understatement to say that the targeting required for such a propellant beam would have to be very accurate.

One physicist Gerald Nordley has examined beamed particle systems in detail [17, 18]. He makes the good point that decoupling the question of interstellar travel from the concerns of rocketry (exhaust velocities, mass ratio) will make such missions more feasible in the future. It does appear to be an efficient way to convert collected energy in the solar system consistent with other propellantless systems and is worth exploring further. Nordley is optimistic about the potential of beam power propulsion:

> When I read that the wind from the Sun could blow magnetic sails toward the stars, I thought nice, but that wind is rather tenuous and is limited to 600 km/s or so. Why not provide our own wind?

10.6   Beamed Microwaves and Starwisp

Instead of using laser beams to propel a spacecraft over long distance one can send out a beam of microwaves using a maser. This is the hypothetical propulsion method for the unmanned probe concept named *Starwisp*. This was a design study conducted in the 1980s and first proposed by Robert Forward [19] and developed further by Geoffrey Landis [20]. The original idea apparently came to Forward from a discussion with the physicist Freeman Dyson who had earlier considered the idea of a perforated solar sail pushed by microwave radiation. Forward combined Dyson's ideas with other ideas from the field of long distance communications using thin wire meshes to produce the microwave sail concept that embodies Starwisp as is shown in Fig. 10.6 – the name being analogous to a thin and nearly invisible spider web.

Like the solar sail and laser sail, the microwave sail or maser-pushed sail offers a rocket-less solution for an unmanned probe with no necessity to carry on board fuels, one of the repeating themes of the work of Robert Forward and the motivation for developing something like Starwisp. The source of the microwave beams could be a 50,000 km circular diameter Fresnel zone lens located close to Earth orbit and having a mass of around 50,000 tons. This would convert solar energy into

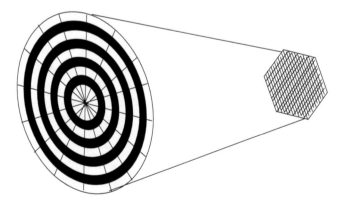

Fig. 10.6 Illustration of starwisp probe concept

electricity and then use that to generate a microwave beam to give an initial momentum impulse to the 1 km diameter spacecraft as well as reflection off the carefully designed wire sail, thereby imparting further momentum. A wire mesh is preferred due to its ability to reflect microwaves well.

The principle of operation is that the radiation wavelength of the beam would be much greater than the holes within the wire sail mesh. The wire mesh has holes, which are larger than the microwave wavelength but small enough to affect the phase of the microwaves passing through them so that the phase shift would be exactly 180°. This is a necessary condition in order to obtain sufficient reflectivity and thereby momentum. As a result any microwaves arriving at the spacecraft will be deflected and thus impart momentum. A microwave beam is thought to be more efficient than a laser beam although it is more difficult to transmit a narrow well-collimated beam over long distance. This means that once the spacecraft moves away from the source of the microwave beam, the photon pressure will begin to reduce, so any rapid acceleration must be accomplished early on in the mission before the beam pressure decays. This acceleration gradient necessitates high g forces and so it is unlikely that a microwave powered sail would be suitable for manned missions.

The spacecraft would be accelerated by a 10–50 GW microwave beam at 1,130 m/s^2 or 115 g and would attain a cruise velocity of 60,000 km/s or 0.2c with a specific impulse of around 6,000 s. Forward speculated that such a probe could even be accelerated to near light speed, due to its low mass. The really clever aspect of Starwisp is that once it arrives at the destination, provided it gets within a few AU of the local planetary bodies, it can form images using the sail mesh as an antenna to pick up reflected microwave radiation from the planetary bodies as well as any other matter in the system from which the microwave flooded area is able to detect in advance from a long distance. This would also allow any other instruments on the spacecraft to power up and perform additional science measurements.

The hexagonal wire mesh sail would be constructed of an ultra-low mass material such as carbon with a wire thickness of micrometer in width each wire spaced at equal distance from each other equivalent to the wavelength of the emitted microwaves, 3 mm. They would have a mass of order tens of grams and have a diameter of around 1 km.

The choice of carbon is to deal with any thermal heating loads due to the absorption of some of the microwave beam, which ultimately reduces the propulsion efficiency. Originally Forward suggested the use of aluminum or beryllium due to their high conductivity and he speculated possible superconductivity at the low temperatures of deep space for maximum reflectivity, but this was later shown not to be true by Geoffrey Landis and that in fact most of the microwaves would be absorbed by the sail and thereby overheating it. One of the difficulties with constructing such a fragile probe is in the manufacturing of the wire sail, perhaps using microprocessor chip technology for its assembly. For maximum reflectivity, the wires need to be light and thin, but if they are not sufficiently massive then structurally they may break and could also absorb too much energy, hence there is a performance trade-off to be understood with this technology.

In the original proposal by Forward the spacecraft would have minimal structural mass as all of the electronics and computer technology would be incorporated into the thickness of each wire located at each intersection. The wire sail would constitute its own payload. Although in the later design by Landis the 100 kg/km^2 density microwave sail would have a mass of around 1 kg as well as pulling an 80 g payload. The wire sail would have to remain of uniform density throughout to ensure a polarized microwave push in the same direction, and mesh distortions would have to be prevented.

Forward proposed that larger Starwisp probes could be launched to more distant locations such as to Epsilon Eridani at 10.7 light years. With much higher velocities of around half the speed of light they would require microwave energy beams of order 10 TW. Such sails would be around 30 km in diameter and have a mass of around 14 kg using a transmitter lens diameter of around 9,000 km, much smaller than the first example he studied. In theory a Starwisp probe could reach the nearest star systems within decades, with the flyby probe shooting through the target star system in a matter of hours. However, the reliability and repair issues associated with such a delicate probe would make it difficult to achieve such missions. The radiation and particle bombardment issues, particularly at the high speeds of 0.2–0.5c, would also present a significant technical obstacle to mission success. To be successful it would also be necessary to ensure that the efficiency of the Fresnel lens, beam and momentum transfer are high. However, it is worth noting that the potential of a maser beam sail shouldn't be ignored, especially if they offer orders of magnitude reductions in the cost and improvements in the power efficiency compared to a laser beam, as some have claimed.

Finally in this chapter we mention some of the exciting experiments that have been conducted into microwave beam driven sails by people such as the physicist James Benford. In published work the author discussed the effect of acceleration from sublimation pressure as mass is ejected downward to force a sail to move upwards [21–23]. The product of the thermal speed of evaporated ultralight carbon-carbon microtruss material and the associated mass ablation rate gives this force. According to the authors the magnitude of this effect can greatly exceed the microwave photon pressure provided the temperature is sufficiently high. In the experiments a 10 kW, 7 GHz microwave beam was sent into a vacuum chamber to impinge on 5–10 g/m^2 sails which were heated to between 1,700 and 2,300 K from absorption and then accelerated by 1.2–132.3 m/s^2 (many times Earth gravity) corresponding to flights of up to 60 cm in the test chamber which represented velocities of 0.3–4.08 m/s. Extensive diagnostics were used including optical, infrared photography, reflected microwave power and gas analysis. It is these sorts of experiments that will help to move propulsion technology to a higher Technology Readiness Level and should be supported, particularly if microwave beam propulsion is ever to become part of the propulsion options for real spacecraft in space.

10.7 Problems

10.1. A solar sail has a mass loading of 0.15 g/m². If the sail efficiency is 0.85 and it is launched from a distance of 0.5 AU, what will be the characteristic acceleration on the spacecraft? Assuming the Lightness number is 1.1 what escape velocity will the sail achieve and how long will it take the spacecraft to reach the outer edge of the Oort cloud at 50,000 AU distance?

10.2. You are designing a sail mission to reach the outer edge of the gravitational lens point at 1,000 AU. It is to arrive at its destination within 20 years from launching within the inner solar system. To escape from the solar system and reach its target within the specified duration it will require an average cruise speed of around 100 AU/year or 474 km/s. What distance in AU from the Sun will correspond to this escape velocity assuming a Lightness number of unity? What will be the solar pressure at this distance? Assuming a total spacecraft + sail mass of 100 kg with a sail loading of 1.2 g/m² what will be the area of the sail required for this mission?

10.3. A giant 0.35 λm laser is assembled in solar orbit with a lens equal to the diameter of the Moon, 3,476 km. Assuming that it accelerates a 1,000 kg spacecraft for 10 days at 0.0001 m/s² what will be the pressure if the reflectivity is 0.8? After the 10 days of acceleration what will be the cruise velocity of the spacecraft at this time and the distance in Astronomical Units where this occurs? If the spacecraft continues onto Barnard's Star 5.9 light years away, what would the hypothetical beam spot size be on the receiving spacecraft?

References

1. Parker, EN (1958) Dynamics of the Interplanetary Gas & Magnetic Fields, The Astrophysical Journal, 128, p664.
2. Zubrin, R (1991) The Use of Magnetic Sails to Escape from Low Earth Orbit, AIAA-91-2533, Presented AIAA/ASME Joint Propulsion Conference, Sacremento, CA, June 1991.
3. Andrews, DG & R Zubrin (1990) Magnetic Sails and Interstellar Travel, JBIS.
4. Wingley, RM et al., (2000) Mini-magnetospheric Plasma Propulsion: Tapping the Energy of the Solar Wind for Spacecraft Propulsion, J.Geophys.Res.,105, 21067–21077.
5. Wingley, RM (2001) Laboratory Testing of the Mini-magnetospheric Plasma Propulsion (M2P2) Prototype, CP552, Space Technology & Applications International Forum.
6. Clarke, AC, (2003) The Wind from the Sun, Galancz.
7. Matloff, GL & E.Mallove (1989) The Starflight Handbook, A Pioneers Guide to Interstellar Travel, Wiley.
8. Matloff, GL (2010) Red Giants and Solar Sails, JBIS, 63.
9. Matloff, GL et al., (2007) Helios And Prometheus: A Solar/Nuclear Outer-Solar System Mission, JBIS, 60, 12, pp439–442.
10. Lyngvi, A, et al., (2005) The Interstellar Heliopause Probe, Acta Astronautica, 57.
11. Liewer, P et al., (2000) NASA'S Interstellar Probe Mission, Presented Space Technology and Application International Forum.

12. Landis, GA (1999) Beamed Energy Propulsion for Practical Interstellar Flight, JBIS, Vol.52.
13. Niven, L & J Pournell (1999) The Mote in Gods Eye, Voyager.
14. Forward, RL (1976) A Programme for Interstellar Exploration, JBIS, Vol.29.
15. Cassenti, BN, (1982) A Comparison of Interstellar Propulsion Methods, JBIS, Vol.35.
16. Jaffe, LD et al., (1980) An Interstellar Precursor Mission, JBIS, Vol. 33.
17. Nordley, GD (1993) Relativistic Particle Beams for Interstellar Propulsion, JBIS, 46.
18. Nordley, GD (2000) Beamriders, Analog Science Fact/Science Fiction.
19. Forward, RL (1985) Starwisp: An Ultra-light Interstellar Probe, J.Spacecraft, Vol 22, No 3.
20. Landis, G (1999) Advanced Solar and Laser Pushed Lightsail Concepts Final Report for NASA Institute for Advanced Concepts, May 31.
21. Benford, G et al., (2006) Power-beaming Concepts for Future Deep Space Exploration, JBIS, Vol.59.
22. Benford, J et al., (2001) Microwave Beam-Driven Sail Flight Experiments, Proc.Space Technology and Applications International Forum, Space Exploration Technology Conf, AIP Conf. Proceedings 552, ISBN 1-56396-980-7 STAIF, p540.
23. Benford, J (2003) Flight and Spin of Microwave-Driven Sails: First Experiments, Proc.Pulsed Power Plasma Science, IEEE 01 CH37251, p548.

Chapter 11
Nuclear Fusion Propulsion

Thermonuclear rocket propulsion will be advantageous for interplanetary missions. The deuterium fuel is inexpensive, and the thermonuclear power planets may well prove to be comparable in size, mass and cost to the power plants presently used in ocean-going vessels. Thermonuclear spaceships, combined with nuclear shuttle rockets, may prove to be the basis of a commercially practical system of interplanetary transportation

J.R. Roth

11.1 Introduction

Long the subject of a promised dream, inertial fusion energy is now receiving more investment than at any time in its history. If the energy generated from a fission reaction is orders of *magnitude* that of a chemical reaction, then a fusion reaction gives a much greater improvement still. We know fusion reactions can be made to work in theory, because this is what powers the Sun every day of its existence. Taming this energy source will surely free up the people of Earth from an over reliance of fossil fuel technology and eventually make the world a better place to live for everyone by improving the quality of life. When the first commercial power producing reactor goes live, scientists and engineers will already be working on ways to make smaller, cheaper and more efficient designs. When this happens, the application to space propulsion will be obvious for all to see and finally the solar system and beyond will be within our grasp.

K.F. Long, *Deep Space Propulsion: A Roadmap to Interstellar Flight,*
DOI 10.1007/978-1-4614-0607-5_11, © Springer Science+Business Media, LLC 2012

11.2   Fusion: The Holy Grail of Physics

In 1953 the former US President Dwight D Eisenhower gave a tremendous speech to the United Nations General Assembly on the peaceful uses of atomic energy. The 'atoms for peace' speech was a call on all world leaders to find ways of fulfilling this goal. In particular, tensions were growing between nations in the West such as the United States with the Soviet Union in the East. Finding ways to work together was a prerequisite for developing a fundamental understanding between nations in the future. One of the projects that the United States first instigated was a civilian program called Project Plowshare which examined the possibility of using nuclear explosives in a safe manner to make harbors, canals and dams or even stimulate natural gas reservoirs. This project was cancelled in 1977 due to public concern over the environmental consequences.

One of the first areas of research for the United States (and countries like Great Britain) to openly discuss with the Soviet Union was magnetic fusion energy as a method of power generation. Scientists had long realized that the development of nuclear reactors on Earth would solve many energy shortage problems across the globe but also unite scientists and nations behind a scientific endeavor that would enhance humanity technologically to an extraordinary level. In particular, the development of reactors based upon the principle of fusion was seen as the key solution to all our problems. It was in 1920 that the astronomer Arthur Eddington had first speculated about what powers the Sun, this enormous plasma engine that burns for ten billion years before it enters retirement. The unsolved question however, was what exactly was the Sun burning to allow it to remain in this careful balance of hydrostatic equilibrium for so long? Because the Sun is so massive, it wants to gravitationally collapse to a smaller size, but energy generated at the core slowly makes its way towards the surface over a tens of thousands of year's timescale as radiation pressure maintaining the balance. It was not until 1939 in a paper titled *Energy production in stars*, that the Hungarian Physicist Hans Bethe first described the network of fusion reactions that allow the Sun to keep producing energy, sustaining it and all objects in orbit around it.

So what exactly is fusion? If fission is the fragmentation of two heavy isotopes to release energy, then fusion is the combination of two light isotopes to release energy. Because we discuss fusion reactions in detail in this chapter it is necessary to use notation. Fusion reactants are usually isotopes of hydrogen such as deuterium (D) and tritium (T) and isotopes of helium, such as helium-3 (He^3) and helium-4 (He^4). Neutrons and protons are usually produced in the reactions as well as we use the notation of n and p to refer to these. The process of sustained fusion reactions taking place is referred to as thermonuclear burning. This is perfect for the Sun, because it is made up of mostly Hydrogen and Helium. It quickly became obvious that the potential exists to generate the same fusion reactions on Earth for use in electrical power generation. A fusion reactor would be much safer than conventional fission reactors.

Although fusion reactions will produce a large neutron flux (for DT reactions expected to be around 100 times larger than for a fission reactor) and thereby activate the surrounding materials with induced neutron radioactivity, the half-life of the radioisotopes produced from these reactions is much less than those from atomic fission reactions. Also, a the materials around a fission reactor have to be carefully chosen for their cross section properties, but this wouldn't be the case for a fusion reactor so you could use materials which had a low activation energy and further minimize any radioactive contamination.

A technical issue for any fusion based space propulsion is the choice of fuel to use. DT reactions will ignite at the lowest temperature, followed by DHe^3 and DD. The main fusion reactions include:

DT: $H^2 + H^3 \rightarrow He^4(3.52MeV) + n(14.06MeV) \Rightarrow 17.58MeV/reaction$
DHe^3: $H^2 + He^3 \rightarrow He^4(3.67MeV) + p(14.67MeV) \Rightarrow 18.34MeV/reaction$
DD: $H_1^2 + H_1^2 \rightarrow H^3(1.01MeV) + p(3.03MeV) \Rightarrow 4.04MeV/reaction$
DD: $H_1^2 + H_1^2 \rightarrow He^3(0.82MeV) + n(2.45MeV) \Rightarrow 3.27MeV/reaction$

In this notation the energy release is given in the units of MeV, this is Mega-electron Volts. Because the energy release is so large from fusion reactions, physicists tend to use the eV notation for convenience, where 1 eV is equal to 1.602×10^{-19} J of energy. But how does one build a star on Earth and what do we know about the Sun? Well, we know that it is very big, very hot and dense at its core. But we can't possibly fit a whole Sun on the Earth, so this leads to the requirement for something that is hot and dense but at the same time small.

The research effort has led to the development of a whole community of scientists across the globe trying to work out how to produce fusion energy on Earth. After several decades of working out the basic theory and conducting numerous experiments, they have realized that there are two main routes to solving fusion known as Magnetic Confinement Fusion (MCF) and Inertial Confinement Fusion (ICF). The key to both is the fusion triple product, which states that a sufficient reaction rate will be sustained provided a criteria is met. This was first described by the British physicist John Lawson in 1955 and so known as the Lawson criteria: [1]

$$n\tau T \geq 5 \times 10^{21} m^{-3} s keV \qquad (11.1)$$

where n is the plasma density, τ is the energy confinement time for self-heating and T is the temperature of the fuel mixture. For a 10 keV plasma:

$$n\tau \geq 5 \times 10^{20} m^{-3} s \qquad (11.2)$$

MCF uses magnetic fields that trap and control charged particles and confine a low density plasma (10^{-6} cm^{-3}) but for a long duration (seconds). ICF uses laser beams to irradiate the surface of a fuel capsule, causing material ablation, and then uses the inertia of the material itself to confine the high-density plasma (10^{23} cm^{-3}) but for only a short duration (nanoseconds). The key to any fusion reactor and designed capsules is the attainment of ignition (defined by the triple product) and gain which is when more energy is released from the capsule that is used to confine it, measured by the ratio of fusion alpha (He-4 particles) heating power to input heating power or otherwise denoted Q. For any commercially viable fusion reactor $Q \gg 10$ is required. A fundamental part of any fusion reactor is the use of a Lithium blanket to convert the energy of fusion neutrons into heat.

It is well known that the Sun is able to confine the fusion plasma by the presence of a massive gravitational field. An experimental Tokomak reactor will confine plasmas by using magnetic fields. The basis of MCF is the heating of a gas contained within a vacuum chamber, by microwaves or electricity, then ionizing it into a plasma state. The gas is then squeezed by super conducting magnets to a state of ignition. The magnetic field produces an efficient Toroid shape for the control of the plasma, enabled by a Tokamak device. Significant progress has been made in the development of MCF technology. In particular, the Joint European Torus (JET) has demonstrated substantial progress in igniting and sustaining fusion reactions for a brief period. It is the world's largest Tokamak facility and construction started as early as 1978 with the first experiments taking place in 1983 and has been managed by United Kingdom Atomic Energy Authority (UKAEA) since 1999. JET uses DT fusion fuel for the fusion reactions. During one campaign using DT fuel JET achieved 40 MW of fusion power for one second equating to a gain $Q = 0.7$ (the ratio of energy in to energy out). Future upgrades may exceed this. The next generation of MCF technology is the International Thermonuclear Reactor (ITER) facility, which aims to prove the technical feasibility of a fusion reactor. ITER is twice as big as JET and began construction in France in 2007 but will not begin operation until 2018. It should produce ~500 MW of fusion power for ~1,000 s and is a true technology demonstrator.

In ICF a fusion gas is contained within a high Z pusher capsule [2]. Laser beams are used to compress the capsule, although instead ion beams or electron beams can be used. The beams will then impede the surface of the capsule and via a 'rocket effect' cause the inner surface to move inwards, compressing the gas. Eventually, when sufficient density and temperature is reached, a central hot spot region will be created and ignites via fusion reactions. This releases alpha particles (He^4), which are trapped within the central hot spot region, and self heats. Eventually, the hot spot region causes a propagating burn wave through the gas, generating fusion energy production for the whole capsule volume, producing more fusion energy, the total of which is equivalent to burning a barrel of oil, assuming 100% efficiency.

The gas would ideally be ignited on the $D(T,He^4)n$ reaction, because this doesn't require as high a temperature as other reactions. However, from a propulsion perspective the neutron has neutral charge, making it difficult to magnetically direct for thrust in an engine. An alternative is to use the $D(He^3,He^4)p$ reaction, the proton having both charge and not being radioactive. Also, a DHe^3 combination provides a more manageable exhaust at greater power than D/T, although the latter is easier to initiate. It is easier to get a DT combination to ignite than DD and similarly this is easier to ignite than DHe^3. The reason for this is in the ratio of the number of neutrons to protons of the reacting products. D has one proton and one neutron; T has one proton and two neutrons; He^3 has two protons and one neutron.

The combined number of neutrons and protons in DT reactions leads to a ratio of $3/2 = 1.5$. Similarly the ratio for a DD reaction is $2/2 = 1$ and for a DHe^3 reaction is $2/3 = 0.66$. The lower the combined neutron to proton ratio the higher the temperature required to react two nuclides. This is because the Coulomb force from the protons will dominate the reaction with a repulsive force (the neutrons are bound by the strong force) and so if the neutron to proton ratio is low then the ignition will be harder to achieve.

In the 1970s it was proposed by the physicist John Nuckolls [3], that to achieve ignition and high gain, laser energies of 1 kJ and 1 MJ would be required respectively. The basis of ICF is employing the inertia of a material to confine the plasma. The key to this technology is the use of laser beams (the driver), where laser light (photons) are optically amplified and transported into a target chamber to irradiate a typically 2 mm diameter capsule containing 10 mg of fusion fuel which is surrounded by a metal cylinder (indirect drive) forming a cavity around the capsule. The beams enter the cavity, heat it to a plasma state and generates thermal x-rays which then irradiate the capsule surface, are absorbed, and cause it to explode outward and produce a 'rocket' effect which accelerates the target inwards. This compresses the capsule sending shock waves into the centre further compressing the fuel until a state of ignition is attained in a central hot spot region; typically characterized by a central fuel density of 1,000 gcm^{-3} ($20 \times \rho_{Pb}$) and ignition temperatures of 100 million K or 10 keV. Expressed in the language of ICF the capsule fuel would ideally have hotspot performance defined by a density of $\rho = 10^3$ gcm^{-3} and an areal density (density $\times$ thickness) of $\rho R = 1$ gcm^{-2}. The surrounding fuel will ideally have a performance of $\rho = 10^2$ gcm^{-3} and $\rho R = 0.3$ gcm^{-2}.

These densities and areal densities are the baseline target for the US National Ignition Facility (NIF) based in the US and there is every confidence that successful ignition will be obtained at high gain. NIF uses 192 beams at 1.8 MJ requiring 500 TW of power. It went online in 2009 and should eventually achieve a 45–100 MJ of fusion power for around 1 nano-second. A laser driven facility is also being built in France called the Laser Mégajoule (LMJ) which plans to deliver 1.8 MJ of power to the target, equivalent to the US NIF. ICF research is focused on several issues today, including a uniform delivery of laser energy to the target, preventing the fuel from heating prematurely prior to maximum compression as well as mixing of the hot and cold fuel instabilities, the symmetry of the imploding capsule and associated shockwave convergence. Lasers also lack efficiency, delivering perhaps 10% to the target.

The physics requirements for ignition of a typical ICF capsule as used at NIF are made difficult by the need to firstly achieve sufficient compression of the fuel and the central hotspot. Secondly, there is the need to achieve hot spot ignition and alpha particle deposition for sustained burn. However, an alternative approach has come to light in recent years known as the *Fast Ignition* method [4]. This relies on the use of a long pulse laser beam to achieve the compression of the cold fuel, but then uses another short pulse laser to ignite the central fuel to fusion conditions. An appropriate analogy to compare the concepts of hot-spot ignition and fast ignition is that of a diesel and petrol combustion engine respectively, where the latter makes use of a sparkplug (i.e. analogous to short pulse laser) to ignite the fuels. A facility known as High Power Laser for Energy Research (HiPER) has been proposed using the fast ignition method, and using a 250 kJ long pulse laser and a 100 kJ ps short pulse laser to produce 30 MJ of power output and attain a gain Q ~ 100. Before such a facility could be constructed numerical modeling of fast ignition problems will help to elucidate fundamental physics issues.

There are really three ways to achieve ignition using the short pulse laser enhancement route. The first method is to first use a long pulse laser to compress the fuel and then use a second short pulse laser of order 100 ps to burn through the plasma. A shorter pulse laser 1–10 ps is then used through the same channel to heat up the central hot spot further and cause ignition. A second method to achieving fast ignition is again to first use a long pulse laser to compress the fuel but then use a short pulse laser to generate a hot electron beam which can penetrate the fuel and heat up the central hot spot to ignition. The problem with this method is that the electrons have to penetrate through all that high-density plasma without being stopped. A third method is to use a long pulse laser to compress the fuel but to have a gold cone in part of the capsule that tapers off at an angle from the axis. A short pulse laser beam is fired down this channel impinging on the gold cone tip and causing the generation of electrons at the tip which go straight into the central hot spot to ignite the fuel.

The problem with this method is that the generated electrons do not leave the tip as a collimated beam but instead diverge widely so that only a fraction of the electrons produced will actually reach the central hot spot. Another problem with this approach is that the act of passing the laser beam down the channel causes plasma infilling, which is an obstacle to the continued laser transport. Some of the laser energy will be scattered reducing the laser-capsule coupling efficiency. Although a novel approach, fast ignition is not without its complications. The Russian physicist Misha Shmatov has proposed that the probability of successful ignition in a cone-guided fast ignition scheme can be increased if the design aim is to generate not one central hot spot, but two – although one would be concerned about the implosion symmetries of such a scheme [5].

A recent proposal called *Shock Ignition ICF* is a form of slow ignition. A laser pulse is first used to slowly compress the fuel and then just as the assembly is about to stagnate (stall and turn around) a second drive pulse is sent in to compress the central fall to an ignition state. One estimate [6] claims that this method is capable of producing high output energies (up to 250 MJ in a typical NIF capsule) with laser driver energy requirements lower than those required for conventional ICF and gains ten times higher in the region of 150 (compared to say 10–12). This scheme would also have less instabilities associated with it, compared to conventional ICF, because the high laser intensity is not applied until late into the implosion. Such a scheme would seem to be a credible approach for a commercial fusion reactor producing around 2,500 MW of fusion power for a 10 Hz pulse frequency.

There are many other schemes that explore fusion ignition and it is worth just mentioning some of them. The first is the use of a *Z-pinch*, where such a facility uses large amounts of electrical energy to heat and vaporize several high atomic number wires (typically tungsten) in a cylindrical arrangement rapidly, producing x-rays to implode a fusion fuel capsule. The arrangement makes use of the magnetic $J \times B$ force to drive the implosion of the plasma. The best-known facility today is the US 'Z-Machine' which is the largest x-ray generator in the world. Using 50 TW power pulse it produces an output of 290 TW and this led to the successful fusing of deuterium in April 2003. But after a substantial refurbishment the machine was able to produce 350 TW of power and generates 390 TW of energy output. Future plans aim for power outputs of up to 30 MJ and plans for the next generation machine talk of a 1,000 TW power facility. It's a real competitor to the conventional inertial fusion and magnetic fusion concepts.

Another approach to obtaining fusion ignition has been proposed by the American Physicist Robert Bussard, the same person who invented the Bussard interstellar ramjet discussed in Chap. 13. This is known as *Inertial Electrostatic Confinement* and is the basis of a Polywell reactor. It relies upon the principle of magnetically trapping negatively charged electrons at the centre of a set of polyhedron shaped electromagnetic coils that forms a cathode. Positively charged ions are then injected into the potential well and are accelerated by the cathode until they are confined to fusion conditions. Bussard was very confident in the potential of the Polywell reactor and believed that a prototype demonstrator could be built for moderate costs enabling the path to clean fusion energy. As with all reactor schemes discussed in this chapter, if it could be made to work there would be clear applications to fusion propulsion.

Despite the tremendous advances that have been made in demonstrating some of the technology of a fusion power plant, the maturity of this technology is still not sufficient for the application of powering the electrical grid of cities. The prototype fusion reactors of the near future, namely ITER and HiPER, should further demonstrate this technology and get us closer to the desired goal. This means that the technology is also not mature for the application of space-based propulsion, but is likely to reach demonstrator levels within decades. These exciting developments in fusion research present an opportunity for deep space propulsion research in future years.

With all this technology, there is also the problem of specific power. Engineering the fusion reactor technology to much smaller dimensions and mass will be a critical problem to occupy researchers in the near future. In particular, current Tokamak designs are very massive and may not be easily engineered for space applications. Hence advancements in alternative approaches (e.g. inertial) are required. Once this is done, space based applications can be appropriately considered. The other difficulty with fusion-based technology is that the maintenance of any power production necessitates more than 1 target capsule. In reality, multiple targets would be required, each detonated in succession (several or tens per second) to generate the sustained heat production for powering a commercial turbine. Although fusion research is moving fast, there are still many problems to solve. So until this technology has matured it remains impractical for space power in the immediate future but inspires great hope for the decades ahead.

11.3   Fusion Power for Space Exploration

Many historical research projects have explored the possibility of nuclear pulse technology for space applications. This includes the external pulse rocket in the guise of Project Orion conducted in the 1950s and discussed in more detail in Chap. 12. This involved the use of nuclear bombs being detonated rearward of a vehicle, the products from which would 'push' the vehicle along and provide thrust. It would obtain exhaust velocities 10,000 km/s (3% of light speed) and reach the nearest stars within a century or so. Although the historical calculations clearly show that external pulse technology can produce a performance appropriate for deep space missions, the existence of a Comprehensive Nuclear Test Ban Treaty rules this technology out.

Instead, designers turned to alternative propulsion schemes, which will give similar performance such as fusion based, or what is also referred to as fusion micro-explosion propulsion. This relies upon the ignition of a target by use of laser light or relativistic electron beams and a magnetic thrust chamber to direct the exhaust products. The main attraction of fusion propulsion is the ability to provide high specific power whilst providing a high exhaust velocity. Alternative propulsion systems such as chemical, fission or nuclear electric cannot provide high values of both. For fusion, specific power levels of 1–10 kW/kg are expected. One of the concerns about such systems is that the mass of any magnets or fusion trigger systems may limit the application of this technology to large vehicle of order Megagrams in size. This means that fusion propulsion systems may not be competitive for missions within the solar system [7].

Let us understand why fusion propulsion continues to generate such interest and consider the energy release from fusion reactions. The four principal reaction combinations were described earlier in this chapter, mainly those due to DT, DHe^3 and DD. DHe^3 reactions have the advantage that they produce fewer neutrons than DT. DD also produce only ~1/4 neutrons of DT reactions, but DD reactions are around ten times harder to ignite. Tritium will also decay to He^3 with a half-life of around 12.5 years. Let us assume a pellet design containing an equimolar mixture of DT propellant at 3 g mass, typical for a planned civilian based ICF reactor. We calculate the molar mass of the combination:

$$H_1^2 + H_1^3 \approx 5g/mole$$

The amount of propellant in moles is given by:

$$N(moles) = \frac{mass(g)}{molar\_mass(g/mole)} \approx \frac{3g}{5g/mole} = 0.6moles \qquad (11.3)$$

It is worthwhile nothing that for other reactions using a 3 g mass the number of moles will be 0.6 moles (DHe^3) and 1.5 moles (DD). For the DT mixture we then

Table 11.1 Fusion reaction energy release

Propellant	Reaction products	Total energy release from primary product (tons TNT equivalent)
DT	$He^4 + n$	49
DHe^3	$He^4 + p$	51
DD	$T + p$	35
DD	$He^3 + n$	28

calculate the number of nuclei in the pellet which is the molar amount multiplied by a constant known as the Avogadro's number N_A.

$$N(nuclei) = N(moles) \times N_A = 0.6 moles \times 6.022 \times 10^{23} atoms/mole$$
$$= 3.613 \times 10^{23} atoms$$

We then assume an energy release per reaction for the He^4 product to be 3.52 MeV which we multiply by the number of atoms in the pellet to give us an estimate for the total energy release in the form of He^4 products which is 1.272×10^{24} MeV or the equivalent of 49 tons TNT. We can do a similar calculation for DHe^3 and we find that the total energy release in the form of He^4 products with energy of 3.67 MeV per reaction is 3.613×10^{23} MeV or the equivalent of 51 tons TNT. This is shown in Table 11.1 along with the energy release from DD reactions. However, it must be noted that the products from each reaction will go on to react with other products and thereby contribute more energy. This is the case for the DD reaction for example which produces tritium, and this tritium will then go on to react with the deuterium further and so add more energy release. The other DD reaction will release He^3 products which will also react with any deuterium and drive the energy release to higher rates. These secondary reactions will enhance the burn quite substantially.

So what is the maximum performance that you can get from a fusion based rocket engine? We can approach this question in terms of exhaust velocity and specific impulse. We shall calculate the maximum performance based upon two methods. Firstly, we shall simply look at the difference in mass between the two reacting particle species and then consider the kinetic energy of the excess mass. Next we shall examine the question from the stand point of total thermodynamic energy content otherwise known as enthalpy. We begin by examining the DHe^3 reaction which produces the products of He^4 and a proton. Looking at the atomic mass unit balance between the reaction products and the released products we have:

$$2.013553 + 3.014932 \rightarrow 4.001506 + 1.007276$$

The difference in mass between the two sides of the reaction is 0.019703. We next consider what kinetic energy is associated with this mass difference. The fractional energy release defined by the symbol ε, as applied to the DHe^3 combination is simply the mass difference divided by the total mass of the reacting products $0.019703//(2.013553 + 3.014932) = 0.00392$. A similar calculation for different

Table 11.2 Fusion reaction performance

Propellant	Reaction products	Exhaust velocity (km/s)	Specific impulse (million s)
DT	$He^4 + p$	26,400 (8.67%c)	2.64
DHe^3	$He^4 + p$	26,500 (8.88%c)	2.65
DD	$T + p$	13,920 (4.64%c)	1.39
DD	$He^3 + n$	12,510 (4.17%c)	1.25

propellant combinations will yield a fractional energy release of 0.00375 (DT), 0.00433 (DD $\rightarrow$ D + p), 0.00351 (DD $\rightarrow$ He^3 + n). We then work out the thermal exhaust velocity V_e by inverting the equation for kinetic energy.

$$V_e = \left(\frac{2E_{kin}}{m}\right)^{1/2} \tag{11.4}$$

Then using $E = \varepsilon mc^2$, the mass cancels and we are left with an equation for the exhaust velocity as a function of the fractional energy release

$$V_e = (2\varepsilon)^{1/2} c \approx 0.088c \tag{11.5}$$

This is nearly 9% of light speed. This corresponds to a velocity of 26,500 km/s and if we divide by $g_o = 10$ m/s^2 we get an estimate for the specific impulse to be 2.65 million seconds. The performance for various propellant combinations is shown in Table 11.2. This is then the maximum performance of a fusion-based engine although subsequent energy from reacting products will also increase this maximum. In reality, efficiency issues will come in such as burn fraction and this will reduce the potential exhaust velocity. Neutron energy losses will also reduce the DT exhaust velocity, not an issue for DHe^3, which produces protons in the reaction (also useful for magnetic thrust directivity due to the charge).

The fractional energy conversion for fusion based propulsion systems was computed to be an average of approximately 0.004. Similar calculations can be performed for other propulsion schemes for comparison. For a chemical based fuel combination (e.g. hydrogen/oxygen) the fractional energy release is of order 10^{-10}. For propulsion systems that rely upon the fission of atomic nuclear the fractional energy release is of order 0.0008. So in terms of energy release fusion reactions are many orders of magnitude better than any chemical based scheme and an order of magnitude better than fission. If we take the exhaust velocity of our best performing propellant combination, DHe^3 at 26,500 km/s, we can then estimate the maximum enthalpy performance for a fusion based engine, where $h_o \approx V_e^2/2$. This computes to 348 million MJ/kg; a number much greater than fission reactions (around 82 million MJ/kg). The exhaust velocity of a rocket can also be obtained by balancing the kinetic energy release with the average thermal energy of the exhaust products

$$\frac{mV_e^2}{2} = \frac{3k_B T}{2} \tag{11.6}$$

Table 11.3 Exhaust velocities assuming different reaction chamber materials

	He^4 (16.0104 g/mol)	He^3 (12.0078 g/mol)	D (2.0136 g/mol)	T (3.0238 g/mol)
Lead = 600 K	0.97 km/s	1.11 km/s	2.73 km/s	2.22 km/s
Titanium = 1,930 K	1.73 km/s	1.99 km/s	4.89 km/s	3.99 km/s
Molybdenum = 2,890 K	2.12 km/s	2.45 km/s	5.99 km/s	4.88 km/s

where k_B is a physics constant known as the Boltzmann constant for a perfect gas kB $= 1.381 \times 10^{-23}$ J/K, and T is the average temperature of the particles. The exhaust velocity is then given by

$$V_e = \left(\frac{3k_B T}{m}\right)^{1/2} \tag{11.7}$$

where m is converted from units of g/mol to kg by using the Avogadro number $N_A = 6.022 \times 10^{23}$ mol^{-1}. It is notable that the exhaust velocity is proportional to the square root of the combustion temperature, so for a higher exhaust velocity a higher combustion temperature is desirable. However, there is a physical limit to how high you can have the temperature based upon the melting temperature of the materials used to contain the combustion products, e.g. the reaction chamber. Table 11.3 shows the results of calculating the exhaust velocities for various exhaust products and assuming different reaction chamber materials, where the temperature of the exhaust products (the heat of reaction) are assumed to be limited to the material temperature. A DT propellant combination with a Molybdenum chamber would seem to be a good choice for maximum exhaust velocity. One solution to increasing the exhaust velocity is to move the combustion process outside of the vehicle so that a higher temperature can be obtained. This is the basis of external propulsion systems.

Before we leave our discussion on potential propellant combinations for fusion propulsion it is worth mentioning the potential for proton-boron-11 reactions. This will only produce a helium-4 charged particle and no neutrons. It does require a higher ignition temperature than say DT but it can be initiated by using a DT trigger at the centre of the pellet, which would reduce the ignition requirements substantially. The interesting thing about this propellant combination is that both hydrogen and boron-11 are easily accessible on Earth, so you don't need to send an expensive mission to the gas giants to mine the materials.

In the 1960s physicists from the NASA Jet Propulsion Laboratory published a paper exploring the feasibility of interstellar travel [8]. Crucially, they identified two major factors that have limited the performance of previous assessments for the feasibility of going to the stars. Firstly, that the available energy corresponds to a fixed fraction of the final vehicle mass instead of the propellant mass, not considering that all of the fuel could be exhausted. The second was the assumption that nuclear-based rockets were limited to a single stage, unlike chemical rockets which are multi-staged. With both of these assumptions removed, the potential performance of nuclear-based engines clearly demonstrates application to interstellar missions.

The authors also considered the effect of the mass ratio depending on the mission. For a mission that decelerates to a stop as it approaches the target the mass ratio would become squared. Similarly, for a mission that required the vehicle to return to the origin and decelerate the mass ratios would be raised to the fourth power. They gave an example of a mission, which was less than a decade in duration, using Deuterium fusion, which accelerated at 1 g to obtain a cruise velocity of between 0.6 and 0.8c using mass ratios of between 10^3 and 10^6. Their analysis showed that a five-stage vehicle would nearly reach the maximum possible velocity increment for a particular payload fraction.

Some analysis was also conducted of the burn up fraction. It was found that the required propulsion time becomes longer with increasing burn-up fraction because the higher specific impulse of the engine produces lower thrust and more vehicle acceleration at the same power level. The Spencer paper is referenced many times in the Project Daedalus study. Indeed, it appears that this paper had a fundamental influence on the design team in deciding on flyby only as well as a multi-stage design configuration. Since the Spencer paper International space agencies have continually reviewed the future use of fusion based space propulsion and kept an eye on developments presumably for eventual use in precursor missions–should it reach technological fruition [9].

The idea of using fusion propulsion has inspired many. The *British Rail Space Vehicle* is one of those gems of a find in the interstellar literature. This proposal was very curious for two reasons. Firstly, a single individual Charles Osmond Frederick produced it. Secondly because that individual was a worker for British Rail and with his employers support even managed to arrange for a patent application and it was granted in March 1973. The actual proposal was for a large nuclear driven passenger craft for use in interplanetary missions. In some ways the inventor was a visionary because he clearly saw the potential emergence of a space tourism market well ahead of its time. The propulsion mechanism was an inertial confinement fusion driver where capsules of fusionable material are imploded internally by impeding laser beams. Detonations would occur at a frequency of around 1,000 Hz to mitigate any potential damage from sudden detonations. The energy generated from the fusion reactions would then be transferred to electrodes for conversion to working electricity to power either electromagnets or superconductors that could then accelerate any fusion products to provide thrust generation.

Because this was a passenger-carrying vehicle the designer realized that he would need a radiation shield and this was located around the reactor core. Since then many scientists have gone on to discredit the basic design saying it is unworkable and inefficient. In fact, if one considers what was accomplished by a British Rail worker one can't help but be impressed by the imagination of the designer and indeed his grasp of physics when his day job involved calculations for the interaction of train wheels with their tracks. Although in a previous occupation he had apparently worked for the UK Atomic Energy Authority calculating stress in nuclear fuel elements. The principle of laser-induced fusion is sound, as is particle acceleration by electromagnetism. What probably is true however is that the designer way under estimated the vehicle mass that is required for any electromagnets as well as the state of the fusion technology today, let along in 1973. The size of the laser technology required was especially way too small.

11.4   The Enzmann Starship

In the 1960s the MIT Professor and Raytheon Corporation employee Robert D Enzmann discussed the idea of a massive Starship, which had the resemblance of an augmented lollipop stick as shown in Fig. 11.1. This was a bare 305 m (1,000 ft) diameter sphere consisting of around 3 million tons of frozen deuterium, a 'snow-ball', with a long 1,000 ft spacecraft stuck on the end of it containing the Starship modules. The original concept had 8 nuclear fusion type engines and later versions had up to 24. The idea would be to fuse deuterium nuclei and so produce He^3 and a neutron as well as Tritium and a proton. The deuterium would be mined from the icy moons of the gas giants. The spacecraft had a modular design so that if part was damaged it could be 'unplugged' and replaced. The spacecraft was mainly made up of a cylindrical habitation unit each 91 m (300 ft) in diameter and 91 m (300 ft) in length, with three of the modules coupled end to end. The modules would be self sufficient with their own nuclear power plant. The propulsion system was contained within the engineering modules. A central load-bearing core 15 m (50 ft) in diameter would run from the fore to the aft of the ship for structural strength. Each of the three engineering compartments would be around 76 m (250 ft) in diameter. The total Enzmann Starship length would be around 610 m (2,000 ft).

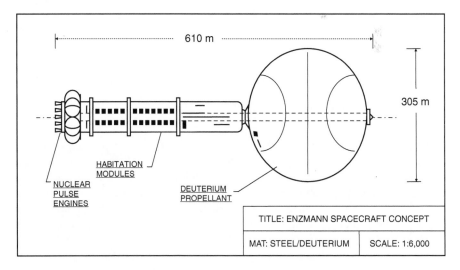

Fig. 11.1 Enzmann starship proposal

A later design also changed the configuration to having a metallic shell around the propellant, as it was recognized that this could present problems for the spacecraft acceleration. The Enzmann Starship was the intended subject of a New York Academy of Sciences article in the 1960s but it never appeared. It would have included illustrations from the renowned space artists Don Dixon and Rick Sternbach. The space artist David Hardy later painted another version. Plate 9 shows a version of this commissioned in 2010. Additions to the design by Enzmann saw that the total vehicle could actually separate into three separate craft with each consisting of a habitation unit, 1–2 Deuterium spheres and 8 engines. The separation maneuver would be performed once the vehicle had reached the destination star so that each sub-ship could explore different parts of the solar system. The starship as designed by Enzmann would have been capable of a cruise speed of 9% of light.

In the 1970s, G.Harry Stein wrote an article in the October edition of Analog titled "*A program for star flight*" [10] which discussed the Enzmann Starship in some detail where he argued that a mission to the stars would have to be completed as part of a full program plan rather than a one-off hope it gets there scenario. The roadmap would involve three phases, which included identification of an astronomical target, the launch of unmanned probes to the destination and finally the launch of a full expedition fleet to the target system. A full program of exploration was examined and he proposed a fleet of 10 Enzmann Starships to be launched, beginning in the year 1990 at a cost of around $100 billion spread out over a couple of decades and based on 1972 costs, one tenth of the then Gross National Product. Each ship was claimed to be around 12 million tons in mass, although most of that was made up of the Deuterium propellant. The Enzmann Starships would be assembled in Earth orbit and capable of reaching (implausible) speeds up to 30% of light. Any shocks generated from the nuclear pulse engines would be mitigated by absorbers on each of the eight propulsion units. Artificial gravity would be provided for any crew by spinning the cylinder on its longitudinal axis. The Enzmann Starship was a truly original idea and an example of large-scale visionary engineering. It may yet represent part of a future human colonization agenda in deep space exploration.

11.5   Project Daedalus

Between the periods 1973 to 1978 members of the British Interplanetary Society designed a fusion based unmanned interstellar probe called *Project Daedalus*. Just like *Project Orion* before it, Project Daedalus has obtained legendary status as one of the most complete interstellar engineering design studies ever undertaken. Members of the study group were volunteers who all shared a romantic vision of humans one day travelling to the stars. These were people like Anthony Martin, Alan Bond and Bob Parkinson – the current President of The British Interplanetary Society. Over a period of 5 years they calculated, discussed and eventually arrived at an extensive set of engineering design reports which were published in the later 1970s in a special supplement to the Journal of the British Interplanetary Society [11–14].

There were two key motivations for instigating *Project Daedalus*. Firstly, there was a need to move discussions of interstellar travel from the speculative to the credible, based upon engineering calculations. This would allow a proper assessment for whether travelling to other stars were even a possibility. Second, the lack of contact with an intelligent species from another world; this was in contradiction to our theoretical expectation given the number, age and spectral type of stars that we observe in the night sky, a problem previously termed the Fermi Paradox, named after the Italian physicist Enrico Fermi who first presented the problem in a scientific context (see Chap. 7). If it could be shown that at the outset of the space age travelling to other stars was feasible in theory, then with future advances in science this would imply that it must be possible in practice in the decades or centuries ahead. Therefore the absence of an encounter with any intelligence species other than from the Earth would require a more sophisticated explanation.

Project Daedalus had three stated guidelines. Firstly, the spacecraft was to be designed using current or near-future technology, extrapolated to a few decades hence. This was to establish whether any form of interstellar space flight could be discussed within reasonable credibility using the existing knowledge of science. Second, to design a simple vehicle that was an unmanned and undecelerated flyby mission of a target destination and return useful scientific data back home to Earth. Third, the mission was to be completed within the working lifetime of a designer. This amounted to 40–50 years duration to a nearby star. The target star was Barnard's star 5.9 light years away, chosen because it was believed to have several planets at the time based on the latest astronomical information. This was later shown not to be true. Two crucial design philosophies were adopted by the team, this included the adoption of conventional technology where possible and when this could not be accomplished it was to be projected until the end of the twentieth century.

The end result of the engineering study was a 190 m long vehicle that was nearly 54,000 tons in mass of which 50,000 tons was the D/He^3 propellant. The science payload had a mass of 450 tons. Daedalus was a two-stage design as shown in Fig. 11.2. A more detailed image of the Daedalus configuration is shown in Plate 10. Plates 11 and 12 show the actual engineering drawings from the original 1978 design study reports. The first stage would burn at 754,000 N thrust for around 2 years up to a speed of around 7% of light. The second stage would then burn at 663,000 N for around 1.8 years up to a speed of just over 12% of light. The vehicle would then cruise for around 46 years at a velocity of around 36,000 km/s and would reach its destination in around 50 years from launch. During the boost phase the propellant tanks would be dropped along the way, this included around 500 tons for the 1st stage and around 50 tons for the 2nd stage. The first stage structure would also be ejected. Table 11.4 shows the nominal performance for the Daedalus design.

The idea for the propulsion system for Daedalus first emerged from work on pulsed micro-explosion of small pellets using laser or electron beams. This was based upon two studies by Los Alamos and Livermore Laboratories giving rise to performance levels of 10^4 s and 10^6 s specific impulse respectively. Then, studies by Friedwardt Winterberg clearly showed the potential of electron driven ICF ignition [15, 16]. Winterberg has been a prolific inventor of proposals for space

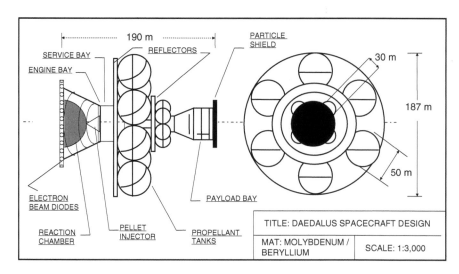

Fig. 11.2 Project Daedalus starship proposal

Table 11.4 Nominal Daedalus mission profile and vehicle configuration

Parameter	First stage value	Second stage value
Propellant mass (tons)	46,000	4,000
Staging mass (tons)	1,690	980
Boost duration (years)	2.05	1.76
Number tanks	6	4
Propellant mass per tank (tons)	7666.6	1,000
Exhaust velocity (km/s)	1.06×10^4	0.921×10^4
Specific impulse (million s)	1.08	0.94
Stage velocity increment (km/s)	2.13×10^4 (0.071c)	1.53×10^4 (0.051c)
Thrust (N)	7.54×10^6	6.63×10^5
Pellet pulse frequency (Hz)	250	250
Pellet mass (kg)	0.00284	0.000288
Number pellets	1.6197×10^{10}	1.3888×10^{10}
Number pellets per tank	2.6995×10^9	7.5213×10^9
Pellet outer radius (cm)	1.97	0.916
Blow-off fraction	0.237	0.261
Burn-up fraction	0.175	0.133
Pellet mean density (kg/m^3)	89.1	89.1
Pellet mass flow rate (kg/s)	0.711	0.072
Driver energy (GJ)	2.7	0.4
Average debris velocity (km/s)	1.1×10^4	0.96×10^4
Neutron production rate (n/pulse)	6×10^{21}	4.5×10^{20}
Neutron production rate (n/s)	1.5×10^{24}	1.1×10^{23}
Energy release (GJ)	171.82	13.271
Q-value	64	33

propulsion over the years, particularly in the field of fusion. Winterberg is one of the scientists to have come out of the golden age of physics discoveries and his PhD supervisor was Wernher Heisenberg. One of Winterberg's ideas was the use of electron beams to drive inertial fusion capsules to ignition. This was later picked up and eventually adopted by the Project Daedalus Study Group. In reviewing relativistic electron beams the report said that present systems were capable of producing energies in excess of 1 MJ per pulse and powers greater than 10^{13} W. Projected machines should be capable of powers 10^{14} per pulse. Winterberg has also proposed using magnetic compression reactors powered by Marx generators as well as suggestions of using GeV proton beams to initiate fusion in deuterium pellets.

The propulsion system for Daedalus used electron beams to detonate 250 ICF pellets per second containing a mixture of D/He^3 fuel. The fusion products would produce He^4 and protons, both of which could be directed for thrust using a magnetic nozzle. The D/He^3 pellets would be injected into the chamber by use of magnetic acceleration, enabled by use of a micron-sized 15 Tesla superconducting shell around the pellet. The complete vehicle was said to require 3×10^{10} pellets, which if manufactured over 1 year would require a production rate of 1,000 pellets per second. Although neutron production would be minimized by the choice of D/He^3 it was estimated that neutron production rates per pellet pulse would be 6×10^{21} neutrons/pulse for the 1st stage and 4.5×10^{20} neutrons/pulse for the second stage. This corresponds to a neutron production rate of 1.5×10^{24} neutrons/second from the 1st stage and 1.1×10^{23} neutrons/second from the second stage.

The problem with using He^3 is where to find it. The acquisition of large quantities of He^3 was said to represent a "*major obstacle to the realization of the project*". Three potential sources for He^3 were examined: First there was artificial breeding in fusion reactors. The different ways of creating He^3 through the D/D and Li^6/n reaction paths were investigated including power production rates. Second, there was the solar wind, where He^3 is known to exist in small abundances and could be collected. In order to collect the small amounts, vast collecting units were said to be required thousands of kilometers across. Thirdly, for Jupiter, around 17% of the atmosphere was said to be made up of helium with an abundance of between 10^{-4} and 10^{-5}. This would be a 20-year mining operation.

The third option of mining from Jupiter was the only one that "*appears practicable*" [17]. Two approaches were discussed to accomplish this. The first was 'sky mining' from Jupiter orbit using large collection scoops, a method thought to be impractical. The second (and chosen method) was the deployment of a collection plant directly into the atmosphere, attaining neutral buoyancy and then launching the collected He^3 back into orbit using a combined rocket-ramjet. One of the interesting observations of the report was that He^3 would also be required by a fusion-based society here on Earth for use in power reactors. To continue this form of power would necessitate a large source of He^3, of which Jupiter does provide requiring a space transport infrastructure for He^3 mining and transfer to Earth.

Another interesting feature of the Daedalus design was the acknowledgement by the study group that such a long duration mission was likely to result in some system failures. To deal with this they included 'warden' repair vehicles to fix any

Fig. 11.3 Illustration of a Daedalus-like fusion probe being prepared for launch in lunar orbit

failed systems. Once the vehicle reached its destination, it would deploy autonomous robotic probes to any nearby planets. These probes would be powered by nuclear ion engines. The Daedalus team maintained that it was necessary to consider the type of society that would attempt such a mission. This led to envisaging the type of society that would exist and therefore the timescale upon which the mission would be launched. They envisaged Daedalus type vehicles being built by a wealthy solar system wide community, sometime in the latter part of the twenty-first century. The community would already be employing nuclear pulse rockets for space flight and transporting He^3 from the outer planets to the inner solar system for Earth commercial power usage on a routine basis. Such a society was also said to be politically stable. Figure 11.3 shows an illustration of the Daedalus spacecraft being assembled in lunar orbit.

In Chap. 5 we considered a two-stage Daedalus design as well as a single stage version in our calculations of velocity increment. The calculations estimated a final ΔV of 12.3%c and 9.9%c respectively, assuming an exhaust velocity of 10,000 km/s. Within the same propellant constraints and a minor addition to the structural mass, it is worthwhile considering what alternative Daedalus staged designs might look like. Figure 11.4 shows such a design (calculated by this author) compared to a single and the reference two-stage design along side a three and four-stage configuration, where the total propellant and structural mass was fixed for the calculations, along with an assumed constant exhaust velocity of 10,000 km/s (for a 250 Hz pulse frequency) and constant boost period of 3.81 years. The configurations were not

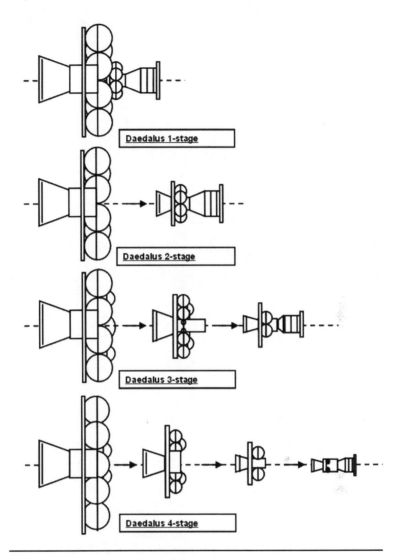

Fig. 11.4 Illustration of multi-stage Daedalus variants

completely optimal but they are useful for illustrating the effect of engine staging.
Table 11.5 shows the breakdown for each engine stage and mission performance.

The 3-stage configuration would achieve velocity increments of 18,137, 12,587
and 10,619 km/s to reach a final cruise velocity of 41,343 km/s or 13.8% of
light speed. Carrying a 170 tons science payload it would reach Barnard's star
(5.9 light years away) in 44 years or Epsilon Eridani (10.7 light years away) in
79 years. The 4-stage configuration would achieve velocity increments of 16,949,
12,179, 9,035 and 7,634 km/s to reach a final cruise velocity of 45,797 km/s or 15.3%

Table 11.5 Multi-stage Daedalus-like configurations and mission performance for a Barnard's star (BS) and Epsilon Eridani (EE) mission

	First-stage concept	Second-stage concept	Third-stage concept	Fourth-stage concept
First stage	50,000 tons propellant 2,670 tons structure deltaV 29,819 km/s (0.099c) Burn time 3.81 years Stage distance 0.107 light years	46,000 tons propellant 1,690 tons structure deltaV 20,664 km/s (0.069c) Burn time 2.05 years Stage distance 0.048 light years	44,081 tons propellant 1,300 tons structure deltaV 18,137 km/s (0.06c) Burn time 1.97 years Stage distance 0.04 light years	44,149 tons propellant 1,150 tons structure deltaV 16,949 km/s (0.056c) Burn time 1.92 years Stage distance 0.039 light years
Second stage	–	4,000 tons propellant 980 tons structure deltaV 16,256 km/s (0.054c) Burn time 1.76 years Stage distance 0.224 light years	6,218 tons propellant 1,000 tons structure deltaV 12,587 km/s (0.042c) Burn time 1.54 years Stage distance 0.187 light years	7,050 tons propellant 1,050 tons structure deltaV 12,179 km/s (0.041c) Burn time 1.45 years Stage distance 0.172 light years
Third stage	–	–	1,070 tons propellant 370 tons structure deltaV 10,619 km/s (0.035c) Burn time 0.3 years Stage distance 0.375 light years	1,235 tons propellant 360 tons structure deltaV 9,035 km/s (0.03c) 0.38 years Stage distance 0.344 light years
Fourth stage	–	–	–	236 tons propellant 110 tons structure deltaV 7,634 km/s (0.025c) Burn time 0.05 years Stage distance 0.524 light years
Final science payload	450 tons	450 tons	170 tons	51 tons
Cruise velocity	29,819 km/s (0.099c)	36,921 km/s (0.123c)	41,343 km/s (0.138c)	45,796 km/s (0.153c)
Total burn time	3.81 years	3.81 years	3.81 years	3.81 years
Boost distance	0.107 light years	0.224 light years	0.375 light years	0.524 light years
Cruise time BS	58.29 years	46.12 years	40.09 years	35.22 years
Total mission BS	62.1 years	49.93 years	43.91 years	39.03 years
Cruise time EE	106.58 years	85.12 years	74.92 years	66.66 years
Total mission EE	110.39 years	88.93 years	78.74 years	70.48 years

of light speed. Carrying a 51 tons science payload it would reach Barnard's star in 39 years or Epsilon Eridani in 70 years. Making the configuration more optimal in design would reduce the final science payload mass for the four-stage configuration to something like 5 tons. By adding more stages slightly higher cruise velocities could be attained but the performance gain has to be traded with the engineering reliability and system complexity. It is doubtful that a Daedalus-like design giving more than around 20% of light speed is attainable, unless the engine is made even more efficient. Enhancing the burn up efficiency by using methods such as fast ignition techniques has the potential to do this or antimatter catalyzed fusion (see Chap. 13).

In 1980 one author extended the work of Project Daedalus and produced the self-Reproducing Probe or REPRO design [18]. It was a variation on the Daedalus design where all of the non-replicating subsystems for Daedalus were redesigned to make them amenable to self-replication. The idea was that the first probe sent out to the target destination would also carry a 443 tons 'seed factory' which could then be used to replicate itself many times, producing versions with masses over ten million tons, mostly consisting of fuel. Each version would be constructed from the local raw materials such as from asteroids, Moons and planets. The original idea of a self-replication machine was proposed by the Hungarian physicist John von Neumann and called 'universal assemblers'. In the context of spacecraft people tend to refer to them as von Neumann probes. The monoliths in Arthur C Clarke's novel "*2001: A Space Odyssey*" is an example of von Neumann probes [19].

Self-replicating probes are an interesting idea because some estimates claim that the entire galaxy could be covered by them in a timeframe of about half a million years. This has obvious implications for the Fermi Paradox where any intelligent life in the Universe that sends out such probes should already be within the vicinity of our solar system – a tantalizing prospect. The fact that we have not seen any such probes, may suggest that there is no other intelligent life in the Universe, who would have exploited such technology by now. However, others have argued that in fact any such probes would have by now consumed all the matter in the galaxy, suggesting that intelligent races would try to avoid the use of such technology. Alternatively such probes may already be here but beyond the detection range of our observations.

That the Project Daedalus study has inspired much research since its publication is a testament to the excellent work produced by the original study group, allowing for a careful balance between being sufficiently bold whilst maintaining credibility. The main objective of the study was to show that interstellar travel *is feasible* and it is the opinion of this author that this was clearly demonstrated. In a 1986 review of the study Alan Bond summed up the important question that the project had further highlighted [20]:

> As it become evident that the conclusion would be that interstellar missions were feasible, it also became evident that Fermi's paradox is still unresolved. To reiterate, if there is a high density of intelligent life in the Galaxy, and if our aspirations to conquest and colonisation are not unique, then there has been sufficient elapsed time since the first habitable star systems formed to permit complete colonisation of our galaxy. There should never have been a time when alien presence on Earth was not commonplace. Project Daedalus has only repeated the question. Where are they?

11.6   Project Longshot

In the late 1980s several members of the US Naval Academy undertook an advanced university design study for an unmanned autonomous probe to be sent to Alpha Centauri, chosen due to its proximity and trinary system [21]. Centauri B was the target star of main interest. The project was titled *Project Longshot* and was a response to a requirement from the US National Commission for a planning program for spacecraft to be sent outside of the solar system to the nearest star. The probe would take around 100 years to reach the destination and the designers optimistically expected sufficient technology advances within 30 years to allow a credible launch. The vehicle configuration is described in Fig. 11.5. Plate 13 shows an artist illustration of the Longshot probe in flight, especially commissioned for this book.

The launch of the Longshot spacecraft involved several phases of assembling the modular components (engine, structure, tanks, fuel, and payload) on the ground and then launching the nearly 400 tons structural mass over many flights. They proposed the use of heavy lift launch vehicles to get to LEO for full assembly at the orbiting space station, but plugging in the different component to enable the full vehicle integration. The vehicle would then be chemically boosted over several velocity change maneuvers to the injection point for interstellar escape at 61° to the ecliptic plane and then fire up the main engines.

The main engine for Longshot would be a pulsed fusion micro-explosion drive similar to Daedalus with a specific impulse of one million seconds. Before they decided on a fusion drive the designers conducted a short feasibility trade study

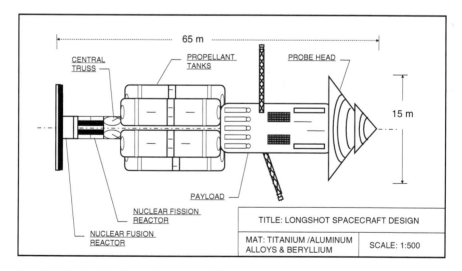

Fig. 11.5 Illustration of longshot spacecraft design

where they considered ion drive (high feasibility), fusion (medium), laser-pumped light sails (low), high temperature thermal gas expansion (very low) and an antimatter rocket (extremely low). The assessment in brackets is the conclusion reported by the Longshot team. Based upon near technological fruition and potential specific impulse requirements fusion was selected as the best candidate. The designers didn't think it would be practical or safe to ignite the fusion drive while in Earth orbit which necessitated a secondary propulsion requirement. These were said to be solid rocket boosters temporarily attached to the rear of the spacecraft and similar in size to those used for the Space Shuttle, but with a larger specific impulse of around 600 s; although this is not possible with conventional chemical propellants.

The design would use D/He^3 fuel, where the He^3 could either be manufactured on Earth using particle accelerators to bombard a lithium blanket, captured from the solar wind or be mined on route from the atmosphere of Jupiter. The team rejected the Jupiter option due to high cost and complications. They also considered the use of alternative fusion reactions using higher Z elements where the nuclides may be more abundant and produce fewer neutrons, although require a higher threshold energy. A magnetic nozzle would be required to direct the charged particle products for thrust. The team did little work on the design of the fusion pellets themselves although stated that an engineering analysis must be done on the proposed engine in order to determine the pellet size and pulse frequency. They speculated that they would need a mass range between 0.005 and 0.085 g and would be detonated at a pulse rate between 14 and 250 Hz.

The fuel would be kept in pairs of cylindrical tanks for ease of construction located on the main vehicle truss that also connects the probe head with the main engine. Once the fuel in the tanks is used they would be discarded to minimize mass. A total of six tanks would be required. Explosive bolts would jettison the first two tanks at 33 years into the mission and a further two at 67 years. By the time the spacecraft reached the turnaround point 71 years into the mission the spacecraft reverses itself, shuts off the main engines and injects into the orbit of the target star. The fission reactor is then used once again to power up the main engine to provide correct orbital insertion speed and thereafter at 100 years the final two tanks are jettisoned along with the main engine and shields. The designers did not define how orbital insertion speed would be achieved which would require some form of deceleration to the target and presumably more propellant. These three periods would be separated by acceleration phases boosting the spacecraft from 0.004 ms^{-2} to 0.006 ms^{-2}, to 0.009 ms^{-2} and then finally up to a maximum of 0.02 ms^{-2} at orbital insertion around Centauri B. This would result in a maximum spacecraft velocity of 14,400 kms^{-1} (0.05c). The massive impulse would induce a current in the surrounding coils which could then be used to power any laser beams and other systems.

The primary purpose of the probe would be to study the properties of the interstellar medium on route, study the target stars at the destination and also take accurate distance measurements using the parallax technique to improve the accuracy of astrometry so that the distances of thousands of star systems could be better known which also improves the determination of their physical properties. As a scientific driver, this was an excellent reason for launching the Longshot probe. A long life fission reactor was also included in the design located at the rear of the spacecraft with a 300 kW power output. This would be used to initially power the fusion reactor and associated laser systems for fuel ignition but also other systems on board, using enriched uranium nitride fuel pellets. A system of hydrazine thrusters would also be employed for attitude control. A particle shield would be located at the back and rear of the spacecraft to protect it from the high velocity incoming particles.

Several instruments were also proposed for the spacecraft including a star tracker that could be powered by the fission reactor. Other instruments all located on the end of an extended boom included visual and infrared imagers, ultra violet telescopes, particle detectors, spectrophotometers, magnetometers and a solar wind plasma analyzer. Communication lasers for data transmission back to Earth would also be included, located on the fuel tanks and probe head for the acceleration and deceleration phases of the mission respectively. Clearly the team recognized the importance of science data return for such a mission. The total payload mass including the instruments, fission reactor, lasers and shields was approximately 30 tons. The structural mass of the vehicle including the fuel tanks and central truss and was around 70 tons. The engine was composed of the chamber, igniter and field coils all coming to 32 tons. Most of the vehicle mass was due to the fuel, taking up a mass of 264 tons hence the need to jettison the fuel tanks throughout different parts of the mission. The team correctly identified that one of the main issues for the spacecraft would be reliability. In particular, the construction of technology that can be sustained for a 100 year mission was a challenge. They had specified a requirement for three duplicates of each instrument for redundancy, coming to a total instrument mass of 3 tons.

Unfortunately, the configuration layout for Longshot doesn't match the claimed performance. With the claimed total wet mass of 396 tons and total propellant mass of 264 tons, ignoring tank drops this would equate to a mass ratio of around 3. However, the authors claimed the vehicle would achieve a cruise speed of 12,900 km/s or 0.043c with a specific impulse of one million seconds, which corresponds to an exhaust velocity of 9,810 km/s and implies a mass ratio of 3.72. The total dry vehicle mass (including tanks) was to be around 132 tons and to attain a mass ratio of 3.72 (for the required performance) this would need an initial mass of 491 tons (propellant + structure) and imply a propellant mass of 360 tons, around ~100 tons extra propellant than specified in the Longshot report. Then to decelerate into the target system would require the square of the mass ratio, which is around 13.8, and with a structure mass of 132 tons would equate to a total propellant mass of around 1,700 tons.

Considering that the designers of Longshot were mainly first class midshipmen performing this study over two academic semesters, a lot of areas were identified which deserves credit. However, most of the research is non-specific with no detailed calculations presented for sub-systems. It was an opportunity for example to theoretically demonstrate fusion-based propulsion, but no real work was done in this area – possibly due to time constraints with the project. The work was a good effort in the design of an interstellar probe from a student team under time constraints, which could be taken forward to a more detailed and credible study that would include correcting some of the fundamental errors in the concept such as an appropriate configuration layout to match the required performance and including deceleration requirements.

11.7   Project VISTA

This proposal was for a manned single stage vehicle named *VISTA* or *Vehicle for Interplanetary Space Transport Applications* and one of the principal designers was the US physicist Charles Orth from Lawrence Livermore National Laboratory [22, 23]. The main vehicle engine was based upon inertial confinement fusion like the Daedalus design, and is in some ways a scaled down version of Daedalus meant only for interplanetary travel. However, VISTA differed from Daedalus in several key ways. Firstly, the choice of fuel was 4,000 tons D/T instead of D/He^3. The authors said that although D/T propellant combination was the best choice for near-term interplanetary applications, the use of the other combinations would be more useful in far off applications, when mining of the gas giants for He^3 became feasible for example. Although other authors have argued that thermonuclear power plants on Earth can serve as He^3 sources [24].

The fusion pellet was imploded with a 5 MJ laser beam instead of an electron beam. The development of the 5 kJ laser was seen as critical to the design and later designs discussed the use of a 150 tons diode-pumped solid state laser which is said to have an efficiency of at least 10%. This laser would impinge on a pellet target positioned at the apex of the conical spacecraft structure. VISTA proposed the adoption of a fast ignition technique for high target gain of from 200 up to 1,500, where the amount of energy released by the capsule exceeds the energy expended to implode it. Also, whereas Daedalus used direct laser impingement on the pellet, VISTA uses mirrors to redirect a dozen lasers onto the pellet. The performance for VISTA was said to be much better than for nuclear-thermal and nuclear-electric rockets because the effective exhaust temperature is around ten times larger at high mass flow rate.

The VISTA spacecraft is illustrated in Fig.11.6. Plate 14 shows an artist illustration of the VISTA spacecraft especially commissioned for this book. The vehicle totaled a launch mass of around 6,000 tons along with a 100 ton payload and a total propellant mass being around 4,165 tons. The spacecraft would be around 100 m (328 ft) high and 170 m (558 ft) wide and the whole vehicle had a cone shaped geometry, being wider at the base. Around half way up the cone, was a 12 T superconducting ring magnet, which would generate the required magnetic fields to contain and magnetically direct the explosion, then ejected through the nozzle. Any particle products would be deflected for exhaust thrust by using the superconducting magnet. Any generated excess heat would be expelled by the system of trusses and supporting heat pipe radiators positioned around the surface.

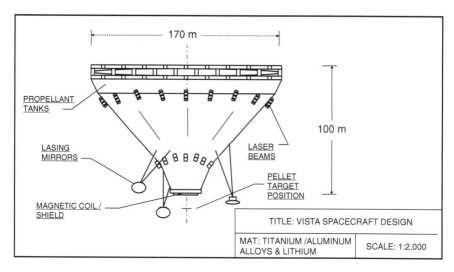

Fig. 11.6 Illustration of vista spacecraft design

The pellet repetition rate would vary for throttling between frequencies of 10 to 30 Hz. The pellet ignition would take place behind the plane of the superconducting magnet so that the radiation shield for the coil would cast a hollow conical shadow, which maximizes the efficiency of the exhaust chamber. The unusual conical shape of the VISTA design is due to a requirement to mitigate the generation of fusion neutrons heating the spacecraft structure, as well as being a structurally efficient design. Because this was a manned vehicle some degree of radiation shielding would be required, and the team proposed that the magnetic fields would perform this role. The vehicle would be assembled in sections at Low Earth Orbit and then the complete vehicle would be moved to a higher orbit above the radiation belts. When going between Earth and Mars it would always remain in space, but deploying its own planetary Lander probes where required.

The mission analysis planned that the vehicle would reach Mars within months and could reach any planet in the solar system within around 7 years. Its high performance engines, which would attain a specific impulse of around 17,000 s, enabled this and a generated thrust of 2.4×10^5 N. For a typical return mission to Mars within 60 days, a 5,800 tons vessel could deliver a 100 tons payload. Missions to Mars of order 100 days were seen as crucial to minimize radiation exposure to any crew as well as to avoid the need for massive amounts of radiation shielding. The whole point behind the design of VISTA was to produce a conceptual configuration that would be a viable, realistic, and defensible spacecraft concept using ICF technology that was projected to be available between 2,100 and 2,150. In this respect the team achieved their goals and identified the critical areas of the design to be the target gain, the engine mass and the laser driver efficiency.

11.8 Discovery II

In Chap. 1 of this book, the *Discovery* spacecraft design from Arthur C. Clarkes novel "*2001: A Space Odyssey*" was mentioned. This was based on ideas developed in the 1950s for a dumbbell like spacecraft with the reactor and engine section separated from any habitation module. Technical engineering accuracy played an important part in the development of the *Discovery*, from both Clarke and the film director Stanley Kubrick. The propulsion system of *Discovery* was a fission gas core nuclear reactor type with specific impulse up to around 6,000 s [25]. Although the design was technically credible, there were some glaring errors such as the omission of radiators for dumping waste heat. Apparently they were in the original design but removed for the film for aesthetic reasons, so Clarke was not to blame. Despite this a 2005 NASA report led by the spacecraft designer Craig Williams, concluded that the attempt to produce a credible engineering design "*appears in retrospect to have been successful*". The same authors set out to produce a concept vehicle design that enabled a fast piloted outer solar system mission that was similar in many respects to *Discovery*. For this reason, the authors decided to call their design *Discovery II*.

The design reference mission was a Jupiter rendezvous with a more demanding alternative mission to Saturn with a crew of between 6 and 12 people and a 5% payload fraction. The total trip time from Earth orbit to Jupiter orbit would be around 1 year mainly due to the presence of a human crew. The mission would include a flyby of Jupiter's Moon Europa and Saturn's Moon Titan. The spacecraft would first be assembled in Earth orbit using Heavy Lift Launch Vehicles (HLLV) and then the crew would follow, possibly via air breathing propulsion-rocket combined horizontal take-off vehicles, where they could be docked with the assembled spacecraft.

On leaving Earth orbit the spacecraft would have a total mass of 1,690 tons and carry a payload mass of around 172 tons. *Discovery* II was 240 m long (Similar to *Discovery* at 213 m) design with 25 m long extended heat rejection radiators coming off the central truss. The 17 m radius (compared to 5.8 m on *Discovery*) human crew compartment would rotate at the front end of the spacecraft so generating a small 0.2 g lunar level artificial gravity field (compared to 1 g on *Discovery*), so as to facilitate a minimum of locomotion and therefore offset bone decay due to a long duration space mission. A radiation shield protected the crew and was a layer of liquid water surrounded by a 51 mm thermal protection layer as well as a half mm aluminum micrometeoroid shield.

Following on from the philosophy adopted by Clarke for a crewed spacecraft, the engine compartment was kept separated from the human crew, with a long hexagonal based central truss linking each end constructed of aluminum Graphite Epoxy. All communication booms would also come out from the central truss with a total communications power of 0.2 MW. A prominent feature of the design were four cylindrical shaped hydrogen propellant tanks as well as tanks containing 11 tons of DHe^3 fuel and the refrigeration system for storing the propellant system.

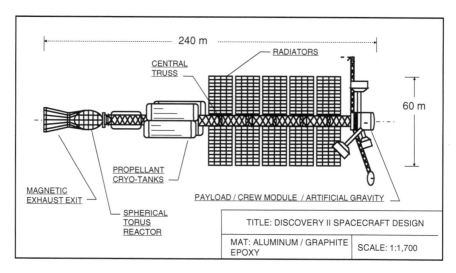

Fig. 11.7 Illustration of Discovery II

Aft of the central tanks was located the shielded spherical torus nuclear fission reactor and the magnetic nozzle. The configuration is illustrated in Fig. 11.7.

The propulsion system for *Discovery II* used a closed magnetic reactor system, with a small aspect ratio (2.0) spherical torus with a radius of 2.48 m. Twelve coils were used to generate the toroidal magnetic field and seven poloidal field coils were used to provide plasma stability, all generating a coil current of 9.2 MA. The engine would produce nearly 8,000 MW of power with 96% being in the form of charged particles for thrust generation. The design had a specific power of 8.62 kW/kg and an exhaust jet efficiency of 0.8 was assumed. The fusion reactor has a total mass of 310 million tons. The 1 g mass pellets were constructed of solid deuterium and liquid He^3 and would be injected into the reaction chamber using an electromagnetic rail gun at a conservative frequency of 1 Hz. This would generate an exhaust velocity of around 345–460 km/s (0.01c) and a specific impulse of around 35,000–47,000 s.

The authors discussed the need for placing nuclear fusion space research on the critical path for enabling order of magnitude improvements in space transportation capability, particularly in the context of human goals for expansion throughout the solar system. They said:

> It is the judgment of the authors that direct nuclear fusion space propulsion is the leading technology that can reasonably be expected to offer this capability.

They claimed that *Discovery II* could accomplish fast interplanetary trip times with significant payload and over a large range of launch window opportunities. The Jupiter mission would take around 118 days and the Saturn mission around 212 days. One of the options left open for the mission design was the potential for refueling at the outer planets; in particular, planets with atmospheres or Moons rich in hydrogen, deuterium and He^3. Other options include sources of water ice such as at the lunar poles or on other moons. They assumed that by the time such a mission could be launched, it is likely that the required technological infrastructure for planetary mining on a solar system wide scale would be available. He^3 atmospheric mining operations would be similar to those proposed for the Daedalus study.

11.9 Problems

11.1. A single stage fusion engine has a specific impulse of 10^5 s and ejects propellant at a mass flow rate of 500 g/s. What is the thrust of this vehicle? If each ICF pellet has a fusion mass of 5 g what is the pulse frequency in Hz? If the pulse frequency is then reduced to half what is the new thrust assuming the same exhaust velocity?

11.2. Calculate the average reaction debris velocity, which is 0.95 of the exhaust velocity, calculated in the previous problem. Then estimate the kinetic energy released from the 5 g pellet assuming that the burn fraction is already accounted for. Assume a laser driver energy of 10 MJ and show that the energy gain from the pellet is a Q value of over 200.

11.3. Write out the four key fusion product equations given in this chapter. Make a list of the main reacting nuclides. Identify various sources for these nuclides within the confines of the solar system or just beyond. Do some research to find out the abundance of each one and on the basis of this research reach you're your own conclusions on the supply of fuel for a fusion-based engine. Re-examine the credibility of different design concepts discussed in this chapter with this new knowledge.

References

1. Lawson, JD (1955) Some Criteria for a Useful Thermonuclear Reactor, Atomic Energy Research Establishment, GP/R 1807.
2. Lindle (1998) Inertial Confinement Fusion – The Quest for Ignition and Energy Gain Using Indirect Drive, Springer.
3. Nuckolls, J (1972) Laser Compression of Matter to Super-High Densities: Thermonuclear (CTR) Applications, Nature, 239.
4. Atzeni, A et al., (2008) Fast Ignitor Target Studies for the HiPER Project, Physics of Plasmas, 15, 056311.
5. Shmatov, ML (2009) Advantages of Fast Ignition Scenarios With Two Hot Spots for Space Propulsion Systems, JBIS, 62, 6, pp219–224.
6. Perkins, LJ et al., (2009) Shock ignition: A New Approach to High Gain Inertial Confinement Fusion on the Nation Ignition Facility, Physical Review Letters, 103, 045004.
7. Garrison, PW (1982) Advanced Propulsion for Future Planetary Spacecraft, J.Spacecraft, 19, 16, pp534–538.
8. Spencer, DF et al., (1963) Feasibility of Interstellar Travel, Astronautica Acta, 9, 49–58.

9. Schulze, NR (1991) Fusion Energy for Space Missions in the twenty-first Century, NASA TM4298.

10. Stine, GH (1973) A Program for Star Flight, Analog Science Fiction Science Fact.

11. Bond, A et al., (1978) Project Daedalus – The Final Report on the BIS Starship Study, JBIS.

12. Bond, A & RM Martin (1978) Project Daedalus: The Mission Profile, JBIS, pp.S37–S42.

13. Martin, RM & A Bond (1978) Project Daedalus: The Propulsion System Part I – Theoretical Considerations and Calculations, JBIS, ppS44–S62, 1978.

14. Bond, A & RM Martin (1978) Project Daedalus: The Propulsion System Part II – Engineering Design Considerations and Calculations, JBIS, ppS63–S82.

15. Winterberg, F (1973) Micro-fission Explosions and Controlled Release of Thermonuclear Energy, Nature Vol.241, pp449–450.

16. Winterberg, F (1977) Rocket Propulsion by Staged Thermonuclear Microexplosions, JBIS, Vol.30, pp333–340.

17. Parkinson, B (1978) Project Daedalus: Propellant Acquisition Techniques, JBIS, ppS83–S89.

18. Freitas, RA, Jr., (1980) A Self-Reproducing Interstellar Probe, JBIS, 33, 7.

19. Clarke, AC, (1968) 2001: A Space Odyssey, New American Library.

20. Bond, A & AR Martin (1986) Project Daedalus Reviewed, 39, 9.

21. Beals, KA et al., (1988) Project LONGSHOT, An Unmanned Probe to Alpha Centauri, N89-16904.

22. Orth, CD (1987) The Vista Spacecraft–Advantages of ICF for Interplanetary Fusion Propulsion Applications, 12th Symposium on Fusion Engineering.

23. Orth, CD (2000) Parameter Studies for the VISTA Spacecraft Concept, UCRL-JC-141513.

24. Shmatov, ML (2007) Some Thermonuclear Power Plants as the Possible Sources of He3 for Space Propulsion Systems, JBIS, 60, 5, PP180–187.

25. Williams, CH et al., (2001) Realizing '2001; A Space Odyssey': Piloted spherical torus nuclear fusion propulsion, NASA/TM-2005–213559.

Chapter 12
External Nuclear Pulse Propulsion

Men have an extraordinary, and perhaps fortunate, ability to tune out of their consciousness the most awesome future possibilities.

Sir Arthur C. Clarke

12.1 Introduction

A supernova explosion of a star will release a colossal amount of energy, around 10^{44} Joules. It is no surprise then, when physicists look to the stars to understand how they operate and what drives them. This curiosity has led to the development of bombs for use in war that utilize the power of nature based upon the processes of fission and fusion. However, there is also a peaceful application of this technology, which is in the propulsion of a space vehicle. This 'game changing' technology has been around for decades and is probably the only technology we could build today, if sufficient political public support was present, for the application of an interstellar mission. In this chapter we discuss external nuclear pulse technology with particular emphasis on the historical Project Orion. A vehicle designed to go to Saturn and back within a few years travel time or even further on to the distant stars.

12.2 Nuclear Pulse Detonation and Project Orion

The field of interstellar flight is filled with many so-called crazy ideas. However, it is often the case that the craziest ideas are also the most promising. The idea of using atomic bombs to propel a spacecraft was first proposed by Stanislaw Ulam in 1946 and then extended further by Ulam and Cornelius Everett in 1955, both physicists working on the Manhattan project during the Second World War. They talked about using low yield detonations located 50 m from the external spacecraft

K.F. Long, *Deep Space Propulsion: A Roadmap to Interstellar Flight*,
DOI 10.1007/978-1-4614-0607-5_12, © Springer Science+Business Media, LLC 2012

at a frequency of 1 per second and producing an exhaust velocity of around 10 km/s. The initial concept called *Helios* was to see nuclear detonations occur within a combustion chamber, but with a performance limit due to the surrounding materials. In Chap. 5 we discussed the fact that exhaust velocity is proportional to the square root of the combustion temperature, but a physical limit is reached when the temperature becomes too high.

An ingenious way to get around this is to move the 'combustion' process outside of the vehicle. Enter the *External Nuclear Pulse Rocket* in analogy with the external combustion engine. The idea here is to detonate many nuclear bombs or 'units' at the rear of the spacecraft of varying yield and with perhaps a few seconds between each detonation. Each unit would contain a solid layer of tungsten propellant that can be accelerated towards the pusher plate similar to a shaped charge over a cone angle of around 22°. Once around 80% of the expanding explosive products reach the pusher plate they then give it a 'kick' and provide the momentum transfer to push the vehicle. This is continued in a succession of such pulses until the required total impulse is achieved. Another name for this concept is the *uncontained pulse system* as opposed to a *contained pulse system* which wouldn't work using bombs due to the high temperatures generated, sufficient to melt any materials. The uncontained system is not power limited, because the detonations occur externally to the vehicle. *Project Orion* was the name given to the research project in the 1950s that was initially studied as a research program at the General Atomic Division of General Dynamics in 1958. The British born Physicist Freeman Dyson was one of the people that worked on this exciting research, along with the founder of Project Orion Theodore Taylor. An Orion-type launch vehicle is illustrated in Fig. 12.1.

The energy source would be displaced some distance, 50–100 m, rear of the spacecraft by ejecting it through a central hole in the pusher. It would then detonate and some fraction of the expanding explosion propellant would be intercepted by a large circular Aluminum or Steel pusher plate over a solid angle. There were several issues, but the main one was whether or not the explosively generated hot plasma would melt the pusher plate through ablative mass loss. As the propellant hits the pusher plate it loses kinetic energy to form a hydrodynamic stagnation layer, which if approximated to reach equilibrium through blackbody radiation into the vacuum of space corresponds to a temperature up to 20 eV, where 1 eV corresponds to 1.602×10^{-19} J of energy or ~11,600 K in temperature (physicists like to work in these units at high temperature).

To capture this material, the large pusher plate was located at the rear of the vehicle, and the subsequent impulse would ablate some of the surface, and huge pneumatic shock absorbers would cushion the acceleration of a few g's from any crew or spacecraft systems located at the front. After the shock absorbers have absorbed the energy of one pulse cycle the pusher plate is returned to its neutral position to take the next pulse seconds (or less) later. The actual velocity increment gained from a series of pulses can be tailored depending on the frequency and yield of each detonation as well as the distance that it occurs from the pusher plate.

Fig. 12.1 Illustration of an Orion ship prepared for a lunar ground launch

The Orion vehicle designs varied in mass and there were several types of space transport vehicles designed throughout its history. The first was the *Recoverable test Orion*. This had a gross mass between 50 and 100 tons and powered by small atomic bombs it would have a specific impulse of up to 3,000 s. The vehicle had a pusher plate diameter of around 12 m (40 ft) and a height of 15 m (50 ft). The acceleration loads would be between 2 and 4 g. Such a test vehicle was incapable of reaching Earth orbit on its own.

Another design was the *Orbital test Orion* which had a gross mass of around 880 tons and powered by atomic units could achieve a specific impulse between 3,000 and 6,000 s. The pusher plate had a diameter of around 24 m (79 ft) and a total vehicle height of 37 m (121 ft). It would undergo accelerations of up to 2 g and was capable of taking a 300 tons payload to 483 km orbit with a velocity of 10 km/s or a 170 tons payload for soft lunar landing with a velocity of 15.5 km/s.

The *Interplanetary Orion* was a much more massive vehicle with a gross mass of 4,000 tons. Also powered by atomic units it had a specific impulse between 1,700 and 4,000 s and would accelerate up to 2 g. The pusher plate diameter was 41 m (135 ft) and it had a height of 61 m (200 ft). It would have been capable of taking a 1,600 tons payload to a 483 km orbit with a velocity of 10 km/s or a 1,200 tons payload to a soft lunar landing with a velocity of 15.5 km/s. A return mission to Mars orbit carrying 800 tons would attain a velocity of 21 km/s and a return mission to Venus carrying 200 tons payload a velocity of 30 km/s. The initial assessments suggest that this could have been a very capable vehicle design and serving our immediate near-term planetary exploration objectives, although the final proof of concept experiments were never done to truly show how the performance would have scaled in a specific design and whether the engineering technology (e.g. shock absorbers) was appropriate for the expected environmental conditions.

The team went further however and designed the *Advanced Interplanetary Orion* concept. With a gross mass of 10,000 tons and also powered by atomic units it would have been capable of a specific impulse of up to 12,000 s and undergone an acceleration of up to 4 g. The design had a pusher plate diameter of 56 m (185 ft). It would have been capable of delivering a 6,100 tons payload to a 483 km Earth orbit with a velocity of 10 km/s or a 5,700 tons payload to a soft lunar landing with a velocity of 15.5 km/s. Alternatively it could deliver a 5,300 tons payload for a lunar or Mars orbit return mission with a velocity of 21 km/s or even a 4,500 tons payload from Earth surface to Venus or Mars orbit and back with a velocity of 30 km/s. Where it really stood out from the other designs was in its capability to deliver a 1,300 tons payload directly from the surface of Earth to one of the inner satellites of Saturn within a 3 year round trip with a velocity of 100 km/s. This was truly an interplanetary spacecraft design.

When Dyson first addressed the *Interstellar Orion* concept in his now famous *Physics Today* article [1] he first considered an energy limited design where the pusher plate would absorb all of the thermal energy from the detonation and not vaporize. However, his initial calculations showed that this concept would need a pusher plate approximately 20 km in diameter and it would require of order 100 s cooling, before the next detonation products would arrive at the pusher plate surface. To mitigate this he considered the design of an *Interstellar Momentum Limited Orion* concept. The large 100 m (328 ft) pusher plate would be coated with a material such as graphite to aid in getting rid of the excess heat through energy absorption of the impinging debris. He calculated that the velocity transferred to the

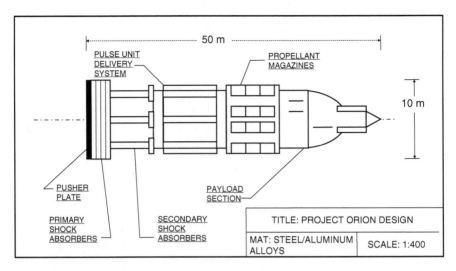

Fig. 12.2   Illustration of project Orion nuclear pulse proposal

pusher plate for every detonation was around 30 m/s so that it accelerates at 1 g with
a pulse rate of one detonation every 3 s.

The gross mass of the momentum limited design was 400,000 tons and the main
difference between this design and all the others was that it was to be powered by
Hydrogen bomb units rather than atomic bombs. The debris velocity from each
detonation would be around 3,000–30,000 km/s but the impact with the pusher plate
would reduce this to between 750 and 15,000 km/s. It would reach a maximum
velocity of 10,000 km/s or 3.3% of light speed, which is equivalent to a specific
impulse of around one million seconds. The maximum theoretical performance was
limited to a cruise speed 10% of light (compared to 5% for the atomic bomb
designs) so that the vehicle could in theory reach the nearest stars in around
44 years. Alternatively it could reach more distant interstellar targets such as
epsilon Eridani at around ten light years distance in about a century travelling at
the same cruise speed. Because the interstellar Orion is so large it can only be used
in space and is the equivalent of a 'Battlestar Galactica' type Starship. It would have
carried around 300,000 bomb units with a mass of around 1 ton each as well as its
20,000 tons payload. An Orion concept is illustrated in Fig. 12.2.

In all of the Orion designs the bombs would be surrounded by inert propellant
material, all comprising the unit. The frequency of each unit detonation would
depend upon the design but varied between 0.4 and 50 s. When a greater thrust was
required the frequency would simply be increased. Project Orion showed that to
reach an exhaust velocity of around 10,000 km/s you would need around 25 million
units. In theory an Orion like design could obtain an exhaust velocity of around
10,000 km/s (0.003c) and a specific impulse of around 1,000,000 s. The realistic
performance would in principle allow a crossing of the solar system in a matter of
months and a trip to Alpha Centauri in around 150 years. The Orion designs would

Table 12.1 Velocity (km/s) for an Orion-type vehicle assuming 5,000 units

g (m/s^2)	V (km/s) (d = 3 m)	V (km/s) (d = 5 m)	V (km/s) (d = 10 m)
1	27.1	35.0	49.5
2	38.4	49.5	70.0
3	46.9	60.6	85.8
4	54.2	70.0	99.0
5	60.6	78.3	110.7
6	66.4	85.8	121.3
7	71.8	92.6	131.0
8	76.7	99.0	140.1
9	81.4	105.0	148.6
10	85.8	110.7	156.6

have a mass ratio of around 10 with around 90% of the original launch mass being propellant. The number of units required for any design was given by:

$$n = \frac{V}{\sqrt{da}} \tag{12.1}$$

In this equation V is the vehicle velocity and a is the acceleration. The longer the shock absorber length d the lower the number of units required. Any manned vehicle would impose acceleration limits on the design but an unmanned interstellar probe would not be so limited. Hence the concept is ideal for an interstellar probe. If a crewed Orion vehicle were limited to only 3 g acceleration, around 29.43 m/s^2, a shock absorber length of 3 m the number of units to be 5,000, leading to a velocity of 47 km/s or 0.001c. Table 12.1 shows a similar calculation but for varying acceleration and shock absorber length.

If an Orion spacecraft is accelerated uniformly at a rate of 1 g with a transfer velocity of detonation to plate, of 30 m/s per detonation then the interval between each detonation will be around 3 s. The shock absorbers would have a stroke length given by:

$$d = \frac{V^2}{4g\omega} \tag{12.2}$$

where in (12.2) ω is the detonation frequency. Depending on the actual velocity delivered from each detonation and transferred to the pusher plate, this would compute to a stroke length of around 50–100 m, approximately the length of the spacecraft.

One of the advantages of an Orion type vehicle is that it can deliver a high specific impulse as well as a high thrust, otherwise known as a high specific thrust. A vehicle such as the US Space Shuttle can deliver high thrust but only at low specific impulse, due to its chemical fuel. Similarly, an electric propulsion system can deliver high specific impulse but only at low thrust. Orion (and nuclear pulse generally) is the only type of propulsion system that can deliver both. The specific impulse of

Table 12.2 Debris velocities specific impulse (s) as function of debris velocity and collimation factor

f/V_d	10 km/s	100 km/s	1,000 km/s	10,000 km/s
0.1	100	1,000	10,000	100,000
0.2	200	2,000	20,000	200,000
0.5	500	5,000	50,000	500,000
1.0	1,000	10,000	100,000	1,000,000

an Orion like vehicle is given by multiplying the collimation factor f (some fraction of an isotropic distribution) by the debris velocity V_d as follows:

$$I_{sp} = \frac{f V_d}{g_o} \qquad (12.3)$$

The collimation factor is the fractional amount of the explosion products that hit the pusher plate. Table 12.2 shows some of the results of calculating this for different debris velocities and with different collimation factors. The collimation factor is also directly related to the surface area of the pusher plate so that the larger the plate diameter the greater the specific impulse, although this is a trade-off with vehicle mass.

The hemispherical pusher plate at the rear of the vehicle was to be constructed of thin steel and designed to intercept the propellant and transfer its momentum and heat through the shock absorber system so as to accelerate the payload section. The thickness of the plate was to be varied to provide a mass distribution which matches the radial distribution of the propellant impulse, a requirement to attain uniform acceleration over this pusher plate area in order to minimize bending stresses. When the heat flux from the propellant interacts with the pusher plate it would ablate some of its surface. A simple ablation model for the pusher plate can be described by a decreasing exponential as follows:

$$\dot{m}(kgs^{-1}) = \frac{\sigma T_s}{E_a} e^{-\alpha_R X} \qquad (12.4)$$

Where T_s is the stagnation temperature, E_a is the ablation energy for the specific material, σ is the Stefan Boltzmann radiation constant which equals 5.67×10^{-8} Wm^{-2} K^{-4} and X is the material thickness and α_R is termed the Rosseland mean opacity (when using this equation ensure that T_s and E_a are in the same units). Assuming a stagnation temperature of 20 eV, an ablation energy of 1.602×10^{-18} J (equivalent to 10 eV), an opacity of 0.01 m^{-1}, this gives an ablation rate of around 0.0001 g/s for a 1 m thick pusher plate.

It was Freeman Dyson's view that the Orion vehicle would work in principle. In the book by his son George, Freeman says NASA approached him in 1999 to discuss what outstanding problems there were with the design [2]. Freeman said that the most useful thing that could be done to improve the design today was a computer simulation of the entire vehicle systems. This included using modern radiation-hydrodynamics codes for the propellant-pusher interaction, modern neutron-transport codes for the internal radiation doses, and modern finite element codes for the mechanical structures. He said that with the large improvement of machines and codes in the last half-century, designers could answer many of the questions that could not be adequately addressed in the 1950s–1960s. Specifically, he saw the use of three dimensional hydrodynamics codes as vital to finding out how much the turbulent mixing of propellant with pusher plate material would increase the ablation, a major unknown in the original project. This would allow designers to answer the basic question of technical feasibility without having to test any of the systems. A NASA illustration of an Orion-type spacecraft is shown in Plate 15.

In their efforts to demonstrate the practicalities of the Orion propulsion system the design team built a scaled down model demonstrator known as the 'Hot Rod' device or the 'Putt-putt' tests. The tests were conducted in 1959 in Point Loma and flew to a height of around 56 m in 23 s using about seven detonations The model was around 1 m in height and was powered by conventional explosive charges. The vehicle was seen to 'put-put' through the air as each unit exploded at the rear of the vehicle pushing it higher. An overall conclusion of the Project Orion study was that it was perfectly feasible for vehicles to take off from the ground and escape the earth using external nuclear pulse propulsion. If the Orion vehicle had ever been built it would have been one of the most awesome sights we had ever witnessed. For those who have had the honor of watching a conventional rocket launch, multiply this experience by an order of magnitude in terms of sound, vibrations, brightness and overall exhilaration. Plate 16 merely gives us a hint of what such a sight may have been in flight. Such a sight may never be witnessed but the $11 million spent over the 7 years of the project certainly allowed a scientific assessment to show that the concept was feasible. Freeman Dyson once said that:

> this is the first time in modern history that a major expansion of human technology has been suppressed for political reasons.

If there ever was an impending asteroid threat, Orion may be our best hope of getting there quickly to deploy whatever deflection method we wish to use. Ironically, the technology that many see as one of sciences worst creations may one day turn out to be our salvation. In a twist of irony, a vehicle designed during the cold war may yet become the technological savior of our world should we be presented with such a crisis. The future will show if we ever need Orion, but at least we have a plan. It's a question of engineering and political will, rather than one of science. The external nuclear pulse rocket is clearly highly viable for the desired goals, however, the controversial nature of using nuclear bomb technology and the existence of a test ban treaty rule this technology out, at least for now. Because the atomic bomb concepts are inherently radioactive, it would probably not be appropriate to launch such a vehicle in the Earths atmosphere and if it had any application it is in outer space only.

In his autobiography *Disturbing the Universe* [3] Freeman Dyson clearly argues his view that today he does not support the propulsion scheme as proposed by Orion:

> Sometimes I am asked by friends who shared the joys and sorrows of Orion whether I would revise the project if by some miracle the necessary funds were suddenly to become available. My answer is an emphatic no.....By its very nature, the Orion ship is a filthy creature and leaves its radioactive mess behind it wherever it goes.....many things that were acceptable in 1958 are no longer acceptable today. My own standards have changed too. History has passed Orion by. There will be no going back.

It is possible that he was just referring to using Orion as an Earth launch vehicle rather than for deep space propulsion, but only Dyson could clarify that. In the television series Cosmos and subsequent book Carl Sagan made the comment that he couldn't think of a better use for the world's stockpile of nuclear weapons [4].

12.3 The Medusa Concept

When reviewing interstellar propulsion concepts, one is tempted to draw the conclusion that solar sails and nuclear pulse detonation are two ends of the spectrum of design types. However, from the late 1990s onwards a US physicist Johndale Solem, published several papers in the Journal of the British Interplanetary Society, on the construction of a nuclear pulse design called *Medusa* [5–7]. It used elements from both propulsion schemes, although the designer was mostly interested in interplanetary missions such as to Mars. The name for Medusa comes from the idea that its motion through space is expected to be analogous to the motion of a jellyfish through the ocean. In contrast, the Daedalus vehicles motion would be analogous to an accordion musical instrument. Figure 12.3 shows an illustration of a Medusa sail.

Medusa uses a version of external nuclear pulse propulsion such as Project Orion. Orion had two main problems. Firstly the pusher plate at the back only subtended a small solid angle so wouldn't capture all the exhaust products from the

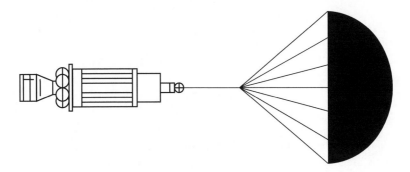

Fig. 12.3 Illustration of a probe utilizing a Medusa sail for stage boost

explosions. Second, the shock absorbers were limited in the pulse amplitude they could take. Medusa proposes to solve both of these problems by utilizing a large sail (or spinnaker) ahead of the vehicle, probably made out of a very high strength polymer such as aligned polyethylene enabled by future nanomaterials technology.

Nuclear explosions would be detonated in front of the vehicle but behind the hemispherical sail so accelerating it and therefore pulling the vehicle along with it. The cable line between the sail and the spacecraft is quite long, so that longer shock absorbers can be used in the vehicle, thereby taking larger amplitude shock pulses, and also nearly a 360° solid angle of the exhaust products is captured by the sail. So in theory the performance would be less limited compared to Orion. Automated winches between the sail and the vehicle control the acceleration pulses. The author claimed good scaling of the design and stated that the mass of the sail is theoretically independent of the size of its canopy or the length of its tethers. So the canopy can be made very large so as to minimize any radiation damage to the crew or payload systems. Similarly, the tethers can also be made very long to mitigate any radiation damage.

Nuclear bombs or 'units' provide the propulsion as with other nuclear pulse detonation schemes, however, the force is provided by the lightweight blast wave sail deployed towards the front of the spacecraft attached via cables. There is no large pusher plate in a Medusa design, unlike Orion. Any units would be sent forward to detonate using either highly elastic tethers or even a servo winch, and then give the required impulse to the sail. This would then pull the attached payload along with it, by transmitting the impulse smoothly to the payload using shock absorbers.

Clearly, the main advantage of a Medusa design, over that of Orion is the potentially enormous reduction in mass, using a sail instead of a pusher plate. The designer also claimed that the shock absorption stroke would be longer for Medusa. The main disadvantages of this type of design is that any crew on board or even the payload systems will have to be dragged through the radioactive detonation cloud of each pulse. The Medusa sail could be used as a first stage boost and then ejected along with all the accompanying mass. It is claimed that a Medusa like spacecraft would attain a specific impulse between 50,000 s and 100,000 s. The thrust for a Medusa type vehicle is given by:

$$T = PA_c \tag{12.5}$$

where P is the pressure applied to the canopy and A_c is the projected canopy area. For any object of mass m (i.e. spacecraft mass) at a distance r from the detonation point the velocity produced by a single detonation will be given by:

$$V = \frac{25A_c}{24\pi m r^2} \left(\frac{2m_b E_b}{5} \right)^{1/2} \tag{12.6}$$

The terms m_b and E_b is the mass and energy of the unit. So obviously the larger the canopy area and the closer it is to the detonation point the higher the velocity. It is no surprise that the velocity is also proportional to the yield of the detonation.

The final velocity attained will be directly proportional to the number of units detonated, where each detonation pulse adding more velocity. This leads us to an equation for the specific impulse of a Medusa design, which is given by:

$$I_{sp} \approx \frac{1}{g} \left(\frac{2E_b}{5m_b} \right)^{1/2} \tag{12.7}$$

Hence the specific impulse is proportional to the square root of the energy to mass ratio. Assuming values given by the original 1993 Medusa paper a unit of mass 25 kg with energy of around 100 GJ (25 tons) subtending a canopy angle of 2π would give a specific impulse of around 4,100 km/s, which is much greater than that available from chemical propulsion systems. This would place the exhaust velocity in the region of 40 km/s.

One way to increase the specific impulse of the Medusa concept is to use a programmable servowinch in place of elastic tethers. Another interesting application of the Medusa sail could be in its use as a deceleration mechanism. As a probe approaches a target destination, the sail is unfurled rearwards with detonations causing a force in the negative thrust direction to the direction of motion. This may only require a small quantity of units to produce this result. Alternatively, ICF capsules could be ignited by laser beams rearwards of the vehicle, giving rise to the same effect, but on a more moderate level. One of the biggest problems for the Medusa sail being used in a deceleration mode however is the reliability of it deploying in deep space after decades of being stored away. If the sail doesn't deploy, then the probe will essentially be a flyby probe with limited observing time of the target star system and potentially left with a lot of unused units, which would have to be destroyed safely in a controlled way by the vehicle main computing system.

The Medusa concept is a clever idea, combining the elegant technology of solar sails with the heavy engineering of Project Orion. It is this sort of 'fusion' of ideas, a hybrid, which will lead the way to the first technical blueprint for an interstellar probe.

12.4 Problems

12.1. An Orion type interstellar vehicle is accelerating at 1.5 g and detonates a pulse unit every 5 s transferring a velocity of 40 m/s to the pusher plate with each detonation. Show that the shock absorber length will exceed 100 m. Then show that the number of pulse units required will exceed 26,000 if the vehicle velocity is of order 10^7 m/s.

12.2. An experiment is performed on a prototype pusher plate, which has a Rosseland mean opacity of 0.08 m^{-1} and material ablation energy of 15 eV. If a stagnation temperature of 30 eV is measured what is the initial thickness of the pusher plate? For the same material properties stagnation conditions what ablation rate would correspond to a material of thickness 2 cm?

12.3. Using the equations given in this chapter for a Medusa sail calculate the specific impulse for yields of 100 tons, 500 tons, 1,000 tons, assuming a constant unit mass of 25 kg. For an interstellar mission specific impulse values of order one million seconds are required. Show that this would correspond to a yield of around 1.4 Mtons assuming the same unit mass. What technical challenges do you foresee with using such a large yield? Finally, again using the equations in this chapter, what can be done to mitigate any technical problems that you identify?

References

1. Dyson, F (1968) Interstellar Transport, Physics Today.
2. Dyson, G (2002) Project Orion: The Atomic Spaceship 1957–1965, Penguin.
3. Dyson, F (1979) Disturbing the Universe, Basic Books.
4. Sagan, C (1980) Cosmos, the Story of Cosmic Evolution, Science and Civilization, Abacus.
5. Solem, JC (1993) Medusa: Nuclear Explosive Propulsion for Interplanetary Travel, JBIS, 46 (1), pp21–26.
6. Solem, JC (2000) Deflection and Disruption of Asteroids on Collision Course with Earth, JBIS, Vol.53, pp.180–196.
7. Solem, JC (2000) The Moon and the Medusa: Use of Lunar Assets in Nuclear-Pulse-Propelled Space Travel, JBIS, Vol.53, pp.362–370.

Chapter 13
Towards Relativistic Propulsion: Antimatter and the Interstellar Ramjet

Successive breakthroughs in techsing speeds. Foreseeable breakthroughs in nuclear fusion and antimatter technology may one day allow us to travel at appreciable fractions of the speed of light. But what about that ever present barrier — the speed of light itself? Special Relativity enforces that no material object can ever reach or exceed this cosmic speed limit. However, within the equations of General Relativity lie tantalizing hints and intriguing suggestions that it may be possible to overcome Einstein's relativistic hurdle; that FTL travel could actually occur, albeit under extreme conditions. Are these mere anomalies? Perhaps mathematical curiosities? Or could they serve to illuminate the path to exotic and revolutionary modes of propulsion?

Richard K. Obousy

13.1 Introduction

In this chapter we consider the implications of the special theory of relativity as discovered by Albert Einstein. These discoveries impose a speed limit on the universe, and ultimately anything that possesses mass is prevented from reaching this limit, although it may approach it. Despite this, the limitations of sub-light speed don't stop us from achieving journey times across the galaxy in times that are acceptable, but this comes with a penalty that will be discussed. We then discuss two schemes, which may have the potential to actually approach the speed of light limit, and although they may seem speculative today they could form an important part of tomorrow's space faring civilization.

K.F. Long, *Deep Space Propulsion: A Roadmap to Interstellar Flight*,
DOI 10.1007/978-1-4614-0607-5_13, © Springer Science+Business Media, LLC 2012

13.2   Relativity in Space Flight

When Albert Einstein published his special theory of relativity in 1905 he changed the way we view the universe. One key element to his theory was the invariance of physical laws – the idea that the laws of physics are the same in every frame of reference. This has profound implications for our ability to understand how galaxies and stars work despite being light years away from us. Another key element was the fact that the speed of light is the same in all inertial frames of reference and is independent of the motion of the source. It is with this postulate that all the trouble starts for space flight. If a spaceship is moving along at 10,000 km/s and fires a probe forward at a speed of 5,000 km/s relative to the spaceship then the actual speed of the probe relative to the distant observer will be 15,000 km/s – this all makes sense.

But what if the spaceship fires a light beam forward that is traveling at 300,000 km/s – how fast is the light beam then traveling relative to the distant observer? If your answer is 305,000 km/s, then you are wrong because 300,000 km/s is the speed of light limit on the universe. We are approximating here that the speed of light is 300,000 km/s for simplicity, but the actual speed of light limit in a vacuum is 299,792 km/s, although it will be less through any material medium due to the effect of slowing light speed via optical refraction.

There are several profound implications concerning the speed of light limit, such as two observers at different locations in the universe observing the same event may measure a different time interval. This can lead to the phenomenon of time dilation and is a direct consequence of the speed of light limit and the motion of the observers. If two events occur at the same point in space in a particular frame of reference and the time interval between them as measured by an observer at rest is Δt_o, then the observer in the second frame of reference moving with a constant velocity v relative to the first frame of reference will measure a different time interval, which we denote Δt. This is summarized neatly in the following relation:

$$\Delta t = \frac{\Delta t_o}{\sqrt{1 - v^2/c^2}} = \frac{\Delta t_o}{\sqrt{1 - \beta^2}} = \gamma \Delta t_o \qquad (13.1)$$

A consequence of this equation is that for any crew on board a spacecraft that is in motion the clocks will run slower from the perspective of a distant observer. So that if one of two twin brothers goes off in a relativistic spaceship for a year as measured by clocks on board the ship with time Δt and then returns to Earth, he will find that the twin he left behind has measured a time interval Δt_o not a year-long but perhaps decades to centuries long and so will be significantly older (or have passed on) than the one who explored the universe. This is illustrated dramatically in Table 13.1. What determines this time dilation factor is the speed of the ship v relative to the speed of light c. As the ships speed approaches the speed of light, the time dilation effect increases.

Table 13.1 Relativistic time dilation for twins starting out at 20 years old

Fraction of light speed	Time measured by earth clock ΔT (years)	Possible events in home solar system during time duration
0.1c	20	5 U.S. presidential administrations pass
0.5c	23	Twin brothers children born
0.9c	46	Twin brothers' grandchildren born
0.95c	64	Twin brother on Earth dies ages 84
0.99c	142	Twin brother great (great) grandchildren born
0.999c	447	Nations collapse and rise
0.9999c	1,414	Empires collapse and rise
0.99999c	4,472	Human colonies on all main Solar System objects
0.999999c	14,142	Miniature ice age comes and goes

One remains on Earth while the other goes off into deep space at relativistic speeds for 20 years' duration

Most of the motion we experience on Earth and indeed in Earth orbit is so slow in comparison to the speed of light that relativistic effects are not noticeable. Indeed, when the relative velocity between two frames of reference is negligible then the denominator inside the square root of equation (13.1) is nearly unity and the laws of physics, as laid down by Isaac Newton, prevail. The effect of time dilation has been demonstrated however by flying atomic clocks around the globe or by measuring the time elapsed by orbiting satellites from the perspectives of both satellites and ground-based observers. Relativistic time dilation effects are used in minor corrections for car satellite navigation aids. If we used computers for a high-speed deep space missions and didn't take account of relativistic time dilation, astronauts could find themselves hundreds or thousands of kilometers off course.

Another consequence of the speed of light limit in the universe is relativistic momentum p. Ordinarily; momentum p is simply the product of the rest mass m_o of a body multiplied by its velocity. In this framework a constant force will give rise to a constant acceleration. However, in the world of relativity the momentum is no longer directly proportional to velocity but to a relativistic version where a constant force no longer gives rise to a constant acceleration, as defined by the ratio of the objects velocity v with respect to the speed of light c:

$$p = \frac{m_o v}{\sqrt{1 - v^2/c^2}} \tag{13.2}$$

This equation shows that as a particle's velocity increases towards the speed of light, the relativistic momentum will tend to zero. Because force, equal to the product of mass multiplied by acceleration, is also equal to the rate of change of momentum then the acceleration will also approach zero. In other words the faster a mass goes the closer it gets to the speed of light limit and the acceleration begins to drop. It is therefore impossible for any object with a non-zero mass to attain a speed equal to that of light and so is a fundamental limitation to the speed of any spaceship. If we re-arrange equation (13.2) we get the following relation for the relativistic mass:

$$m_{rel} = \frac{m_o}{\sqrt{1 - v^2/c^2}} \tag{13.3}$$

This shows that the relativistic speed limit is the equivalent to saying that any spacecraft that rapidly approaches the speed of light limit will undergo an effective increase in mass. This means that a 1 ton object moving at a speed of 0.1c gets an effective 5 kg mass increase and the same object moving at a speed of 0.5c gets an effective 150 kg mass increase. With higher speeds still the effective mass rises rapidly to over 7 tons at 0.99c and over 700 tons for 0.999999c. In the context of a relativistic rest energy $m_o c^2$ this can be expressed as a relativistic energy equation of the form:

$$E_{rel} = \frac{m_o c^2}{\sqrt{1 - v^2/c^2}} - m_o c^2 \tag{13.4}$$

So that as a spacecraft approaches the speed of light the amount of energy that is required in order to accelerate it further increases infinitely. In the limit of Newtonian, non-relativistic, physics (i.e., $v \ll c$) this equation and the others shown above will approximate to the classical form:

$$\Delta t = \Delta t_o; \ m = m_o; p = m_o v; \ E = \frac{m_o v^2}{2} \tag{13.5}$$

At the speeds of today's spacecraft technology, it would take many thousands of years to reach the nearest stars as measured by clocks on Earth. However, if one could build a spacecraft that could accelerate to relativistic speeds, then it is possible to reduce trip times to the nearest stars to a few years. This astonishing fact is a consequence of Einstein's special relativity theory discussed above. To handle the relativity of space flight special equations have been developed and using these results it has previously been demonstrated originally by the astronomer Carl Sagan [1] and discussed later by others [2] that for a spacecraft traveling a distance in meters of S/2 under an acceleration a_n of 1 g = 9.81 ms^{-2} and then followed by an equivalent deceleration for the same remainder distance, that the time duration of the mission as measured by clocks on board the spacecraft is given by:

$$t = \frac{2c}{a_n} \cosh^{-1}\left(1 + \frac{a_n S}{2c^2}\right) \tag{13.6}$$

where the hyperbolic cosine function can be written as the following provided $x = 1 + a_n S/2c^2$ is positive:

$$\cosh^{-1}(x) = \ln\left(x + \sqrt{x^2 - 1}\right) \tag{13.7}$$

Accelerating and decelerating at 1 g would bring the spacecraft to the nearest star of Alpha Centauri 4.3 light years away to within around 3.5 years. Similarly a trip to Barnard's Star 5.9 ly away would take around 4 years and a trip to Epsilon Eridani 10.7 ly years away would take around 5 years. These trip times become even more mind boggling when applied to the scale of galaxies. A trip to the center of the Milky Way at 30,000 light years distance would take around 20 years and to the edge of the Andromeda Galaxy a mere 28 years. How can it be possible that such vastly distant locations in the universe can be reached within such a short transit time? Well, it comes at a penalty, due to the time dilation effect. What may only seem like years or decades to the crew aboard the ship will be the equivalent of thousands or hundreds of thousands of years back home on Earth. If the crew were ever able to turn the ship back towards home, they would return to a very different planet, a very different civilization and a species that may not even be aware of their existence. Einstein's relativity changed our view of the universe in a radical way. If we can ever build relativistic spacecraft they will come at a price. Two potential options for relativistic spacecraft include the interstellar ramjet and the antimatter-based engine.

13.3   The Interstellar Ramjet

The name is due to the similarities with the atmospheric ramjet (discussed in Chap. 4), which intakes air for combustion but only at high speed. It was used as the main spacecraft in the science fiction novel *Tau Zero* [3] by Poul Anderson and is often referred to as the great hope of interstellar travel because it has the potential to reach such enormous speeds. The idea behind the interstellar ramjet was simple; instead of carrying huge quantities of propellant mass along the journey, you use interstellar gas as the source of energy to power a nuclear fusion-based engine at high exhaust velocity. The gas of particular interest is interstellar hydrogen or protons. The diffuse hydrogen would be collected by use of a huge 10,000 km^2 collector with the total structure and spacecraft having a mass of several 100 tons. To collect just 1 g of interstellar hydrogen with an abundance of 10^{-21} kg/m^3 the spacecraft would have to sweep out the volume of a cone given by the mass of the products m and the density ρ:

$$R^3 = \frac{3m}{\pi\rho} \approx 9.5 \times 10^{17} m^3 \tag{13.8}$$

Because the abundance of interstellar hydrogen may be at low levels, a laser beam would be shot out ahead of the vehicle, ionizing the space ahead and thereby creating more charged protons to collect with a large magnetic field. This electro-magnetic field is provided by large electromagnets, assisted by a 1 mm thick Mylar mesh structure to guide the field direction. The generation of an enormous magnetic

Fig. 13.1 Illustration of a fusion ramjet-powered spacecraft entering another solar system

field would require a correspondingly large power source. This could be large capacitor banks, for example, but all increasing the mass of the vehicle. An interstellar ramjet is illustrated in Fig. 13.1.

The main problem with the interstellar ramjet is the choice of the fuel, interstellar hydrogen. This has a much smaller cross section for fusion reactions and a slow energy release rate compared to other low-density isotopes such as helium-3, deuterium or tritium. The fusion of two protons will produce deuterium as well as a neutrino and a positron. However, the necessity for beta decay of a proton leads to the low reaction cross section. Hence, alternative fusion reactions for the interstellar ramjet includes deuterium + deuterium, which either leads to helum-3 + neutron or to tritium + proton. Another option is proton + deuterium, which leads to the products helium-3 + gamma ray. These reactions are much better because their cross sections are so many orders of magnitude larger than that for proton + proton. However, the problem is reaction rate, which is given by the product of the average cross section, $\langle \sigma \rangle$ and the number density squared of the two reacting nuclides n_1 and n_2 as follows:

$$R = n_1^2 n_2^2 \langle \sigma \rangle \tag{13.9}$$

Although the cross section may be much smaller for proton + proton, the number density of these atoms is so much larger in space than for deuterium. The second problem with the interstellar ramjet is that in order for it to collect sufficiently useful

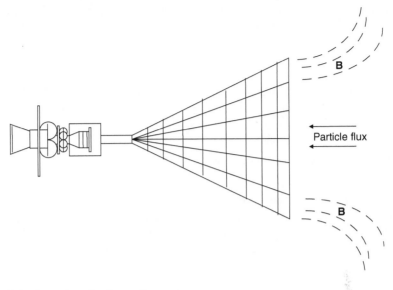

Fig. 13.2 Illustration of an interstellar ramjet concept

quantities of fuel, it must already be moving at a high velocity, of around several percent of the speed of light. In order to do this it must first use its own (and separate) propulsion engine to achieve this speed. This could also be a fusion engine with a single tank of propellant, sufficient to get the vehicle up to the velocities required before ram scooping starts to occur. Alternatively, it could be boosted up to moderate velocity using sails, lasers or even electromagnetic pellet stream systems.

Because the interstellar ramjet will be moving so fast, it will pick up a lot of drag from the medium from which it is collecting the hydrogen, which is proportional to its speed squared. Some think that this may be a killer for the concept. As the interstellar hydrogen is collected into the scoop and momentum transferred, it will also begin to heat up as it is compressed, resulting in thermal energy loss. This will act as a form of drag and decrease the overall engine efficiency. Figure 13.2 shows an illustration of the scooping process at work.

One of the ways that you could still use the collected interstellar hydrogen is not to use those protons directly for causing fusion. Instead, the incoming proton stream is decelerated to moderate energy levels and then impacted onto a lithium or boron-based target. The clever thing about this scheme is that although some breaking of the proton stream will be required to decelerate it to manageable levels, the energy generated from the subsequent reactions is given back, provided it is all utilized in the exhaust stream. This scheme is known as the Ram Augmented Interstellar Rocket, or RAIR [4]. It is more efficient than a nuclear pulse engine although not as efficient as the full interstellar ramjet proposal [5]. The scheme can be further improved by utilizing antimatter to catalyze the ram-augmented reaction although it is likely that large amounts of antimatter would be required to catalyze the large proton stream being collected. The key point to know about the interstellar ramjet is that because it doesn't carry its own fuel, it is possible in theory to continue to accelerate the spacecraft very near to the speed of light depending on the rate of fuel collection and fusion reactions. This means that excluding faster than light travel schemes and the potential for interstellar drag, it is the highest performer of the interstellar ideas proposed. It may be large and require a mammoth construction project to assemble it, but if you can get it to work the stars are, literally, the limit!

13.4   The Bussard Ramjet Spacecraft Concept

A brilliant physicist named Robert Bussard published the idea for the interstellar ramjet in 1960; hence in its original form it is known as the Bussard ramjet [6]. The original scheme proposed that the spacecraft would pass through dense regions of space with a hydrogen abundance of around 1 atoms/cm^3 using a frontal collecting area. In the original 1960 paper Bussard didn't set out a design concept as such, although gave some guiding principles for it. This included three main sections: the electric or magnetic field configuration for charged particle collection and directivity, a focal point within the field configuration for the particles to converge upon, a fusion reactor section where the reactions occur, and an exhaust section that fundamentally delivers thrust to the vehicle. This basic schematic, shown in Fig. 13.3, has laid the foundations for all later versions of the interstellar ramjet. Most subsequent schematics depict a frontal magnetic torus structure converging on a collector head. The charged particles are then channeled through the internal section and reaction chamber, where they are eventually exhausted.

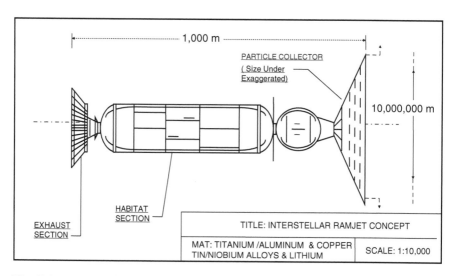

Fig. 13.3 Illustration of Bussard ramjet

In between the exhaust shield assembly and the collector head will be located any scientific payload or habitation section for a crew.

The entire 1,000 tons vehicle would be between 500 and 1,000 m in length, with the magnetic field collecting area being many times larger than this, perhaps up to 10,000–10 million km^2 assuming a low interstellar medium density. The collection area is required to be large because the field strength will decline with an inverse distance from the central point. It has since been found that dense areas exist in the interstellar medium with a hydrogen abundance of between 100 and 1,000 atoms/cm^3. This would require a collecting area of only 1,000 km^2 compared to the original Bussard proposal.

The Bussard ramjet assumed the use of the proton-proton chain reactions with a vehicle that was large in size and tenuous in construction, unless it could be made to pass through high-density regions of space, which tended to be congregated in large 10–40 parsec-sized clouds that were filamentary in structure (where 1 parsec is equal to 3.26 light years). Bussard referred to observations of the abundance of interstellar hydrogen at around 1–2 atoms/cm^3, which made up 90% of interstellar gas – much lower than current estimates. He also referred to the possible use of interstellar deuterium but admitted that not much was known about its abundance. The actual Bussard design was for an Earth-like acceleration, 1 g, in order to get to destinations within a matter of decades. The proposal suggested using a frontal area loading density per unit interstellar gas density of less than 10^{-8} g/cm^2/nucleon/cm^3. This would be required in order for the vehicle to withstand the large accelerations required for short trip times.

The Bussard ramjet appears to be a very large spacecraft design that, if ever built, would likely be in our distant future. However, Bussard himself commented that although it was not thought possible to build anything like this propulsion system in the 1960s, there was no reason to assume such a device is forever impossible since known physical laws are sufficient to describe its performance and engineering requirements. One such technology is the ability to magnetically trap particles. The Joint European Torus fusion experimental reactor has done this on numerous occasions over the last two decades, suggesting that in principle at least the scheme could be made to work. Plate 17 shows an artist's illustration of what the Bussard interstellar ramjet may have looked like in flight.

13.5 Matter-Antimatter Annihilation

One of the most potentially exciting developments for space propulsion is the idea of an antimatter rocket. The mathematician Paul Dirac first predicted antimatter in 1928. Antimatter particles have identical mass to matter particles but reversed electrical charge and magnetic field. The positron (anti-electron) was first discovered in 1932. The antiproton was discovered in 1955, followed shortly by the antineutron in 1956.

Antimatter doesn't exist in large quantities today, although it is thought to have existed in the early universe, with most of it being destroyed in matter-antimatter annihilation reactions. It may also be created in solar flares; a flare event in July 2002 was estimated to have produced half a kg of antimatter.

The collision of a matter-antimatter particle pair results in the annihilation of both into energy proportioned approximately as 1/3 gamma rays and 2/3 charged pions. The pions move at ~94% of speed of light but they exist for just long enough to travel around 20 m and be redirected for thrust by a magnetic nozzle. The reaction of a proton and antiproton will also release electrons and anti-electrons as well as high-energy neutrons and gamma rays. The gamma rays are particularly difficult to direct for thrust generation and given that they make up most of the energy release it makes antimatter propulsion nearly impossible to do practically. The beauty of antimatter is that the potential energy release is around 1,000 times that of fission and around 100 times that of fusion. Around 1 g of antimatter has an equivalent energy release of around 20,000 tons of chemical fuel.

In previous chapters we learned that a fusion reaction will have an associated fractional energy release of order 0.004. For a matter-antimatter based scheme, this will be of order unity and is in the form of pure gamma radiation with an exhaust velocity equal to the speed of light. In fact the total enthalpy release available from the matter-antimatter annihilation reaction is 90 billion MJ/kg. In terms of space propulsion, antimatter may be very efficient, releasing 100% of the energy available from the reaction. This means that very low mass ratios can be attained for antimatter fuels.

The problem is that antimatter production is difficult and you have to contain it, although research into antimatter traps is progressing fast and appears highly practical. Scientists at the NASA's Marshall Spaceflight Center have been developing a High Performance Antiproton Trap (HiPART) for some time, which aims to deal with the issues of production quantity, material transport and thrust production. Antimatter containment is enabled by Penning traps, using electrical and magnetic fields to hold charged particles. The Penn State University 'Mark I' trap has a mass of around 91 kg (200 1b) and can store 10 billion antiprotons for up to a week. Other ideas exist for controlling antimatter along the lines of electric and magnetic fields, radio frequency fields and using laser beams. Also antimatter production is expensive, with estimated costs of $100 billion /mg. However, any space propulsion schemes employing antimatter-based physics will certainly open up the stellar neighborhood to humankind, if enough fuel can be produced. Current global production of antiprotons is 1–10 ng/year [7], and the way to make antiprotons is to use a large accelerator to collide a proton beam at a flux rate of around 10^{15} protons per second into a target. This produces antiprotons on the other side of the target.

There is a range of techniques for enabling an antimatter-based rocket. A solid core engine could get up to a specific impulse of around 1,000 s and works on the principle of antiprotons annihilating protons in a Tungsten or graphite heat exchanger, which absorbs any pions and gamma rays, produced from the reaction. Any hydrogen fuel pumped through the heat exchanger is then warmed by the emitted reaction energy and these gases are then used for thrust generation. Instead of annihilating the particles separately, antiprotons can be injected directly into a hydrogen fuel stream, and although this process is less efficient the use of magnetic fields to control the charged particles means that the performance is as high as around 3,000 s specific impulse. The performance of these sorts of engines can be improved by using diverging magnetic fields to directly focus the relativistic charged particles towards the exhaust. The hydrogen fuel can also be ionized to a plasma state prior to injecting the antiprotons. These ideas are known as beam core and plasma core reactor designs.

Another option for combining antimatter particles with fusion reactions is the Magnetically Insulated Inertial Confinement Fusion (MICF) approach. In this scheme a beam of antiprotons react with the inside wall of a DT fusion pellet, annihilating the protons and giving rise to hot plasma conditions. If a metal shell surrounds the pellet then this apparently produces a self-generated magnetic field that thermally insulates the metal from the hot plasma. The quantity of antiprotons required would be in the range 30–130 mg. In one particular mission scenario the vehicle, a 220 tons (dry mass) spacecraft, conducts a flyby mission to 10,000 AU in 47 years and uses 166 g of antiprotons to accomplish the mission [8]. The antiprotons do not really 'catalyze' the fusion reaction in the chemical sense of the meaning but has the effect of providing an initial energy source to get the ignition going. In particular, when the antiprotons interact with the protons in the nuclei of the pellet material, they immediately annihilate, releasing a shower of neutrons and heat. It is claimed that with this sort of propulsion scheme a specific impulse of order a million seconds and thrusts up to 10^5 N using a pellet frequency of 10 Hz is possible.

13.6   The AIMStar Spacecraft Concept

This was the proposal for an Antimatter Initiated Microfusion Starship or AIMStar [9] and is the name of a spacecraft design that utilizes antiprotons to catalyze fusion reactions in an inertially confined pellet. Researchers at Penn State University are developing this. In the AIMStar design the key propellant reaction is D/He^3 that produces a 14.7 MeV proton and a 3.6 MeV He^4 particle, giving no neutrons provided the burn temperature is high.

The idea behind the AIMStar engine operation is to gravity inject small quantities, around 42 ng, of D/He^3 fuel droplets into a cloud of 10^{11} antiprotons that have been confined within a 10 kg potential well using a Penning trap, a magnetic bottle with a field configuration sufficient for trapping antiprotons. The particles are confined radially within a 0.8 cm diameter container by use of a 20 T axial magnetic field. The antiprotons are trapped axially using a 10 keV electric charge potential. A small amount of a fissile material such as U^{238} is then also injected into the antiproton cloud; the annihilation of the antiprotons with the U^{238} then induces fission, which produces charged particle products with a power density of around 5×10^{13} W/cm^3 and energy sufficient to fully ionize the D/He^3 droplet.

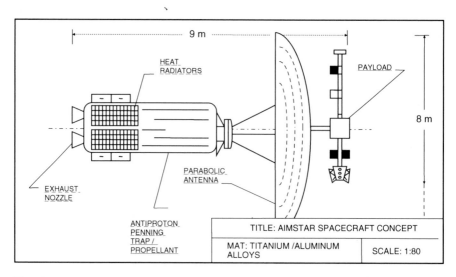

Fig. 13.4 Illustration of AIMStar spacecraft

As the droplet continues to fall into the potential well it is compressed further to sufficient density and temperature, where it eventually achieves fusion ignition and the generation of energy for around 20 milliseconds. This is the equivalent of a 1 kJ in 1 ns short pulse laser, depositing energy over a 200 μm pellet but for much smaller scale. In one pulse around 4×10^8 antiprotons are annihilated on the periphery of the cloud. For the next pulse cycle the potential of the cloud is first relaxed and then a further cloud of antiprotons are injected into the potential well. This is repeated at a frequency of around 200 Hz. Before the D/He^3 droplets are injected into the cloud they are first compressed to high-density 6×10^{17} ion/cm^3 and a temperature of 100 keV using a strong potential well of order 600 kV. This process should if it works produce 0.75 MW of continuous power consisting of the protons and He^4 particle products.

The AIMStar mission objective is to carry a payload of 100 kg to a distance of 10,000 AU, which is deep into the Oort Cloud. The AIMStar reference design using DHe^3 fuel would undergo continuous acceleration for 4–5 years at around 2 N thrust and then coast at around 1,040 km/s (0.003c) assuming complete energy release, reaching Mars in 1 month and 10,000 AU (i.e., the Oort Cloud) within 50 years. This propulsion system would have a specific impulse of around 67,000 s with an exhaust velocity of around 650 km/s, producing a power output of around 0.75 MW from the proton and He^4 particle deposition. A similar mission but using DT fuel would have a thrust of around 76 N, an exhaust velocity 598 km/s, specific impulse of 61,000 s and a power output of 33 MW from He^4 particle deposition. The number of antiprotons required for the two fuel scenarios would be 5.7 μg (for D/He^3) and 26 μg (for D/T). Along the route a flyby of Pluto would be on the plans, revealing information about that mostly forgotten world.

In theory, antimatter-catalyzed fusion spacecraft could attain an exhaust velocity of 297,000 km/s (0.99c) and a specific impulse of around a million seconds. This would allow it to reach Alpha Centauri in a matter of decades. Because the fission fragments from the antiproton-induced fission are apparently not radioactive this would allow for a minimal shielding requirement. Figure 13.4 shows an illustration of what the AIMStar concept may look like just prior to main engine separation and deployment of the payload section. All of the antimatter related machinery is located towards the rear of the spacecraft. This includes the Penning trap and storage tanks. The payload is located at the front of the spacecraft and would contain several instruments including a magnetometer and spectrometer, all located around the 8 m diameter parabolic antenna dish. All navigation and guidance control systems would be located in the unit forward of the antenna and this includes reaction jets for orbital maneuvering.

13.7   The ICAN-II Spacecraft Concept

In an effort to justify the utility of moderate amounts of antimatter for use in space propulsion, the same Penn State University team that created AIMStar has also created the Ion Compressed Antimatter Nuclear Rocket, or ICAN-II spacecraft concept [10]. This works upon similar principles to AIMStar except that it utilizes antiproton-catalyzed micro-fission (ACMF) reactions. In particular, the released neutrons from the proton-antiproton annihilation reaction go on to cause fission within the fissionable material that coats the pellet. This is a somewhat less ambitious objective and is appropriate for a propulsion system designed for interplanetary missions such as to Mars within 30 days and a return mission within 120 days.

The ICAN-II propulsion system works on the principle of using an ICF pellet that contains a mixture of both a fissile isotope such as U^{238} and D/T fuel in a mixture of 1:9. Lasers or ion beams are first used to compress the pellet, and then a beam of antiprotons is injected, causing fission reactions and the claimed release of around 16 neutrons per fission reaction (compared to around 2 neutrons for normal fission). This deposits energy into the fuel and ignites it, releasing fusion energy. ICAN-II is illustrated in Fig. 13.5.

The quantity of antiprotons would be very low, of order 140 ng, which are much smaller than for AIMStar. Although this quantity is small, it would still require many years of production for a facility like CERN or Fermilab at their low rate of 10 ng/year, and so a dedicated antimatter-making factory would need to be designed and built. For ICAN-II any antimatter would be stored in specially designed traps with a mass of 5 tons. As with many internal combustion machines, the propulsion would be generated from the expansion of the hot gases and particle products against the reaction chamber shell, which could then be directed for thrust generation at 180 kN with a specific impulse of 13,500 s from the first few days' acceleration. This would enable a cruise velocity to be reached of around 100 km/s. The detonation frequency would be at a rate of 1 Hz. Some of the generated energy

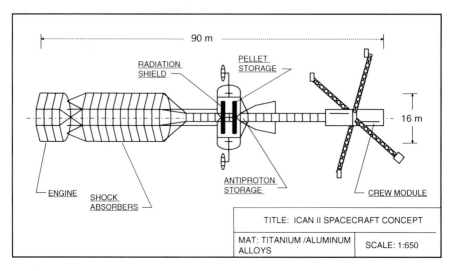

Fig. 13.5 Illustration of ICAN-II spacecraft

would be bootstrapped for electrical power generation to the 10 MW level. This could then be used to drive other spacecraft systems.

ICAN-II would be a 700 tons spacecraft, of which 362 tons would constitute the propellant. The vehicle would also be protected by a several meters of silicon carbide shielding with a mass of around 45 tons to mitigate radiation damage from high-energy neutrons. Any excess heat generated would be radiated away into space. Both antimatter fission and fusion may be the way for powering starships of the future. Such a fuel source would produce lower radioactivity than a pure nuclear reactor drive, and arguably may also be easier to ignite than any of the currently proposed inertially confined fusion methods and only require moderate amounts of antiprotons. Faced with such fantastic potential, we have no reason not to try.

13.8   Practice Exercises

13.1. Equation (13.3) is the relativistic energy result. However, in the limit of low velocities this will approximate to the usual expression for energy $E = m_o v^2/2$. This can be shown by performing a binomial expansion of $\gamma = (1 - v^2/c^2)^{-1/2}$. Confirm this result by performing the expansion so that the first order term approximates to the classical equation. You can consult a college textbook on mathematics to understand how to perform a binomial expansion.

13.2. Assume that a Bussard ramjet or antimatter spacecraft has been built and can travel at the relativistic speeds of 0.9c, 0.95c, 0.99c, 0.999c, 0.9999c. It launches from Earth orbit, accelerates for only 1 year and then cruises at the defined speed to a distance of 100 AU when upon arrival it sends a radio signal back to Earth. What is the time duration measured from the perspective of the crew and observers on Earth? From the perspective of both observers, when will the radio signal arrive at the receiver on Earth? Examine (13.6) for different accelerations and mission distances and to understand the implications of this equation from the perspective of ship crew and those left on Earth.

13.3 Do some research to find out the current and 20-, 30-, and 50-year projected amount of antimatter as well as its production rate. Find out the cost and compare it to diamond, gold and helium-3. Based on your research estimates, when will sufficient quantities of antimatter become available to power a small fleet of AIMStar and ICAN-II-type spacecraft?

References

1. Sagan, C (1963) Direct Contact Among Galactic Civilizations by Relativistic Interstellar Spaceflight, Planet.Space Sci, Vol, 11, pp.485–498
2. Crawford, IA (1990) Interstellar Travel: A Review for Astronomers, Q.JL R.astr.Soc, 31, 377–400.
3. Anderson, Poul (1970) Tau Zero, Orion books.
4. Bond, A (1974) The Ram Augmented Interstellar Ramscoop
5. Cassenti, BN (1082) Design Considerations for Relativistic Antimatter Rockets, JBIS, Vol.35, p396–404.
6. Bussard, RW (1960) Galactic Matter and Interstellar Flight, Acta Astronautica, Vol6, Fasc4.
7. Schmidt, GR & GA Smith et al., (1998) Antimatter Production for Near-Term Propulsion Applications. Penn State University Department of Physics Internal Report.
8. Kamash, T (2002) Antiproton Driven Magnetically Insulated Inertial Confinement Fusion (MICF) Propulsion System. NIAC 98002 Final Report.
9. Gaidos, G et al., (1999) AIMStar: Antimatter Initiated Microfusion for Pre-cursor Interstellar Missions, Acta Astronautica Vol.44, Nos.2–4, pp.183–186.
10. Gaidos, G & RA Lewis et al., (1997) Antiproton-Catalyzed Microfission/Fusion Propulsion Systems for Exploration of the Outer Solar System and Beyond, Penn State University Department of Physics, Internal Publication.

Chapter 14
Aerospace Design Principles in Interstellar Flight

It is recommended that a subsequent study address the possibility of a star mission starting in 2025, 2050, or later, and the long lead-time technology developments that will be needed to permit this mission.

L.D. Jaffe

14.1 Introduction

In this chapter we consider how one goes about designing a spacecraft. Because this book is focused on interstellar probes, we only consider the design process and do not consider the phases which follow the requirement definition such as assembly, qualification, manufacture and in service. These elements would be important for a real spacecraft. The design process begins with an initial requirement for a vehicle to perform a stated mission and then evolves from approximate through to detailed calculations where a solution emerges with a performance that meets the requirements. This is the process known as systems engineering, which we discuss. Finally, an example is given for an interstellar precursor mission probe to aid the reader in developing concepts of his or her own.

14.2 Principles of Aerospace Design

The object of any design process is to produce the optimum solution to a given requirement within predetermined constraints. How does one go about designing either a spacecraft or an aircraft? The principles are essentially the same, which is why they are mentioned in the same sentence. How they differ however is in the engineering requirements. For one, the vehicle must survive in the vacuum of space subject to levels of radiation and undergo significant g loadings during launch and

K.F. Long, *Deep Space Propulsion: A Roadmap to Interstellar Flight*,
DOI 10.1007/978-1-4614-0607-5_14, © Springer Science+Business Media, LLC 2012

Table 14.1 Aviation & aerospace requirements comparison

Parameter	Aircraft	Spacecraft
Propulsion	Turbo-Jet	Chemical rocket + electric
Materials	Aluminum alloys & Carbon Fiber composites	Metal alloys
Fuel	Kerosene	LOX/H2O (launch) + Argon/ Xenon (electric)
Flight medium	Air	Air + vacuum
Environment	Temperature ~300 K	Temperature ~2.7 K (neglecting solar heat)
Loads	Acceleration forces during take off and landing, ~1 g.	Acceleration forces during launch/orbital breaking, $\gg 1$ g
Speeds	~0.3 km/s	~12 km/s
Total redundancy	Low-Medium	Medium-High
R&D time	10 years	10 years
Mission time	Discrete ~hours over 20 years	10–20 year continuous mission
Cost	~$10's million	~$100's million
Reusability	~5,000–10,000	none

eventually arrive at its target destination somewhere within a solar system. For the other the vehicle must survive transport only through the atmosphere within a contained pressure environment and then safely land. For deep space missions the dominant driver for the spacecraft is the propulsion and mass. For the aircraft the dominant driver is also the propulsion and mass, but also includes the aerodynamic lift forces generated. Let us compare both types of vehicles for a moment. Let us consider a typical passenger aircraft that must cross the Atlantic and compare this to a robotic spacecraft that must enter the orbit of Mars. We ignore the human carrying element of both given that the overall objective is to deliver a payload to the destination (Table. 14.1).

We can see from the table that both vehicles are just objects with similar requirements but different levels of these requirements. In particular, the spacecraft will be subject to a more extreme environment than an aircraft. But it is because of the similarities, rather than the differences, that this allows us to use the same approach to evaluate both. This approach is fundamentally based upon the laws of physics and applied to engineering. This includes fluid mechanics, structure mechanics, propulsion science, thermodynamics and material science. These topics encompassed together and applied to a machine is what is meant by aerospace engineering. Before engineers delve into all of the detail, one can perform a top-level study from an initial requirement. For an aircraft we can consider the design by using a driving equation for the vehicle range, which is given by:

$$R = \frac{V(L/D)}{g\mu} Ln\left(\frac{m_o}{m_o - m_f}\right) \qquad (14.1)$$

Where V is the velocity, L/D is the Lift to Drag ratio, g is acceleration due to gravity, μ is the engine efficiency, m_o is the take off mass, m_f is the final mass when all the fuel has been burned. For a spacecraft the driving equation for the vehicle is simply the ideal rocket equation, which was discussed in Chap. 5:

$$\Delta v = g_o I_{sp} Ln \left(\frac{m_o}{m_o - m_{prop}} \right) \tag{14.2}$$

Designing a vehicle is a bit like solving a College textbook problem. You firstly identify the problem and understand what it is you are being asked to do. The next step is to accumulate the necessary data. You then select the appropriate physical principles that you need to solve the problem. Next some assumptions are made. Finally a solution is attempted. If the results do not seem sensible, then one cycles back through this routine until a sensible solution is arrived at based upon an appropriate assumption or model.

The design process is a highly creative one and often requires brainstorming sessions if teams of people are involved and can also involve lateral thinking techniques to think around problems. This has the added benefit of ensuring that all team members are invested in the design and then it will have more chance of succeeding. When presented with several options a morphology chart can be quickly drawn up and the group can discuss the advantages and disadvantage of each option from the perspective of design, manufacturing or performance. Comparisons of the performance to similar products are often required to place the design option into a proper context. To assist a design team visualization of the components is often desirable. This could be in the form of simple sketches, computer models or even a full prototype model.

It is good to have a design team that is a mix of youth and experience. This will allow a good mix of innovative ideas but with a robust design. All interactions and decisions need to be appropriately managed to ensure that those with more experience don't indirectly determine the design solution by inadvertently suppressing ideas. Otherwise a key opportunity to do things better may be missed. Each system and sub-system should be appropriately allocated to a designer so that the responsibility and system interfaces are clearly delineated. The worst thing that could happen is for something to go wrong and then for a designer to say 'I thought Bob was taking care of that'. At some point during the design process all the systems will have to be integrated and so a clear interface definition and a specification of the interactions will be required.

All potential design solutions should start from an approximate or 'back of envelope' basis. There is no point spending valuable time conducting detailed calculations when the design may change significantly later. It is better to 'scope' or 'size' the design space first with appropriate (explicit) assumptions and then later conduct detailed calculations. It then becomes appropriate to also optimize the performance within the design constraints and chosen envelope. For optimization studies it is important to clearly define what parameter is being optimized. E.g. the highest exhaust speed possible.

It is tempting to add an aesthetic element to any design to make it look nice. However, this has little to do with engineering. An exception to this is with passenger liner aeroplanes where comfort and personal confidence in the machine often depends upon an aesthetic quality. However, for a spacecraft designed to go into deep space where every kilogram of mass has large implications, aesthetics are not relevant and should be avoided. Designers should strive for a design solution based solely on the engineering calculations. An interstellar probe that looks like a Concorde has little basis in practical reality. Part of this aesthetic expectation is probably due to science fiction films where rockets were pretty and pointed, despite the fact that the rocket was designed as a lunar Lander where aerodynamics is not an issue. Despite this it is worth noting that often design decisions are based on personal experience and judgment so can contain an element of subjectivity. This is the human element in engineering design and often the origin of aesthetic appearance.

Once designers are given the engineering requirement, it is appropriate and useful to adopt so called watchwords. These help to drive the solution in a particular direction and continuously remind the designers to stay within certain constraints. An obvious watchword could be 'Cost', where a customer has limited financing for the project. For a deep space mission another is 'Redundancy' and this was adopted for the Voyager and Pioneer probes. Another watchword frequently adopted for interstellar design studies is 'Near-term technology', which is a stipulation that only linear extrapolation should be used for a few decades hence based upon currently available technology. Extrapolating technology to say a century would arguably not be in accordance with this watchword.

14.3   Systems Engineering Approach to Concept Design Studies

The way a spacecraft scientist approaches design today is to use a systems engineering practice. Systems engineering is a way of viewing a whole as separated into individual units so that each part can be placed together in a systematic and consistent way. According to the guidelines as laid down by the European Cooperation for Space Standardization, the systems engineering approach to aerospace design can be described as:

> A mechanism for proceeding from interpretation of the customers requirements to an optimized product by steadily applying a wide ranging attention to product requirements, extending to all details of the users needs, produceability constraints and life cycle aspects, essentially through an organized concurrent engineering practice. [1]

Systems engineering begins with the customer needs. Often, the customer is unsure of his own needs and so will discuss this with the potential contractor prior to agreeing on a formal requirements definition with clear constraints on what is not possible. The requirements should be stated at a general level to avoid predetermining the design solutions possible. Often some of the requirements can lead to inconsistencies. This is particularly the case when examining interstellar missions, which use current or near future technology. Somehow a designer must integrate these two technologies in a consistent manner. This should lead to a likely specification, which covers issues such as the definition of systems, the mission and lifetime requirements, environmental assessment, definition of systems functions.

For an interstellar spacecraft consideration will also need to be given to the lifetime maintenance and repair. The end solution will be a balance between the requirements and the constraints. Ultimately, engineering design is about compromise. The ability of any design solutions to meet the requirements can be met by a performance parameter such as cruise velocity or power. In the systems engineering approach each system has an interface so it is clear where one responsibility starts and another begins and this ensures quality in the design process.

For a spacecraft the various systems would include propulsion, payload, communications, power, thermal, structure, guidance & control and any mechanisms on board. Each system will also have various sub-systems. For the propulsion this would be fuel, fuel tanks, pumps, valves, pipes. All of the major systems can be captured in a system requirements document and also in a transparent spreadsheet system model from which the team can then make changes and assess the effect on the overall design. The system definition will then evolve through different versions as a more detailed design is created. This is known as configuration management and it should be centrally controlled. All systems contain sub-systems and these must also be considered in the design process. There are also external elements to be considered which may affect the design solution possible. This includes the operating environment, manufacturing, disposal, ground station & monitoring and for a spacecraft the launch vehicle.

A design project will begin from the initial requirements. The customer who defined these requirements should not then determine the solution path. This should be left to the independence of the designers. From the initial requirements the key constraints on the design should be defined although not all of these will be apparent from the beginning. It could be due to a lack of observational data for example or a lack of capability. All of these should be recorded in a formal document and will act as a constant reference throughout the design study. Once the requirements have been defined they should be reviewed and fixed for the duration of the project. It will then progress through three key phases in the design. All important decisions should be carefully debated as they may have an effect on the design solution later on. This includes the allocation of mass and power budgets, which are distributed among the different systems. This will act as a system constraint so should allow for appropriate margins.

The first part of the design process is known as the *Concept Design Phase* and is really a feasibility study phase to see if solutions are possible for the specified requirement. This addresses issues such as the likely size and weight of the vehicle, the possible configuration layout and will begin a set of trade studies to 'scope' the design space around the stated requirements. Crucially, the results from this phase should address the question of whether a solution is even possible for the specific requirement. If it is not, then designers will have to go back to the customer who stated the initial requirement to see if there is room for modifications. If however several solutions are possible, then the design study may proceed to the next stage having defined several options known as pathways, each option having its own attributes and characteristics. These attributes may be based upon functionality, reliability, cost, weight, geometry and performance for example. The concept design phase should have produced conceptual sketches of the likely vehicle performance, with dimensions and possible performance. Often a design will be a continuation of a previous study, in which case a conceptual sketch already exists.

Calculations should be performed using a variety of different engineering and physics tools. This could be fluid modeling codes, finite element structural analysis, parametric scaling studies or simple hand written physics calculations. This will build up a clear picture of the design space performance for each configuration being assessed and allow for optimization and performance comparison. Margins will also need to be considered in the design. Safety margins are usually important but for unmanned robotic spacecraft one of the dominant margins is mass. Changes in one system margin may propagate through the system and affect other systems; hence the definition of each system interface needs to be correct. At the end of this process it is expected that some level of down select will be performed – this means a gradual selection over an increasingly narrower choice of options until the final design is chosen.

Next there is the *Preliminary Design Phase*. This continues with further studies of the design options and then at some point a *Preliminary Design Review* is held. This involves down selecting to the preferred option and proceeding to more detailed calculations. At the end of the phase it is expected that all of the major systems would have been frozen and a configuration layout should exist. Considerable optimization studies of the design would have been conducted by this stage. It may be appropriate to build prototype models for thermal or wind tunnel tests for example. This will help to produce an optimum design solution.

Finally there is the *Detailed Design Phase*. For an actual flight ready vehicle, this phase will involve detailed design calculations, which lead to the manufacturing requirements and tolerances. However, for most interstellar vehicle studies they will be constrained to a theoretical basis, so instead the detailed design phase will represent detailed sub-systems design calculations and qualification of the likely performance. At the end of this phase a *Detailed Design Review* will be conducted of the final solution and all the calculations. This should demonstrate that the design does meet the initial customer requirements and is optimized to the specified performance.

All of the above will be scheduled in a fully documented program plan from the initial requirement definition through to the final design solution. This way the effect of any slippage on the overall program plans can be fed back to the customer. Between each design phase there may be several (~2–5) key milestones that have to be met. Some of these will also be represented by Stage Gates, which must be passed with a deliverable output, in order for the design work to proceed. There will also be several preliminary design reviews of the work and of course a final *Critical Design Review* towards the end which demonstrates that the solution does meet the initial requirements and is compliant with the claimed specification.

The final design solution will also have to be risk assessed including the effect of one component or system failure of the rest of the systems. This may require a level of redundancy (e.g. multiple sensors) to ensure a high reliability of mission success. For an interstellar probe that has a long duration mission profile, this places harsh constraints on the components to be able to perform decades after launch. One of the ways that designers can assess reliability for mission success with interstellar probe design is to perform a 'nominal' versus 'worst case' assessment. The nominal performance is assuming an all optimistic performance. A worst case assessment involves an ultra conservative assumption on the performance and this approach was used in underwriting the performance of the Apollo Lunar Lander landing gears [2]. Scientists were unsure of the lunar soil properties or the speed and angle of vehicle descent, so a worst-case assessment had to be made.

A 'worst case' assessment may involve the effect on total vehicle performance of say only a 20% burn up fraction for an ICF capsule or sub-MW power levels to the computer. In the limit of all systems performance well below nominal, the spacecraft mission will fail. But it should be designed so that one system failure does not fail the whole mission, hence the need for system level redundancy. Performing these two types of assessments will help bound the likely performance and give a measure of the overall robustness of the design. The estimation of where the failure threshold lies is heavily dependent upon the assumption of inherent uncertainties (e.g. how well do we know the cross section for a DHe^3 fusion reaction?). Once the worst-case performance assessment has been made, an appropriate performance margin should be added to ensure that the failure threshold is avoided.

The above process is illustrated in Fig. 14.1 for a Lunar Lander requirement. For an interstellar spacecraft design that is not likely to be built in the near future, the major deliverable will be a final design report. This should show all the calculations, the configuration layout and expected performance specification. When published in an external peer reviewed Journal this will allow others to critically evaluate the credibility of the design solution and any lessons learned from this study will be fed into any future studies. In this manner, interstellar research is characterized by incremental progress – *ad astra incrementis*.

One of the amazing things about the modern age is that the assembly of teams of people within one location is no longer required in order to undertake a theoretical design study for an interstellar probe. This is all thanks to the power of the internet. A team of people can instead get behind one idea and work towards the fruition of that idea, with some of the team having never met. Meetings can even be held over the internet via phone links such as Skype and the use of web cams. This probably shouldn't be done for an actual spacecraft where safety and financing are at stake. Instead, one to one discussions are required to ensure maximum reliability and remove any ambiguities in the design. There is also the element of feeling part of a team, which all helps to improve the design solution so that all the members can take pride in the achievement. However, it is certainly worth noting that the availability of the World Wide Web provides a unique opportunity to advance interstellar studies by utilizing an international capability. This has implications for proposals such as the 'Alpha Centauri Prize' discussed in Chap. 17.

Fig. 14.1 The design process

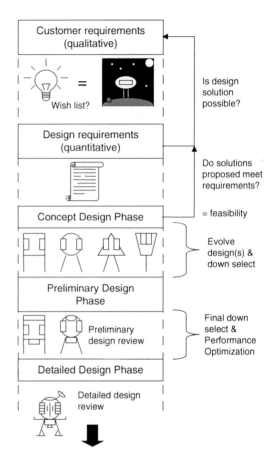

For *Project Icarus* discussed in Chap. 17 an international volunteer team of designers had to decide how to proceed with an interstellar design study. To constrain the scope the team determined the propulsion system in advance as part of the engineering requirement, as well as other parameters such as getting the probe to its destination within 100 years. These starting points are what are termed the engineering requirements or terms of reference for the study and they are fundamental for any team hoping to evolve a design from a wish list to a real engineered application. Once the team begins a basic system breakdown of the spacecraft must be performed ranging from propulsion to communications to science instruments, and any sub-systems. It's a complicated process that requires a watchful eye on the system interfaces, margins, mass budgets and performance. This is where project management is required and a team directed towards the end goal of a successful design solution – a satisfying result for all those involved.

14.4   Technology Readiness Levels

To distinguish between different propulsion schemes we need a way of 'measuring' whether a technology is near to actual use. For spaceflight this is ever more useful given that many of the ideas are highly speculative and require analysis by some consistent manner. Any idea starts with a *conjecture*. This is when you know what you would like to accomplish but do not know if it is possible in theory. To move the idea forward one must 'scope' the issues and understand the problem better. This leads to a basic description of the problem on a physics and engineering basis and the idea then becomes *scientific speculation* where several potential solutions are apparent, within existing knowledge. Further analysis will lead to a detailed description of the problem and potential component functionality.

One can then clearly distinguish between the different methods based purely on *science* and it will be clear how to turn the idea into a real product. The next stage is to build a working prototype or test rig to test out elements of the design systems. This *technology* will clearly show whether the product will give the required performance to meet the original idea. After all the prototype testing has been completed the knowledge gained can be used in an *application* and the actual device can then be built. These five phases of technological development from an idea are crucial steps to building an actual product. All machines, cars, aeroplanes, spacecraft, must go through these phases before they evolve into a real product meeting an initial requirement.

In this book we have only been discussing some of the most popular proposals for interstellar flight but there are many others, perhaps many thousands of specific proposals when one considers that various combinations of schemes that can be combined for optimum performance and specific missions. The selection of any technology for an interstellar precursor mission will depend upon a valid performance comparison. The best way to distinguish between those propulsion schemes that are realistic and practical and those that are speculative fantasy is to consider the maturity of the relevant technology application. A simple assessment of the different proposals can lead to categorization of the technology, where the trend is for a quicker transit time.

- *Accessible (2010 AD)*: chemical; electric, plasma drive; nuclear electric.
- *Near future (+50 years)*: solar sailing, microwave; internal/external nuclear pulse.
- *Far future (+100 years)*: Laser sailing; interstellar ramjet; antimatter.
- *Speculative (>100 years)*: warp drive; space drive; worm holes.

The problem with this approach is that predicting the future and associated timescales is risky. The uncertainty in any prediction becomes larger, the more speculative the concept and the further into the future one may attempt to predict. However, the aerospace industry has already thought of this problem and has devised so called Technology Readiness Levels [3], which runs from conjecture

Table 14.2 Technology readiness levels for various propulsion schemes

TRL	Status	Definition	Method
9	Application tested	Actual system 'flight proven' through successful missions	Chemical, electric
8	Application proven	Actual system completed & 'flight qualified' through test & demonstration (ground or space)	Solar sails
7	System proof	System prototype demonstration in a space environment	Nuclear-electric
6	Prototype proof	System/subsystem model/ prototype demonstration in relevant environment (ground or space)	External nuclear pulse
5	Component proof	Component &/or validation in relevant environment	Plasma drive
4	Physics proof	Component &/or validation in laboratory environment	Laser sail microwave
3	Science	Analytical & experimental critical function &/or characteristic proof of concept	Internal nuclear pulse (fusion)
2	Speculation	Concept and/or application formulated	Antimatter, interstellar ramjet
1	Conjecture	Basic principles observed/reported	Warp drive, worm hole

at TRL1 to the application being fully tested at TRL9. Table 14.2 shows the TRLs written as appropriate for an interstellar probe. It is only by the rigorous and quantitative application of this sort of analysis to the assortment of propulsion schemes that a true reality check is enabled on which schemes are viable for near term missions. If we are to make progress towards the true vision of interstellar travel, then future efforts of the interstellar propulsion community should be directed towards increasing the TRLs of specific design schemes and optimizing them for specific missions. We should be asking what can we do to increase the TRL of a specific scheme and that should be the driver for all related academic research. It is the hope of this author that the TRL of fusion based propulsion systems may increase by around 3 levels within the next two decades.

14.5   A Concept Design Problem for a Precursor Mission Proposal

In this section we work through an example of a concept design study for an interstellar precursor probe. We do this to illustrate how one begins to think about concept sketching. We call this $\mathbf{XiP^2}$, which stands for the *10,000 AU interstellar Precursor Probe*. We only present part of the initial concept phase as

Table 14.3 Engine comparison morphology chart

	V_e (km/s)	I_{sp} (s)	TRL (2010)	TRL (2100)
Chemical	5	500	9	9
Nuclear thermal	60	6,000	6	9
Ion thruster	100	10,000	9	9
Fusion	10,000	1,000,000	2–3	9
Solar sails	–	–	5	9

a feasibility study and the defined options are deliberately limited for the purposes of illustration. This study begins with the following requirements.

A government space agency decides that they want to plan a mission out into the Oort cloud. A review panel of scientists meets to decide on the following system requirements.

- Design an unmanned interstellar flyby precursor probe that is designed to go to a distance of 10,000 AU in a duration that is ideally much less than 100 years and to be launched by 2100.
- The total spacecraft mass should not exceed 100 tons, including propellant.
- The mission must achieve the science goals of optical and spectroscopic observations of the distance dwarf planets en route and distance star system as well as obtain measurements of the interstellar wind particle speed and density at the solar heliosphere and interstellar medium.
- Assume orbital assembly and launch, delivered in several Space Shuttle missions. Although the Space Shuttle fleet has been retired we use this architecture because it's a current baseline upon which to compare future missions.

14.5.1   Scoping Concept Design Space

Now that we have our initial requirements we must fold in any constraints and turn these into quantitative system requirements. For the first concept analysis we are simply interested in finding a concept, which meets the upper bound requirements. We have a mass constraint of 100 tons to 10,000 AU in 100 years. Now 10,000 AU = 1.496×10^{15} m = 0.158 light years and <100 years = 3.1536 billion seconds. This implies a minimum mission cruise velocity to reach this distance of 100 AU/year or 474 km/s (0.0016c).

We then do a quick trade study on the potential engine options, where a trade study refers to a short piece of work usually involving calculations of various performance parameters in order to obtain an estimate for the optimum solution, given the constraints and assumptions of the study such as a 2100 launch. Table 14.3 shows the results, including a comparison of exhaust velocity v_e, specific impulse I_{sp} and TRL for the years 2010 and 2100.

We briefly discuss each engine design in turn by considering the implications to the propellant mass using the ideal rocket equation presented in Chap. 5. The chemical

Table 14.4 Morphology table for down select

Parameter	Meet ΔV requirement	TRL 2100 requirement
Chemical	No	Yes
Nuclear thermal	No	Yes
Ion thruster	Maybe	Yes
Fusion	Yes	Yes
Solar sail	Yes	Yes

Fig. 14.2 Final concept
options for design study

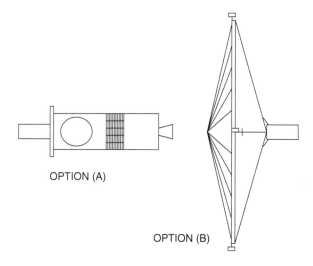

OPTION (A)

OPTION (B)

engine type has been well proven. However to achieve a Δv of 474 km/s with an exhaust velocity of only 5 km/s would require an initial mass that is of order $\sim e^{95}$ times the final vehicle mass, an impractical amount. The final vehicle + payload mass left over would be a very small fraction of the initial mass. Ion thruster technology has also been well proven but to achieve a Δv of 474 km/s with an exhaust velocity of 100 km/s would still require an initial mass of order $\sim e^{5}$ times the final vehicle mass. Nuclear thermal will also supply an insufficient exhaust velocity with a 60 km/s exhaust velocity resulting in an initial mass of order $\sim e^{8}$ times the final vehicle mass.

With nuclear fusion propulsion there is hope with a high exhaust velocity of $\sim$10,000 km/s but we assume here a more moderate 500 km/s, so to achieve the Δv requirement we would need an initial mass of order $\sim e^{0.95}$ times the final vehicle mass. Solar sails potentially offer a high Δv as no propellant is required. This is because the source of the energy is derived directly from the solar intensity flux. All of the spacecraft mass would be close to the 100 tons allocation assuming the entire allocation was used. On the basis of this quick trade study comparison a morphology chart is produced as shown in the Table 14.4. After some consideration from a hypothetical design team this then leads to the down select of two possible candidates for the mission known as Option A (fusion engine) and Option B (solar sail). These are depicted as concept sketches in Fig. 14.2.

Table 14.5 Intensity of solar flux with distance

Solar flux intensity (W/m^2)
1 AU = 1,400
10 AU = 14
100 AU = 0.14
1,000 AU = 0.0014
5,000 AU = 0.000056
10,000 AU = 0.000014

Table 14.6 Sail loading with area

Sail area (m^2)	Sail loading (g/m^2)
100	10^6
1,000	10^5
10,000	10^4
100,000	1,000
1,000,000	100
10,000,000	10

Some further work is then performed to understand the sub-options. In particular, there are serious concerns about the ability to use a solar sail to accelerate a 100 tons vehicle using only solar photons, which reduces inversely with distances. For solar sails the solar flux incident upon a surface normal to the Sun can be calculated assuming that the solar flux at the orbit of the Earth from the Sun is 1,400 Wm^{-2}. This means that at different distance from the Sun to 10,000 AU it will follow an inverse square law trend with the computed intensity shown in Table 14.5 and computed using (14.3):

$$F = \frac{1400 Wm^{-2}}{R(AU)^2} \qquad (14.3)$$

We must also consider the sail loading within a 100 tons total vehicle constraint. Table 14.6 is an example of sail loadings for our proposed vehicle using (14.4). It is clear that a sail loading of around ~10–100 is only attained with massive sail areas with lengths of order 30–100 km in size. On the basis of these numbers Option A (fusion propulsion) is chosen as the candidate propulsion system for the required mission. A more detailed scoping study would then be performed.

$$\sigma = \frac{M_{tot}}{A_s} \qquad (14.4)$$

14.5.2   Concept Solution

From an evaluation of the science requirements of the mission the hypothetical team decides that a payload mass of around 4 tons is required. The team allocates two of each instrument to ensure redundancy, which consists of:

- 2 × Optical imaging telescope
- 2 × Spectrometer
- 2 × Particle analyzer
- Computers for instrument operation and data analysis (in built redundancy)
- Boom with communications antenna

Next we begin by assessing the minimum thrust required as an order of magnitude for a 100 tons spacecraft and assuming a low acceleration of 0.0002 g from deep space determined by science instrument detection sensitivities. The minimum thrust for the maximum mass budget is given by:

$$T_{min} = M_{s/c} \times a = 100{,}000 kg \times 0.0002 \times 9.81 ms^{-2} \approx 196 N \qquad (14.5)$$

We next consider the total jet power. Although an exhaust velocity of 10,000 km/s seems entirely possible with a fusion engine, we assume a more conservative number of 500 km/s which should be sufficient for the mission requirement.

We can next get a rough estimate of the minimum jet power required. This is the power of the rocket exhaust and gives an estimate for the energy expenditure per unit of time. It is given by:

$$P_j = \dot{m} v_{ex}^2 = \frac{1}{2} \times v_{ex} \times T_{min} = \frac{1}{2} 500{,}000 ms^{-1} \times 196 N \approx 49 MW \qquad (14.6)$$

We can then use the concept of specific power to estimate the masses of the key vehicle systems, namely the power supply, engine and thermal management system – usually represented by the radiators which dominate the system mass. The specific power gives an estimate for the mass utilization of the propulsion system. We begin by estimating the mass of the power supply. We assume a specific power to be 5 kW$_e$/kg. The power supply mass is then given by:

$$m_{pow} = \frac{49 \times 10^6 W}{5 \times 10^3 W_e kg^{-1}} \approx 9.8 tons \qquad (14.7)$$

Next we estimate the mass of the engine assuming a specific power to be 10 kW/ kg. This is given by:

$$m_{eng} = \frac{49 \times 10^6 W}{10 \times 10^3 W kg^{-1}} \approx 4.9 tons \qquad (14.8)$$

Finally, we must estimate the mass of the radiators, which will dump any excess heat into space. We assume 25% conversion efficiency then the total power generated will be given by:

$$P_{tot} = \frac{49 \times 10^6 W}{0.25} \approx 196 MW \tag{14.9}$$

The remaining 75% of the total power will be dumped as waste energy into space, which is given by 0.75×196 MW ≈ 147 MW. In the design we assume the use of aluminum bonded honeycomb heat radiators mounted on the inner surfaces. These are to reject excess heat from the fusion reactor. We assume a specific power for the radiators to be 100 kW/kg. The radiator mass is then given by:

$$m_{rad} = \frac{147 \times 10^6 W}{100 \times 10^3 Wkg^{-1}} \approx 1.47 tons \tag{14.10}$$

To allow for additional structure such as propellant tanks and spacecraft casing we allocate an additional 7 tons. This then allows us to produce an initial mass sizing for the spacecraft concept.

- Power supply mass = 9.8 tons
- Engine mass = 4.9 tons
- Radiator mass = 1.47 tons
- Payload mass = 2 tons
- Additional structure mass = 7 tons
- Total ≈ 25 tons

For the estimated 474 km/s velocity increment and an assumed 500 km/s exhaust velocity this gave a mass ratio of $e^{0.95} \approx 2.58$. Assuming no engine staging, the propellant mass is then given by the rocket equation:

$$m_{prop} = 25 tons(2.58 - 1) \approx 40 tons \tag{14.11}$$

The total vehicle wet mass is then ≈ 65 tons. For the final payload mass of 2 tons and final structure mass of 5 tons we can then compute the dry payload mass fraction as follows:

$$f = \frac{2 tons}{25 tons} \approx 8\% \tag{14.12}$$

This is actually quite reasonable compared to conventional spacecraft which are typically between ~5–20%. With this configuration we now compute the performance. Firstly we examine a single staged performance:

$$\Delta v = 500 kms^{-1} Ln\left(\frac{40 + 23 + 2}{23 + 2}\right) = 500 kms^{-1} Ln(2.6) = 478 kms^{-1} \tag{14.13}$$

This can be improved by making this a two stage design, with the assumption that both stages produce an exhaust velocity of 500 km/s. Initial work (by a hypothetical design team) leads to a propellant mass split of 30 tons (first stage) and 10 tons (second stage); a 75% + 25% split. The structure is split with 14 tons (first stage) and 9 tons (second stage), where the second stage mass would also include an additional 2 tons science payload mass equating to 9 tons so comes to a total of 11 tons. The velocity increments are then:

$$\Delta v_1 = 500kms^{-1}Ln\left(\frac{40+23+2}{10+23+2}\right) = 500kms^{-1}Ln(1.86) \approx 309.5kms^{-1} \quad (14.14)$$

$$\Delta v_2 = 500kms^{-1}Ln\left(\frac{10+9+2}{9+2}\right) = 500kms^{-1}Ln(1.91) = 323.3kms^{-1} \quad (14.15)$$

$$\Delta v_{tot} = 632.8 \text{ kms}^{-1} \text{ (0.002c)}$$

A probe travelling at this velocity would reach 10,000 AU in around 75 years. However, the probe must first accelerate to this velocity hence the two boost phases. We can use simple linear theory to estimate the distance and duration of the two boost phases. We assume an acceleration of 0.0002 g for both phases. The two durations are then given by:

$$t_{b,1} = \frac{\Delta v_1}{a} = \frac{309,000ms^{-1}}{(0.0002 \times 9.81ms^{-2})} \approx 5 \text{ years} \quad (14.16)$$

$$t_{b,2} = \frac{\Delta v_{fin} - \Delta v_1}{a} = \frac{632.8,000ms^{-1} - 309.5,000ms^{-1}}{(0.0002 \times 9.81ms^{-2})} \approx 5.2 \text{ years} \quad (14.17)$$

This means that the second stage burn will be completed 10.2 years into the mission. We can then compute the distance achieved under each boost phase as follows:

$$S_{b,1} = \frac{1}{2}at_{b,1}^2 = \frac{1}{2}(0.0002 \times 9.81ms^{-2})(5years \times 3.1536 \times 10^7s)^2$$
$$\approx 163AU \quad (14.18)$$

$$S_{b,2} = S_{b,1} + \frac{1}{2}at_{b2}^2$$
$$= S_{b,1} + \frac{1}{2}(0.0002 \times 9.81ms^{-2})(5.2years \times 3.1536 \times 10^7s)$$
$$\approx 339.3AU \quad (14.19)$$

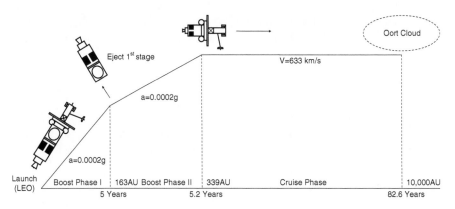

Fig. 14.3 Initial mission profile for XiP^2

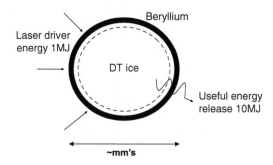

Fig. 14.4 Typical pellet configuration

The remaining distance for the cruise phase is then 10,000 AU − 339.3 AU = 9,661 AU. Then at the cruise velocity of 632.8 km/s it will take the following time to complete the cruise phase:

$$t_c = \frac{S_c}{\Delta v_{tot}} = \frac{9,661AU \times 1.496 \times 10^{11}m}{632,800ms^{-1}} \approx 72.4 years. \qquad (14.20)$$

Hence the total mission duration will be: 5 years + 5.2 years + 69.9 years = 82.6 years. The mission profile is illustrated in Fig. 14.3.

From the distance of 10,000 AU it would take a radio signal around 58 days to reach the Earth based communications. Now in the mission analysis we have assumed linear equations for constant motion to make the concept sketching simple for the reader. However, it is worth noting that the distance achieved during any boost phase will have a logarithmic dependence on the mass ratio given (3.16) of Chap. 3. But we do not consider this level of detail for this example.

The fusion energy release is to be obtained by the laser implosion of multiple inertial confinement fusion pellets. The ~ mm sized radius capsule contains DT ice surrounded by a layer of Beryllium as shown in Fig. 14.4. The total capsule fuel mass is around 3×10^{-6} kg. The efficiency of the pellet energy conversion would be less than 100% and this would have to be factored into the calculations. However, it is assumed that 10 MJ is produced from each capsule, not unreasonable for the design geometry (and similar to NIF targets). This is an energy gain of around $Q = 10$ assuming a 1 MJ laser driver energy. The principal fusion reaction is $D(T,He^4)n$, where the He^4 has an energy of 3.5 MeV and the neutron 14.1 MeV. The choice of DT fuel is due to the large reaction cross section at low temperature. However, a re-analysis would have to examine the neutron yield from the DT reactions and allow for adequate radiation shielding of the science payload. A reassessment may suggest moving to a different propellant combination such as DHe^3 but this has acquisition issues.

The propellant tanks are pre-filled with the DT capsules just prior to launch from the ground. The Tritium will decay to He^4 in around 12 years so this would have to be considered in more detailed studies along with any heat generated. Two problems are presented for the emitted particle products (1) He^4 is a charged particle, so the use of a magnetic field to deflect it towards the nozzle exit for thrust generation is required (2) neutrons do not have charge so cannot be magnetically directed. They will also activate the walls of the reaction chamber and cause structural thermal loads. This necessitates some degree of radiation shielding to protect the payload. The reactions would also emit gamma rays and x-rays which have to be accounted for in any detailed analysis. It is assumed that a radiation shield is made of Beryllium and several millimeters thick.

We can now estimate the number of ICF pellets in each engine stage by dividing the pellet mass into the propellant mass:

$$N_{pel,1} = \frac{m_{prop,1}}{m_{pell}} = \frac{30,000kg}{3 \times 10^{-6}kg} \approx 10billion \qquad (14.21)$$

$$N_{pel,2} = \frac{m_{prop,2}}{m_{pell}} = \frac{10,000kg}{3 \times 10^{-6}kg} \approx 3.33billion \qquad (14.22)$$

We can then divide this number into the boost phase duration to get an estimate for the required pulse frequency for each engine stage:

$$f_{b,1} = \frac{N_{pel,1}}{t_{b,1}} = \frac{10billion}{(5years \times 3.1536 \times 10^7 s)} \approx 63Hz \qquad (14.23)$$

$$f_{b,2} = \frac{N_{pel,2}}{t_{b,2}} = \frac{3.33billion}{(5.2years \times 3.1536 \times 10^7 s)} \approx 20Hz \qquad (14.24)$$

Figure 14.5 shows the results of a calculation for the first and second stage propellant masses with pulse frequency plotted against burn time. We immediately see that the pulse frequency is proportional to 1/burn duration. We require a minimum pulse frequency so as to minimize the propulsion engineering constraints. However, the results show that a minimum pulse frequency will have a longer burn

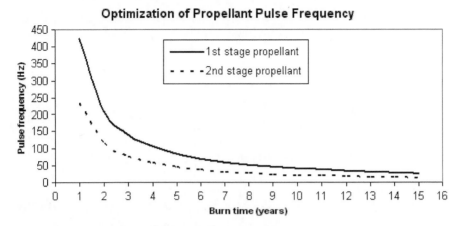

Fig. 14.5 Optimization of pellet performance

time which also places constraints on the technology in terms of the need to operate in a continuous pulse phase for many numbers of years. The optimum pulse frequency and burn time is therefore a trade-off and would be located in the bottom left corner of the graph. We note that we have computed pulse frequencies of 63 Hz for the first stage and 20 Hz for the second stage. A more detailed analysis would attempt to optimize this trade-off.

We assume that each pellet releases 10 MJ of energy. However, only a fraction of this energy will be useful in propulsive thrust. This is due to issues such as pellet burn fraction, exhaust collimation and other effects. To account for all of these effects we assume an efficiency of energy conversion to useful thrust of 0.04 for both stages. This then gives a total energy output to be used in thrust generation as follows:

$$E_{b,1} = 0.04 \times 10MJ \times 10 \times 10^9 pellets \approx 4 billion MJ \qquad (14.25)$$

$$E_{b,2} = 0.04 \times 10MJ \times 3.33 \times 10^9 pellets \approx 1.33 billion MJ \qquad (14.26)$$

This equates to a total energy release of 5.33 billion MJ over the entire boost phase. We can also compare the total energy release to a kinetic energy estimate for the final 11 t structure + payload mass to arrive at the target distance of 10,000 AU. This is obtained as follows:

$$E = \frac{1}{2}mv^2 = \frac{1}{2}(11,000kg)(632.8,000ms^{-1})^2 \approx 2.2 billion MJ \qquad (14.27)$$

We conclude from this that the actual energy produced in the current configuration layout is a factor ~2.4 times larger than the minimum required, so the thrust generation is more than adequate. However, it should be trimmed back on further iteration. We can also divide the staged energy release into the boost duration to get an estimate for the total power as follows:

$$P_{b,1} = \frac{E_{b,1}}{t_{b,1}} = \frac{4 \times 10^{15} J}{(5 years \times 3.1536 \times 10^7 s)} \approx 25.37 MW \tag{14.28}$$

$$P_{b,2} = \frac{E_{b,2}}{t_{b,2}} = \frac{1.33 \times 10^{15} J}{(5.2 years \times 3.1536 \times 10^7 s)} \approx 8.11 MW \tag{14.29}$$

This equates to a total power of 33.48 MW. The actual mass flow rate per stage will be given by:

$$\dot{m}_{b,1} = f_{b,1} \times m_{pell} = 63 Hz \times 3 \times 10^{-6} kg \approx 18.9 \times 10^{-5} kgs^{-1} \tag{14.30}$$

$$\dot{m}_{b,2} = f_{b,2} \times m_{pell} = 20 Hz \times 3 \times 10^{-6} kg \approx 6.1 \times 10^{-5} kgs^{-1} \tag{14.31}$$

The thrusts for each boost stage are determined by the product of the mass flow rate and exhaust velocity as follows:

$$T_{b,1} = \dot{m}_{b,1} \times v_{ex} = 18.9 \times 10^{-5} kgs^{-1} \times 500,000 ms^{-1} \approx 94.5 N \tag{14.32}$$

$$T_{b,2} = \dot{m}_{b,2} \times v_{ex} = 3 \times 10^{-5} kgs^{-1} \times 500,000 ms^{-1} \approx 30.5 N \tag{14.33}$$

As a check we can re-calculate the exhaust velocity for both stages be using the staged power, pulse frequency and mass flow rate as follows:

$$v_{ex,1} = \sqrt{\frac{2P_{b,1}}{\dot{m}_1}} = \sqrt{\frac{2 \times 25.37 MW}{63 Hz \times 3 \times 10^{-6} kg}} \approx 500 km/s \tag{14.34}$$

$$v_{ex,2} = \sqrt{\frac{2P_{b,2}}{\dot{m}_2}} = \sqrt{\frac{2 \times 8.22 MW}{20 Hz \times 3 \times 10^{-6} kg}} \approx 500 km/s \tag{14.35}$$

The minimum specific power for the mission is found by examining each engine stage separately. This specific power should include the mass of the power supply (9.8 tons), radiators (1.47 tons) and engine (4.9 tons), which collective make up the propulsion system and for the total vehicle comes to 16 tons. The masses are split as two third to the first stage and one third to the second stage. So the first stage propulsion mass is 6.54 tons, 0.98 tons and 3.26 tons for the power supply, radiators and engine respectively, coming to a total of 10.78 tons. The second stage propulsion mass 3.27 tons, 0.49 tons and 1.63 tons for the power supply, radiators and engine respectively, coming to a total of 5.39 tons. The staged specific powers are found as follows:

$$P_{min,1} = \frac{P_{,b,1}}{m_{tot}} = \frac{25.37 MW}{10,780 kg} \approx 2.35 kW/kg^{-1} \tag{14.36}$$

$$P_{min,2} = \frac{P_{,b,2}}{m_{tot}} = \frac{8.11MW}{5,390kg} \approx 1.51kWkg^{-1} \tag{14.37}$$

We note that this is the net specific power which is after all the efficiencies have been accounted for and is the useful work as performed by the rocket engines. The raw power is the energy per unit time originating from the exploding ICF hot fusion plasma. To find the raw power we divide by our generic efficiency factor of 0.04 as follows:

$$P_{raw,1} = \frac{P_{min,1}}{0.04} = \frac{2.35kW/kg^{-1}}{0.04} \approx 58.75kW/kg^{-1} \tag{14.38}$$

$$P_{raw,2} = \frac{P_{min,2}}{0.04} = \frac{1.5kW/kg^{-1}}{0.04} \approx 37.62kW/kg^{-1} \tag{14.39}$$

We next consider the likely spacecraft systems. These include: propulsion; structure; communications; navigation & guidance; thermal; power and payload. The spacecraft is made of molybdenum alloy, which has a density of around 1,028 kg/m^3. The first stage has a length of 15 m and a radius of 2.5 m and the second stage has a length of 6 m and a radius of 2.3 m. Both cylinders are assumed to have a shell thickness of 5 mm. So calculating the mass of each cylinder we find:

$$m_{cyl}(1^{st}) = 1,029kgm^{-3}\pi \times 15m\left(2.5^2_{outer} - 2.495^2_{inner}\right) \approx 1.2tons \tag{14.40}$$

$$m_{cyl}(2^{nd}) = 1,029kgm^{-3}\pi \times 6m\left(2.3^2_{outer} - 2.295^2_{inner}\right) \approx 0.45tons \tag{14.41}$$

The engine will comprise components such as nozzle, manifold for exhaust flow control, pumps and pipes, which are to be quantified. The propellant tanks are constructed of a Titanium alloy (density 4,500 kg/m^3) due to its high strength to weight ratio. Titanium also has a high resistance to corrosion, and fatigue. The 1st stage tanks have a radius of 1 m and the second stage tanks a radius of 0.5 m. Both are 2 mm thick giving a total spherical shell mass of:

$$m_{tank} = \frac{4\pi\rho}{3}\left(R_2^3 - R_1^3\right) \tag{14.42}$$

$$m_{tank}(1^{st}) = \frac{4\pi \times 4500kgm^{-3}}{3}\left(1^3_2 - 0.998^3_1\right) \approx 0.11tons \tag{14.43}$$

$$m_{tank}(2^{nd}) = \frac{4\pi \times 4500kgm^{-3}}{3}\left(0.5^3_2 - 0.498^3_1\right) \approx 0.028tons \tag{14.44}$$

There are two tanks for both stages so this means that the total (dry) tank mass is 0.22 tons (first stage) and 0.056 tons (seconf stage). The reaction chamber is to be constructed of Molybdenum (density 1,028 kg/m^3) and 10 mm thick. It has a radius of 0.5 m for the 1st stage and 0.25 m for the second stage. This gives a mass of:

$$m_{chamber}(1st) = \frac{4\pi \times 1028kgm^{-3}}{3}\left(0.5_2^3 - 0.49_1^3\right) \approx 0.032 tons \qquad (14.45)$$

$$m_{chamber}(2nd) = \frac{4\pi \times 1028kgm^{-3}}{3}\left(0.25_2^3 - 0.24_1^3\right) \approx 0.0075 tons \qquad (14.46)$$

There is a load stiffener half way across the first stage is to provide extra rigidity during maneuvers which involve some torque to the vehicle structure. This is made of Molybdenum and has a thickness of 1 cm at an outer radius of 2.5 m. This gives it a mass of 0.2 tons. Energy is supplied to the laser beams by a Marx bank current generator delivering pulse energies in the TW range. A compact nuclear thermal reactor is positioned on the second stage to deliver power to the superconducting magnets and main drive ignition systems. An induction loop then 'bootstraps' the energy from each fusion ignition pulse to run the next pulse. Such technology is expected to be available with these power levels in the coming decades.

A shock absorber is required on the vehicle to protect the payload from the g forces produced from multiple capsule explosions. To deal with this the radiation shield is mounted on a plate of steel (density 7,480 kgm^{-3}) springs around 0.1 m in length and 1 m in radius. Because the second stage cylinder has a radius of 2.3 m, the radiation shield must at least cover this length. The assumption is made of a 3 m radius 1 mm thick Beryllium (density 1,850 kgm^{-3}) radiation shield. Assuming that $m = \rho\pi LR^3$ we find that the shield has a mass of 0.15 tons.

Superconducting magnets are included for magnetic field control and directivity of the charged particle products. There are assumed to be two large magnets on the first stage with a mass of 0.05 tons each. There are two smaller magnets on the second stage with a mass of 0.025 tons each. The primary ignition mechanism is laser irradiation of a foil target which then generates a total combined ion (proton) beam energy of around 1 MJ in 1 ns to the target from 6 ion lasers, thought to be sufficient to give a symmetrical pellet implosion. These produce an beam intensity of around 10^{15} W/cm^2 consistent with a long pulse laser and have a mass of 0.3 tons in the frist stage and 0.2 tons in the second stage. Attitude control corrections are enabled by moderate gimballing of the main exhaust nozzle.

The entire structural component masses discussed above comes to around 3 tons and this should be added to the total propulsion mass of 16 tons which comes to 19 tons. Adding on the science payload of 2 tons we have 21 tons. This fits within the allocated dry mass budget of 25 tons, leaving a 4 tons margin for any additional systems identified above or not mentioned. Figure 14.6 shows the vehicle configuration layout.

We also must consider the launch requirements from a Space Shuttle in terms of payload capacity. The Space Shuttle has a payload bay length of 18.3 m (60 ft); width of 5.2 m (17 ft); height of 4 m (13 ft); approximate payload bay volume = 381 m^3. It has a maximum payload capacity of around 25,000 kg (55,000 1b).

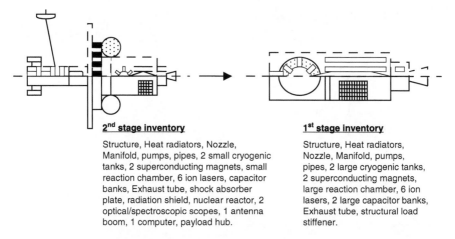

2nd stage inventory

Structure, Heat radiators, Nozzle, Manifold, pumps, pipes, 2 small cryogenic tanks, 2 superconducting magnets, small reaction chamber, 6 ion lasers, capacitor banks, Exhaust tube, shock absorber plate, radiation shield, nuclear reactor, 2 optical/spectroscopic scopes, 1 antenna boom, 1 computer, payload hub.

1st stage inventory

Structure, Heat radiators, Nozzle, Manifold, pumps, pipes, 2 large cryogenic tanks, 2 superconducting magnets, large reaction chamber, 6 ion lasers, 2 large capacitor banks, Exhaust tube, structural load stiffener.

Fig. 14.6   Final concept internal configurations layout

We can use these numbers to work out approximately what the launch costs would be to assemble the vehicle in orbit. The cost is also of interest and we can assume a cost per launch of $450 million. For a vehicle mass of 65 tons (payload + structure + propellant) and a total Space Shuttle payload capacity of 25 tons this would require a minimum of 3 Space Shuttles to place the spacecraft in Low Earth Orbit. It is unlikely that the geometry of the spacecraft would fit within two Space Shuttle bays so we allocate a forth mission. However, there will also be a need to put together the supporting assembly structures and tools that allow in-orbit assembly, which would require a fifth Space Shuttle. We therefore allocate a total of 5 Space Shuttles for the in-orbit assembly so that $5 \times \$450$ million is a launch cost of $2.25 billion. We can also add this to the research and development costs which we must determine.

A typical NASA mission within our solar system (~100 AU), along the lines of a Flagship program would cost up to £3 billion to develop. If we multiply this by the difference to 10,000 AU we arrive at a cost for the $\dot{X}iP^2$ mission to be around $300 billion. We add to this the launch costs and we get a total of $302.25 billion, although it would likely be a lot less than this but this gives us an upper bound. This is around 2.16% of the US Gross Domestic Product (~$14 tonsrillion in 2009) or around ~0.5% of the worlds Gross Domestic Product (~$61 tonsrillion in 2009), as estimated from the International Monetary Fund. The above analysis is rather crude and several major assumptions have been made. But remember, the point of this exercise is merely to illustrate how one goes about constructing a theoretical concept. The final performance specification and conceptual layout is shown below in Table 14.7 and Fig. 14.7. This would then allow a starting point from which further calculations can proceed and a more realistic engineering concept developed.

The next objective would be to cycle back through the above analysis and optimize the design for improved performance. After completing any design

Table 14.7 Preliminary spacecraft performance

Parameter	First stage	Second stage
Propellant mass (tons)	30	10
Number propellant tanks	2	2
Total Mass at stage burn (tons)	65	35
Mass ratio	1.86	1.91
Frequency (Hz)	63	20
Total energy release (billion MJ)	4	1.33
Total power (MW)	25.37	8.11
Jet power (MW)	223.62	7.62
Net specific power (kW/kg)	2.35	1.51
Raw power (kW/kg)	58.75	37.62
Pellet mass (grams)	0.003	0.003
Number pellets (billions)	10	3.33
Mass flow rate (kg/s)	0.00019	0.00006
Thrust (N)	94.5	30.5
Distance at stage burn (AU)	163	339
		Final = 10,000
Duration of stage burn (years)	5	5.2
		Final = 82.6
Exhaust velocity (km/s)	500	500
Specific Impulse (s)	50,000	50,000
Science payload mass (tons)	–	2
Dry payload mass fraction (%)		8
Delta-V (km/s)	309.5	323.3
		Final = 632.8 km/s
		(0.002c)

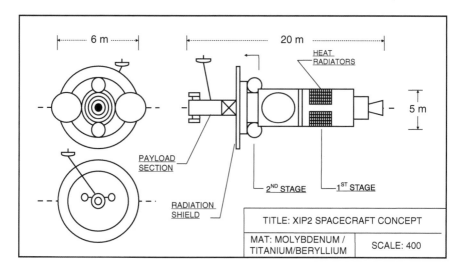

Fig. 14.7 Final concept external configurations layout

work engineers must perform a post design review and seek to improve the concept. This should be done until all system designers are satisfied and then proceed to a more detailed design. Looking at the concept sketch derived here, there are several areas for potential improvement upon which to derive a more credible engineering concept. Some of these include:

- Reduced acceleration burn times. This will relax the burn duration of the engine operation and lead to a more reliable design, if the probe can be accelerated up to the cruise velocity sooner. One can optimize the burn time with the pulse frequency.
- The assumed exhaust velocities of 500 km/s are probably quite low for a fusion engine. Something in the region of 8,000–9,000 km/s may be a more appropriate assumption although this would affect the pulse frequency and may require larger mass engines.
- The pulse frequency between the first and second stage engines is asymmetric, which could present problems in terms of system interfacing and integration. It may be preferable to have similar pulse frequencies of the order 40 Hz each. Similar mass flow rates may also be desirable.
- The fuel pellets consisted of deuterium and tritium. However, the tritium will decay with a half-life of around 12 years. Further iteration of the concept would require consideration of this decay in terms of reduced propellant to use throughout the boost phase of the mission and a new propellant requirement would have to be computed. Also, as the tritium decays it will release heat and this will have to be considered. Neutron production effects on the structure should also be considered.
- Shifting mass between the two stages could optimize the staged performance. In theory, a design is well optimized when the stage mass ratios are equal. The current mass ratios are very similar suggesting the performance is near-optimal.
- The derived structural component masses were chosen optimistically to fit within the mass budget allocation. However, it is likely that the quoted masses under estimate what would actually be required. In particular how a fusion engine scales is not well known. For the Daedalus project the first stage and second stage engine masses were in the ratio 3.1 but there would be a limit to this scaling. It is quite possible that when more realistic figures are used that the total vehicle structure mass increases based upon current technology assessments, which would also affect the performance. Re-visit each component and derive a more reasonable total structure mass.
- When reassessing the concept the exact masses for all systems would have to be computed. A full system and sub-system component definition would be required.
- Analysis of other spacecraft systems such as the communications and data transmission requirements. This should include an estimate for the diameter of any high gain antennas to be employed.

The various items listed above are left to the reader to cycle back through to improve the design concept as a good learning experience. Anyone attempting this should make one change at a time and then re-compute the configuration and performance with each cycle. A spreadsheet would be a useful tool to do this with. With each iteration should emerge a more credible, efficient and optimized design. Only when this is done will a design concept have been achieved and designers can advance to the *Preliminary Design Phase*.

This process to produce a conceptual layout can be applied to numerous concepts by using the following rules. In theory, the concept layout can be started at any point but the other parameters must also be computed. The *Twenty Step Method* to a concept layout is as follows:

1. From initial requirement, conduct trade studies to formulate mission and engineering options. Down select to chosen systems and mission architecture.
2. Assume a payload mass per the mission requirement.
3. Choose the engine type and determine the exhaust velocity and specific impulse to be expected.
4. Determine likely mission cruise speed.
5. Calculate the mass ratio for single or multi-staged performance as well as payload mass fraction.
6. Calculate propellant mass and structure mass budget.
7. Calculate total vehicle wet and dry mass.
8. Derive approximate vehicle configuration including length, shape then construct basic layout drawings.
9. Calculate minimum acceleration and minimum thrust to achieve mission.
10. Estimate total jet power.
11. Estimate likely power supply mass from jet power and assumed supply kWe/kg.
12. Estimate engine mass from jet power and assumed engine kW/kg.
13. Estimate radiator mass from jet power and assumed conversion efficiency and radiator kW/kg.
14. Compute total structure mass of engine, radiators and power supply.
15. Computer mass margin remaining, allocate or throw away.
16. Compute engine performance, mass flow rate, thrust, burn time.
17. Calculate mission profile to include boost, cruise and deceleration phases.
18. After further research repeat the above, resize configuration and performance.
19. After further research, optimize performance and vehicle configuration for required mission.
20. Publish conceptual layout as engineering drawing and final mission specification – move onto the preliminary design stage.

Finally, it is worth noting a comment from the physicist Bob Forward regarding spacecraft design [4]. He said that space missions usually start with a launch vehicle, then the spacecraft, the mission distance, the weight deductions for propulsion fuel to orbit/reach target from the total weight and then any remaining weight is left for payload. But he recommended that for interstellar exploration the design should start at the other end of the rocket ship, the design of the probe itself. Such a probe would probably be of a large physical size to give the transmissibility and receiving apertures required and at high power for active data transmission, all combined with lightweight. The computing power of such a probe would also not be centralized in one location but would be distributed so that if any section was damaged, the probe could still function.

14.6 Problems

14.1. For the design example rearrange the stage masses until the mass ratios are identical, but using the same systems as listed. You may swap components between the stages but no components may be added or omitted and the total mass should stay the same. Construct a new mass table and diagrams for the configuration layout.

14.2. Having completed the first problem optimize the pulse frequency as a trade off with boost phase duration by constructing a chart similar to the one shown in Fig. 14.5. Now re-compute the performance and draw out the revised mission profile.

14.3. Having completed both of the above problems, re-assess the assumptions made in the design calculations such as relatively low exhaust velocity, complete burn-up fraction and 100% efficiency. Choose more realistic numbers and re-compute the performance. Ultimately, derive a credible concept sketch for a 10,000 AU mission according to the requirements specified. For a sensible reality check, you should compare you concept to other spacecraft design like the Voyager and Pioneer probes or Cassini and Galileo. Although note the substantial performance difference due to the fusion propulsion method.

References

1. ECSS-E-10A, (1996) Space Engineering, System Engineering.
2. Rogers, WF (1972) Apollo Experience Report – Lunar Module Landing Gear Subsystem, NASA TND6850.
3. Mankins, JC (1995) Technology Readiness Levels A White Paper, NASA.
4. Forward, RL (1976) A Programme for Interstellar Exploration, JBIS, Vol.29, pp611–632.

Chapter 15
The Scientific, Cultural and Economic Costs of Interstellar Flight

> *It will not be we who reach Alpha Centauri and the other nearby stars. It will be a species very like us, but with more of our strengths and fewer of our weaknesses, a species returned to circumstances more like those for which it was originally evolved, more confident, farseeing, capable, and prudent – the sorts of being we would want to represent us in a Universe that, for all we know, is filled with species much older, much more powerful, and very different. The vast distances that separate the stars are providential. Beings and worlds are quarantined from one another. The quarantine is lifted only for those with sufficient self-knowledge and judgment to have safely traveled from star to star.*
>
> Carl Sagan

15.1 Introduction

This chapter will examine the benefits of investing in interstellar exploration. Emphasis will be placed on the potential astronomical discoveries and the robotic exploration of exoplanets. Other scientific advances will also be considered. General cultural issues will be discussed in the context of understanding our place in the universe and the impact on human civilization. This will include the prospects for encountering other intelligent life. The economic and industrial benefits will also be discussed.

15.2 The Advance of Science

It was Nicholas Copernicus who in 1543 first suggested that Earth was not the center of the Solar System. This idea, thereafter known as the Copernican view, was later adopted and proven convincingly by the Italian astronomer Galileo

K.F. Long, *Deep Space Propulsion: A Roadmap to Interstellar Flight*,
DOI 10.1007/978-1-4614-0607-5_15, © Springer Science+Business Media, LLC 2012

Galilee – we all move around the Sun. Then in the 1600s the English scientist Isaac Newton discovered the laws of gravity and demonstrated how this orbital dance worked according to his classical descriptions. This basic model was accepted and proven by observations over many centuries afterwards, until the arrival of the physicist Albert Einstein and his paper on General Relativity published in 1915. He was the first to demonstrate in theory why the orbit of Mercury was different from all of the other planets – due to its closeness to the Sun. This evidence, along with solar eclipse observations by the English astronomer Arthur Eddington, led to an extension to the Newtonian view of the world and forever after space would be more correctly known by the obscure name of space-time.

Although he did not believe it, Einstein's equations of gravity – predicted the expansion of the universe, a colossal achievement had the prediction been believed. However, in 1929 the American astronomer Edwin Hubble proved that all of the galaxy clusters in the universe were expanding away from each other – due to the expansion of space itself. The fact that space was expanding implied that at some point in the distant past the universe had been in a smaller and hotter condensed state. We now refer to this initial state as being described by the 'Big Bang Theory' and it forms our modern view of the universe. This theory is supported by ample evidence, including measurements of nuclear synthesis of the elements and the cosmic microwave background radiation – the leftover heat emission from the beginning. This led then to our modern view of the universe.

The fact that humans have been able to construct this model of reality through observations and intellect is an astounding achievement. It is no surprise then that such discoveries are inspiring and bring about the idea of visiting the far off places we see twinkling in the dark sky at night. However, the development of machines capable of going into space is a fairly recent phenomenon. The Moon landing and the subsequent achievements of the Apollo program were an inspirational story that shall be told for thousands of years, but we must also face facts. These men of Earth traveled 400,000 km to the Moon. In the vast ocean of space, this distance is barely a step. For example, if the distance from Earth to the Sun is 1 AU, then the Earth-Moon distance amounts to 0.3% of that. If we assume that the Solar System is around 100 AU in size, then it amounts to a tiny 0.003% of that distance. So far, humans have barely left the atmosphere and gravitational influence of Earth. Until we are ready to advance further such as on to Mars, we are fortunate to have other options – robotic probes.

So what can robotic probes tell us about our place in the universe? In the same way that humans have a sense of smell, touch, hearing, sight and taste, robotic probes can act as an extension of our bodies, and tell us what is out there. This can be done relatively cheaply and with no risk to humans, who are not on board. If the mission fails, all that is lost is the immediate opportunity to advance science and some humiliated designers back home who perhaps will find themselves out of work.

What astronomical advances can be gained from space probes that cannot be learned from long distance observations? Certainly it is arguable that many features of a distant planetary system can in the future be observed with high resolution telescopes located in Earth orbit. So any interstellar probe must give science return that at least exceeds these observations in order to justify the cost of the mission. We can distinguish several areas of scientific interest that we wish to address using an interstellar probe, which we now briefly discuss.

15.2.1 Planetary Physics, Including Terrestrial and Giant Planets

It is by no means fully understood how planetary bodies form in a solar system. The general picture goes that a circumstellar disk forms over time, and through numerous collisions become small planetismals that eventually coalesce into planets. However, from observations of planets around other stars it appears that our own Solar System is not necessarily typical in structure. Most of the planets discovered around other stars are giant Jupiter-like gas giants, but this is likely due to the limitations of our technological resolution. At first sight we know that in our own Solar System Earth is at the right (optimum) place within what is called the habitable zone. If it were too close to the Sun it would be too hot; if it were too far from the Sun it would be too cold. Hence we enjoy a temperate climate and a reasonably stable geology. Gaining further knowledge about other planets around distant stars is fundamental to our knowledge of our own planet and how it formed.

15.2.2 Stellar Physics, Including Different Spectral Types

Our Sun is a yellow dwarf star and is not typical. Indeed many stars in the galaxy are part of a binary system. We can learn a lot about stars from observing them through telescopes from Earth; however, the greater the distance of the star from us the greater the uncertainty in its position, speed and properties. Sending a probe close to another star will give us an important second data point from which to calibrate all other measurements. This also adds to our general knowledge of how stars function and evolve, important for us in predicting events such as coronal mass ejections, which can disrupt orbiting satellites and pose a real threat to astronauts.

15.2.3 The Presence and Evolution of Life

As of today there is no evidence that life exists or has ever existed elsewhere in the universe. There are indications that life may have existed once on Mars, but the jury is out subject to further information. However, as discussed in Chap. 7, given the number of likely solar systems in our galaxy, it is highly improbable that this is the only solar system with a planet that has the right conditions for life to begin and evolve. Many astronomers believe in fact that the universe is probably teeming with life, which evolves purely as a function of chemistry. The only way to know this for sure is to embark on missions of scientific discovery and go find out for ourselves. There is only so much we can find out by staying at home. Humankind must be bold, venture forth and discover. An interstellar probe offers this opportunity. When proof is finally found it will be one of the most remarkable moments in human history as the realization dawns on us that we are not alone. Similarly, if we do not find such life then this will be a deep lesson in humility for the entire human race.

15.2.4   The Physics of the Interstellar Medium

Our Solar System is surrounded by a region of space dominated by intergalactic winds and charged particles. If all solar systems are like islands in space then the interstellar medium is the ocean of space. The exact abundance of particles in the interstellar medium is not well known and neither is the ionization state of the many gases. Dust grains are also thought to be abundant, but the quantity is not well known and they could provide a hazard for any spacecraft moving at appreciable fractions of light speed. Finally, the strength of the intergalactic magnetic field is of large scientific interest.

15.2.5   The Physics of the Kuiper Belt and Oort Cloud Layers of Different Stars

The Kuiper Belt is a region of rocky objects that extends from the outer planets out to around 500 AU. It also contains many objects now referred to as dwarf planets, and this includes Pluto. The Oort Cloud is a region of objects extending out to 50,000 AU and contains long-period comets in particular. Little is known about the structure, content and formation of these regions around our own star and how they were involved in the formation of our own Solar System. Other stars will also have similar outer layers, and comparing these to the Kuiper Belt and Oort Cloud will add greatly to our knowledge of solar system formation.

15.2.6   Galactic Structure, Cosmology and the Global Picture

The Milky Way Galaxy is around 100,000 light years in diameter by 1,000 light years thick. As part of a local supercluster it is but one of possibly 100 billion galaxies in the universe. How do galaxies form and produce spiral, elliptical or irregular shapes? Under what time scales does this take place and are supermassive black holes fundamental to the formation of any galaxy? These sorts of important questions will allow us to gain a greater understanding of galactic formation and the evolution of the universe as a whole. An interstellar probe flying at galactic escape speeds will help us to gain greater insights into the structure of the galaxy. As the interstellar engineer Adam Crowl recently said: [1] "The immensity of the Cosmos challenges us to reply with similar sized ambitions."

15.2.7 Exotic Physics, Including Gravitational Issues

Arguably, modern science is five centuries old and much has been learned about the laws of physics from Copernicus to Newton to Einstein. But despite these intellectual achievements there are a large number of missing pieces in the puzzle of knowledge. The first and most important missing piece is so-called dark energy, the fundamental ingredient that is thought to be responsible for the currently observed accelerated expansion of the universe. The next missing piece is dark matter, a name given in response to our failure to correctly measure the mass of galaxies purely from the observations of luminous objects. Then there is the missing Higgs Boson particle in the standard model of particle physics, thought to be responsible for the inertia of an object.

Although we can't see back to the very early universe due to the presence of the cosmic microwave background, there is a possibility of detecting gravitational waves, which can penetrate this layer. All mass in the universe (stars, black holes) that undergoes acceleration will exhibit gravitational waves, although they are yet to be observed. Finally, it is worth mentioning the possibility that our understanding of gravity may be incorrect, particularly over large distances and with large accelerations. This possibility has arisen because of the measurements of the Pioneer 10 and 11 probes, where the spacecraft underwent an unexpected (from conventional theories of celestial mechanics) acceleration of around 8.7×10^{-10} ms^{-2} towards the Sun. Any space probe launched out past our Solar System, into the depths of the interstellar medium and towards the nearest stars will undergo high acceleration, high velocity and cover a large distance – a perfect recipe for addressing these exotic physics issues.

15.2.8 Engineering Design Issues Such as Spacecraft Reliability

A typical satellite launched into Earth orbit today will have a requirement to survive for years. Most planetary probes will also need to survive for at least 5–15 years. There have been exceptions to this and this includes the Voyager and Pioneer probes, which have been in space for around three decades and in theory could have been made to continue to operate for this duration and longer. Pioneer 6 is an example of a spacecraft that was launched in 1965 and is still communicating with ground control despite the fact that it was designed to last for around 6 months.

It probably is the oldest operating space probe in existence. Using solar power it is expected to continue running until the 2030s, around 70 years after launch. To send a spacecraft on a high velocity long-distance mission of many decades or even centuries raises fundamental problems with systems reliability. This includes the survivability of materials in the vacuum of space, the protection of electronics from radiation and the operation of equipment after having been stored for such long periods in the spacecraft payload.

Given these problems, it is not an unreasonable view to assert that an interstellar probe launched on, say, a 100-year mission simply won't survive the journey, and even if it does reach the target destination, the number of system failures will be so great that the actual science return will be dramatically minimized. It is for this reason that spacecraft designers have to factor in system redundancy and this requirement would be even greater for an interstellar probe. Launching precursor probes today to 1,000 AU, 10,000 AU and even further will help us to understand these issues so that when we do finally launch our first interstellar probe it will arrive safely and be worth the effort.

****

Humankind has evolved on two important levels. The first is our biological evolution according to the laws of natural selection. This was first described in Charles Darwin's *Origin of Species* published in 1859. In this respect, humans are no different from the rest of the animal kingdom and we have no right to consider ourselves unique. We are merely a result of nature seeking an efficient form, which maximizes our survival potential. The second way that we have evolved is technologically, and this is a direct consequence of our biological evolution. However, the fact that a biologically intelligent entity has come across the use of technology to maximize our survival potential would imply that we are likely to continue to evolve in such a way that we become more strongly coupled to technology – not less. In the limit of our ultimate evolution, the attainment of a cyborg and eventually artificial intelligence state is not an unreasonable proposition. In fact, some may even say it is inevitable. Arthur C. Clarke believed this – that mankind was merely a stepping stone in the evolutionary advance of intelligence and that ultimately the pinnacle of that intelligence would be expressed in a manner which we would today call a computer form. In the authorized biography by Neil McAleer [2], Clarke expressed the view that the purpose of humans was the processing of information and that ultimately we would evolve to what he called *Homo Electronicus*. Here is a sample of Clarke's thinking on the subject: [3].

> Even machines less intelligent than men might escape from our control by sheer speed of operation. And in fact, there is every reason to suppose that machines will become much more intelligent than their builders, as well as incomparably faster...I suppose I could call a man in an iron lung a cyborg, but the concept has far wider implications than this...Can the synthesis of man and machine ever be stable, or will the purely organics component become such a hindrance that it has to be discarded? If this eventually happens – and I have given good reasons for thinking that it must – we have nothing to regret, and certainly nothing to fear...Yet however friendly and helpful the machines of the future may be, most people will feel that it is a rather bleak prospect for humanity if it ends up as a pampered specimen

in some biological museum – even if that museum is the whole planet Earth. This, however, is an attitude I find impossible to share. No individual exists forever; why should we expect our species to be immortal? Man, said Nietzsche, is a rope stretched between the animal and the superhuman – a rope across the abyss. That will be a noble purpose to have served.

This is stirring stuff from Clarke. What Clarke speaks of is a complete synthesis between humans and machines but where the less efficient partner, the human, will become more and more redundant. Many humans would not share this viewpoint, but it is at least a credible proposition. The interstellar engineer Adam Crowl takes the following view on the subject: [4].

The distinction between organic and machine is a temporary one. Eventually the two will merge or interchange or co-exist in some unimaginable way. Thus our AI descendents will be US in some new way. I think we have something of that with these smart phones and gadgets, which are a breed of AI-mosquito level intelligence – yet already a part of what being human is about.

It is quite possible that already our continued use of technology, such as desktop computers, is influencing the path that our future evolution may take. If performing calculations faster on a computer leads to the increased survival prospects of our species, then it is possible that nature will continue to evolve us in this direction. We are no longer subject to evolutionary changes taking place over many millions of years. With technology, we can skip biological evolution altogether and race towards the future at a pace of our own choosing. Biological evolution is becoming less important for humans as they begin to technologically master the world around them. The discovery of technology therefore represents a breaking free from the laws of biological evolution – an evolutionary divergence. This is what truly separates humans from the rest of the animal kingdom. In essence, our destiny is now determined by the rules of a different reality from the animal kingdom and we dance to the beat of a different and more efficient drum – technology.

Whether we believe this ultimate state is our destiny or not, this does not detract from the reality that we are a technological race and we achieve this by maximizing the efficiency by which we extract information from observations. This is the process we call science or the scientific method. An observation is made, a hypothesis is constructed, further observations are made and then the hypothesis is shown to be either true or false. If it is true science advances to the next level. The true father of the modern scientific method is probably the Italian Galileo Galilee. His observations of the 'Heavens' and Earth were carefully recorded, and Galileo showed a willingness to change his point of view in the light of new information when a contradiction was clearly shown to exist – the mark of the first true scientist. The simple argument went: Earth is the center of the universe, but all the planets appear to go around the Sun, therefore Earth cannot be the center of the universe. Another great observer was the Italian Renaissance man Leonardo Da Vinci. He was a real practical investigator who designed and built many wonderful machines, some centuries ahead of his time. He, too, dreamed of flying like the birds:

"When once you have tasted flight, you will forever walk the Earth with your eyes turned skywards, for there you have been and there you will always long to return."

Fundamentally, humans have only one means by which to expand past the horizon that surrounds our reality, and this is technology – technology to go faster, further, higher, and more efficiently. It is with these drivers that we have expanded out across the lands and oceans of our world, soared upwards, walked on the Moon and then looked towards the distant stars.

So how does humankind continue this push outwards, this expansion towards the rest of the universe? It is simple; in order to continue to grow outwards, we must gain knowledge. The act of gaining knowledge leads to the discovery of more science and the invention of further technology. This then leads to further expansion. This is why science is so important and why it stands above all of the other wonderful topics of interest to our species. It does so because science is about one thing – the truth. Yes, scientists may push a particular dogma for decades or even centuries and all will believe that is how the natural world really works. But ultimately, if science continues to push forward based upon empirical observations alone this dogma will be overturned if it is not truly how nature works. There is only one set of physical laws upon which this universe is bound (excepting the possibility of a multiverse), and no matter how hard we argue we cannot change those laws to fit our view of the world indefinitely. As the American astronomer Carl Sagan once said: [5].

> The Universe is not required to be in perfect harmony with human ambition.

If we continue to expand outwards our scientific and technological growth will continue. This is why space exploration is so important. Any civilization that does not rapidly realize this will surely come to destruction. The human race has been a scientific-technological one for around 500 years – half a millennium. In this time we have also used technology to develop weapons of war to destroy each other, and therein is the heart of the problem. This is the side effect of technological advancement – that with the knowledge to grow technologically comes also the knowledge to destroy technologically.

With the particular application of space propulsion it is often said that any space propulsion is also usable as a weapon. This careful balance has already tested us on several occasions in recent times, especially with World War II. How did we escape then from the precipice of human expansion versus human destruction? It was simply the development of science – the recognition that technology could help win a war led directly to a rapid growth in technology, which also led to an increase in our knowledge of the laws of nature. This appreciation led to activities such as a peaceful competition in the form of Project Apollo and the United States winning the 'space race' against its former adversary, the Soviet Union. The landing on the Moon was a crowning achievement by humankind, although as Arthur C. Clarke once commented, there is the danger that this achievement will be remembered as the biggest ever attained by the human race – a sad indictment of our state of affairs.

To ensure that this will not be the case, humanity needs to continue its quest out into space, increasing our knowledge, making new discoveries, settling new frontiers. What activities in the near term would be a significant move towards achieving these goals? A permanent return to the Moon would be a good next step. This should include a permanent lunar station at the south pole with turnaround crews about once or twice per year. Indeed, with low cost access to space just around the corner in the form of Single Stage To Orbit (SSTO), there is every possibility that space tourists may have the same opportunity in the decades ahead. Then there is the eventual move onto the Red Planet – Mars. This has been a long time coming, and we possess the technology and knowledge to achieve this today with proper planning. Once we have achieved these things, the slow expansion of our species out to the rest of the Solar System will begin followed by the establishment of a Solar System-wide infrastructure and economy. Moving on from this position the stars will be ours for the taking.

15.3   Cultural Growth and the Rise and Fall of Civilization

From the year 1368 China was ruled by the Ming dynasty, considered by many to be one of the high points of Chinese history and most likely at the time the most powerful civilization on the planet. Historians consider this dynasty to be one of the most successful in ancient times. During this dynasty the Great Wall was re-constructed along the northern border in an attempt to maintain control over nomadic tribes. This stupendous achievement has a maximum width of around 9 m (30 ft) and is over a distance covering 8,800 km along the ground. The same dynasty also saw the restoration of the nearly 1,800 km long Grand Canal, one of the longest artificial river systems in the world. This was a civilization that did things big. The emperors in this period created a self-sufficient society consisting of up to 200 million people with agriculture at its heart. The success of this society led to the creation of a large army and navy that saw some 2,000 specially built ships that set sail from 1,405 to distant parts of the world such as to the Muslim countries and parts of Europe. These ships were used to sell agricultural surplus as well as silk and porcelain to other countries and so begin a period of trade, which also saw a large influx of silver into China from European countries, including the Netherlands, Portugal, Spain, and Japan. This period saw the emergence of China into the world and all the glorious opportunities it represented.

However, the generation of wealth in China due to the silver influx and subsequent tax rises that led to the undermining of the Ming economy, coupled with natural disasters such as floods, famine and widespread epidemic were some of the contributing factors leading to the dynasty being directly challenged. In particular, tribes united in a campaign to renounce the authority of the Ming dynasty. This led to the final Ming emperor committing suicide under invasion of a peasant revolt.

Although the power of the Ming was continued for a while they were unable to cooperate and ensure the survival of the Ming throne. The ships that had been sent out to sail around the world were recalled, and no further voyagers were sent forth to undertake trade with other countries. Eventually, it was forbidden for people from China to travel to other countries or indeed build new ships. In fact, it was worse than this, when in around 1,525 all of the ships built for trade with other countries were destroyed. China's contact with the rest of the world was to be cut off for several centuries. Another factor in leading to this self-isolating behavior by a nation was a fear of strange ideas, a fear of cultural values espoused by the Western European countries in particular, which appeared to the Chinese rulers to have nothing to offer them, trade or knowledge. This looking inwards was to be the downfall of a once great civilization. As China moved aside, European countries dominated the world and its ships sailed the high seas in search of new lands and new trade opportunities.

It is worth noting that the interpretation of historical events has not been settled, and the above may be an oversimplification of history. There is a view that the claimed inward turning of China has been over exaggerated and that instead several factors led to the slowing down of the Chinese trading industry. This included the Ming-Manchu transition, the flooding of silver from the New World onto the market and slowness on the part of China, for cultural reasons, to realize the scientific revolution that was taking place in Europe. Also, it is claimed that the lack of availability of coal, due to a lack of railroads, was a major factor in preventing China from being part of the Industrial Revolution. All of this serves to illustrate that the history of civilizations is complex and that only a multiple set of reasons are likely to bear some approximation to the truth. But in any case the above account does illustrate the dangers of cultural stagnation, an important lesson to keep in mind when considering the justifications for continued human expansion into space.

Another great empire was the one whose center was Rome. The Roman Empire existed in some form or another from about 27 B.C. Most people think of Julius Caesar when they think of the Roman Empire, as he effectively seized control of it, but at its height it was under the rule of Emperor Marcus Traianus and it occupied territory throughout southern Europe and up as far as England, spanning a land area of over six million kilometers. Its influence on Europe and the rest of the world – culturally, economically, legally, philosophically and of course with its language of Latin – was significant. Its navy patrolled the ocean trade routes of the Mediterranean and parts of the North Atlantic. The emperor was a powerful person protected by his lethal Praetorian Guard. He could direct the army, negotiate treaties and control the Senate. The Senate was intended to be a separate branch of the government, but in reality it had little authority under the emperor and was often ridiculed. If it had worked as planned, the Senate would have been a constructive check and balance on the rule of the emperor, allowing all powers to be scrutinized and debated before being passed down to the people.

One of the most remarkable aspects of the Roman Empire was its willingness to tolerate other religions in those countries that it conquered, such as the belief in the God Osiris from the Egyptian culture. This included often adapting the Roman way of doing things as the empire expanded to suit a local culture. In this manner the empire became more integrated and eventually accepted. There were exceptions to this, however, such as the persecution of Christians on several occasions and the Druids of Western Europe. Rome was seen as the center of the empire and was also a cultural location for baths and theatres. The Roman Empire was founded on solid legal principles. This includes documents such as the Pandects (or Digest), which were compiled around the sixth century. Much of what was written in the Roman law is thought to have influenced many of the laws of modern civilization today. The influence of the Roman Empire on our society can be seen particularly all over the Western world.

However, at some point during its history, the empire divided into the Western Empire and the Eastern Empire, which included separate emperors. Historians generally consider that the Western Empire collapsed around A.D. 476 and the Eastern Empire around AD 1453. The collapse of the Western Empire was precipitated by a Germanic revolt overland in Italy that resulted in the fall of the Emperor Romulus Augustus. This led to several more revolts and eventually a decreasingly Romanized Western Europe. Over time the Eastern Empire gradually changed, including the adoption of Greek as the national language of the empire as it came under Greek influence, now known as the Byzantine Empire. Subsequent events such as the Muslim conquests of the seventh century led to the loss of further lands and eventually the siege of Constantinople (modern Istanbul) in AD 1204 by the Christian crusaders, leading to a Catholic controlled Eastern Empire. The Roman Empire was to never recover from this and its slide into the history books was to continue.

Although it is a matter of debate among academics, many people consider the fall of the Roman Empire to be due to two fundamental reasons. Firstly, there was the gradual disintegration of the structures of the Empire, those pertaining to the military, political and economic systems. In particular a departure from the principles of its founding laws, including an increased decadence and loss of moral authority, appears to have played a major part. The second reason is the increased number of barbarian invasions over a wide frontier from which the empire could not sustain a defense, especially with incompetent and unstable governance. Finally, it is worth noting that some people don't believe that the Roman civilization ever actually fell – it just evolved under collapse and conquer, but Rome does still exist. It all depends on one's definition of an empire.

In 1951 the science fiction author Isaac Asimov published his first book in the *Foundation* series [6]. The main character was a mathematician called Hari Seldon who had developed his own theories of psychohistory, which assigns probabilities to large events in a society and thereby the potential to predict the future. Although this book is fiction it has some important lessons. In particular, the book focuses on the preservation of knowledge for future generations in the event of a civilization collapsing or regressing into the dark ages. It is the nature of the universe and the nature of humans, too, that change will always occur; nothing stays the same forever. Any civilization that does not evolve with this change must therefore be resigned to eventual collapse. Other authors have written about this inevitable change and how it links to technology [7]. A species will develop its technology in order to alter its environment to its own requirements, but this implies the rate of this technological development must exceed biological evolution and thereby change the conditions under which the population has to exist – a risky strategy unless the population has a full understanding for the environmental system it inhabits and more importantly has evolved in.

The best way for the people of Earth to sustain our civilization for the generations of the future and to avoid a technological-environmental catastrophe is to ensure continued expansion outwards and the progression of knowledge. The society that is best able to achieve this, arguably, is one that is progressive with a forward-looking outlook and a positive view of our future. We can consider how a civilization bests sustains itself once it has expanded into space. Let us address these issues in turn by defining some guiding principles for a human civilization expanding out into space.

Initial human expansion likely requires a guiding legal and moral framework to get it started for the first few centuries. This is prior to becoming a truly independent civilization from the home of origin. The human impulse to explore new horizons will likely be a primary driver in this regard. So once the technology is established people may embark on interstellar colonization without some guiding framework but instead under their own motivations. Whatever the reason, for people moving outwards into space, some guiding principles to help maintain the link with humanity will be a necessary precursor.

Any framework should include the possibility that we may encounter intelligent life in the galaxy. This refers to how we will react when we first meet another intelligent species. One of the benefits that will come from preparing for this is the actual process of thinking through the implications, problems and solutions. No doubt when we finally do meet another intelligent race, it is possible that we may be shocked by their values. Putting some time into thinking about the issues today will at least prepare us for how to deal with an awkward circumstance that we did not necessarily predict. This planning would be an evolving process. Charles Cockell has discussed the ethics of extraterrestrial life and how we should treat it [8]. He argues that the most likely source of an ethical difference is the inability to find a coherent definition of 'life,' which is necessary in order to assess moral relevance in the first place. His conclusion is that this inability to derive a definition should force us to take a position of highest moral relevance when assessing any life we encounter in order to avoid erroneously assigning it a lower moral value.

The nature of humans is to tend towards a central dominant power, which can become corrupt over time. This refers to the need for a separation of powers in terms of the executive, legislature and judiciary. In order to prevent the growth of corruption laws will need to be agreed that enforce the division of powers. However, one could argue that the adoption of a closely coupled system, including a lack of a fixed constitution, actually allows for society more flexibility in evolving to rapidly changing circumstances, not requiring reinterpretations of text perhaps written centuries ago. Charles Cockell has written extensively on the subject of extraterrestrial liberty [9]. He argues that the lethal environmental conditions in outer space (harsher than polar environments) will force a need for hard regulations to maintain safety, and without the traditional buffers to tyranny that exist on Earth despotism is likely to form within permanent human settlements.

The culture that makes up the expanding colony must remain outward looking, knowledge seeking in order to advance. This requires a system of learning, through science, and integrating that knowledge back into the society. The system would be based upon some educational process, although it is worth noting that the American science fiction writer Charlie Stross has argued in personal blog articles that space colonization is implicitly incompatible with both libertarian ideology and the myth of the American frontier. He believes that the organization required for space exploration is an anathema to the ideology of the space colonization movement, which arose from nostalgia to the open frontier and the activities of self-autonomy, self-reliance and advancement.

If two colonies come into contact with one another it will likely result in the eventual weakening of one colony in preference for the dominance of the other. This may not be a bad thing to occur if one of the colonies is practicing ethics we may consider unacceptable. However, the problem with ethics is that they are relative. But whatever merger occurs when two colonies interact, it is unlikely that either colony will not be changed in some fundamental way as a result of this interaction.

The dominant mechanism of interaction among colonies will be trade and so each must have products to trade with. The slow expansion of any colony will therefore depend upon the material wealth of the solar systems and beyond. Stopping off at one place over another will be necessitated by needs. The interaction of two colonies through trade may also necessitate the establishment of a commodity such as an agreed currency to trade with. The existence of free trade between colonies allows for the exploitation of the economic principle of Comparative Advantage, which creates opportunities in wealth for both colonies; this refers to the ability of one colony to produce a product at much lower cost than another colony, thereby allowing trade between colonies despite the fact that one may have fewer resources than the other. The trading of information, transmitted at the speed of light, can be achieved very rapidly, although over light years distance will take an order of years to send or receive information. The trading of material things however presents different problems, easy to accomplish for two colonies within the same solar system, but difficult to accomplish if the two colonies are not.

The knowledge and lessons of history must be communicated to each colony and from one generation to the next to ensure the past is not repeated. This requires a system of information storage and education, probably secured by advanced computing technology. The requirement to be more efficient in the use of materials may lead to advanced artificial intelligence, if it doesn't exist already. The problem with learning from history is that we do not always recognize that one situation is identical or even similar to another one. Indeed, the causes or effects may be very different, and the colony could be faced with entirely new challenges. This is to be expected as we expand into the space frontier and colonize other worlds. Perhaps what is important is the establishment of a process for dealing with such challenges based upon an examination of history so that when something new comes up, as it likely will, the colony is at least prepared in how to respond and seek a solution. Ensuring that knowledge is passed along as humans move ever further apart via space colonies will likely be a major challenge for the human race, preserving knowledge over millennia and ensuring that the core of that information survives into each colony.

Continued expansion into space requires the construction of vessels, which thereby require materials. This necessitates an outwards expansion based upon a live off the land policy. We should be practicing this philosophy on Earth today, reusing materials for example and not wasting our vital energy sources. Perhaps when we have addressed this on Earth we can transfer those lessons to any space-dwelling colony. Alternatively, it may actually be the case that any colony being forced to build efficient power generation equipment or hydroponic systems will transfer that knowledge back to Earth and help us to become a more sustainable world; living within our means is a quality that the human race should aspire to. Examples of adopting a 'live off the land' policy in space exploration may include mining lunar ice for water, mining lunar soil for brick material, mining Jupiter for helium-3 fuel reserves, and mining asteroids for precious metals. The idea that we will just build vessels on Earth and then continually send them out into space is not likely to be how the future will be written. Instead, there will be a drive towards more efficient methods and as we expand our infrastructures out into the Solar System we will be able to use the local material sources and construct such vessels, say, on Mars or in the orbit of the Saturnian moon Callisto. From there we will see crewed vessels moving outwards into the Kuiper Belt, Oort Cloud and eventually into interstellar space and begin the colonization of other solar systems.

For manned flight, it has been estimated by Cassenti [10] that the amount of mass required to support an actual space colony M_c is equal to the product of the number of people P and the amount of mass per person K measured in units of tons/person and the additional equipment, probes, landers required to support the mission A. This is given by

$$M_c = KP + A \qquad (15.1)$$

So for a crew of 20 people with an assumed 100 tons per person the total support mass would be KP = 2,000 tons. For a human colony of 1,000 people at 100 tons per person, this would require a total mass of 100,000 tons. However, if we include a value for A then for the first example of KP = 2,000 tons, according to Cassenti there is an additional 3,000. This would give a total mass of 5,000 tons. Note that A = 1.5 × KP. The numbers used for this calculation as well as the additional mass A then leads to a modified equation where

$$M_c = \frac{5}{2} KP \qquad (15.2)$$

Table 15.1 shows the results for different crew numbers and with a different assumption for K but assuming A = 0. To get the full mass would require a multiplication by the 1.5 factor. As an interesting exercise, if we take a total payload mass similar to that for the Daedalus spacecraft design discussed in Chap. 11, say 450–500 tons, if this was a crewed vehicle this would be an amount of mass sufficient to support a crew of two people with a further addition of the engine mass and propellant mass – an indication perhaps for what a manned mission would really take.

Table 15.1 Estimate for the amount of mass (tons) to support a human colony using (15.1)

K (tons/P)	P = 5	P = 10	P = 20	P = 50	P =100	P = 500	P = 1,000
10	50	100	200	500	1,000	5,000	10,000
100	500	1,000	2,000	5,000	10,000	50,000	100,000
1,000	5,000	10,000	20,000	50,000	100,000	500,000	1,000,000

15.4 The Economics of Space Exploration

Let us consider this question from the basis of an attempt to launch a single interstellar probe, assuming space-based architecture similar to or moderately advanced of the current position, e.g., we assume that no massive space-based infrastructure is in place. Consider that the Cassini mission cost roughly around $3.26 billion. Cassini only went to Saturn, which is located at approximately 10 AU from Earth. The nearest star Alpha Centauri is located ~272,000 AU away, so if costs extrapolate linearly to distance at a rate of $0.326 billion/AU, then the first robotic mission would cost of order of $89 trillion. Similarly a Barnard's Star mission to ~ 373,000 AU would cost $122 trillion. An alternative calculation could be based upon time instead of distance. For example, suppose that it took around 20 years to design, develop and complete the mission for Cassini, this would give a rate of around $0.16 billion/year that, when extrapolated to a 100-year design of a flight interstellar mission would put the cost at around $16 billion. Of course, if we are looking at a probe that employed a fusion-based propulsion system, then perhaps you would want to factor in the time it has taken to develop a workable fusion reactor for space applications, from the 1950s to perhaps the 2050s, of order a century. In this case the estimated time cost would double to around $30 billion. For such a system, the main parameter driving up the cost is the propellant and its acquisition, although the acquisition cost (e.g., mining operations) may be very different from the cost for an end user such as a national space agency, assuming that the propellant is acquired for the purposes of a larger infrastructure agenda than just for the use in deep space probes such as Earth-based fusion reactors for generating electricity on the national grid.

These estimates put the cost for a manned interstellar probe somewhere in between $0.016 and $122 trillion (depending on an Alpha Centauri or Barnard's Star target) and is a staggering amount of money as well as a large cost range. However, with a space mission mass is everything, and probes like Cassini were heavily fitted with science experiments. So perhaps the costs could be reduced significantly by minimizing the number of experiments on board. This would also affect the research and development costs, too. There is also the difference in propulsion technology, which would certainly impact the cost scaling, especially if any breakthrough physics methods can be identified. The current focus is on the use of conventional Earth to orbit architectures, but using single stage to orbit vehicles or even technology such as a space elevator may dramatically alter the cost of access to space. This all illustrates that the cost of an interstellar mission probably requires a more sophisticated calculation than the linear scaling methods

Table 15.2 Cost of different interplanetary probes

Mission	Furthest distance (AU)	Spacecraft mass (tons)	Cost $ million	Cost/Mass ($ million/ton)	Cost/AU ($ million/AU)
Galileo	5	2.38	1,600	672	320
Cassini	10	2.5	3,260	1,304	326
Pioneer 10	80	0.258	350 (P10/11)	1,354 (av)	4.32 (av)
Pioneer 11	82	0.259			
New horizons	55	0.478	650	1,360	11.82
Voyager 1	114	0.722	865 (V1/2)	1,198	7.02 (av)
Voyager 2	134	0.722			

presented above. However, it is likely that the costs of launching an interstellar probe to the stars will be an expensive endeavor.

Let us consider the costs based on some other spacecraft that have been launched from Earth. We consider the probes Cassini, Galileo, Voyager 1 and 2, Pioneer 10 and 11 and New Horizons. These are all missions that were launched by the United States, although both Cassini and Galileo had significant European input. We shall then apply linear scaling to the Daedalus design (see Chap. 11) to estimate the cost of such a mission. Table 15.2 shows some basic data assembled for the historical interplanetary probes. In this simple analysis we have ignored the depreciation of money over time, which if measured in terms of fixed assets would be equivalent to a Net Domestic Product, or NDP. The cost of the Pioneer 10 and 11 together was $350 million. Similarly, the costs of the Voyager 1 and 2 missions were accomplished for $865 million. If we were to build just one of the Voyager probes the cost would be approximately the same as the majority of the costs are in the research and development and setting up of any manufacturing facilities where they don't already exist. The cost estimates represent the mission cost from design and development through to flight and mission completion. For Cassini this was assumed to be around the orbit of Saturn and for Galileo around the orbit of Jupiter. The specified distance in AU is the (rounded up) distance achieved (or expected) at the end of the mission when the power fails or as a result of a complete loss of telemetry. The mission costs range from $350–$3,260 million or $672–$1,360 million/ton or $4.32–326 million/AU.

We can apply some of these numbers to the Daedalus spacecraft. We can apply the numbers separately to just the structure + payload mass (2,670 tons) or to also include the propellant mass (50,000 tons + 2,670 tons = 52,670 tons). Scaling by the $672–$1,360 million/ton range we get costs of $1.8–$3.6 trillion (for 2,670 tons) and costs of $35.4–$71.6 trillion (for 52,670 tons). Scaling by the $6.11–$326 million/AU range for the distance to the Barnard's Star target of 373,000 AU we get costs of $1.6–$121.6 trillion. So taking our total cost estimate for the Daedalus mission is somewhere in the likely range of $1–$122 trillion.

Another way to estimate the cost of an interstellar mission (but without thinking about inflation and having to factor in annual changes) is to base such a cost on the number of man-hours. We can estimate the number of man hours dedicated to such projects based upon 1970s pay ($20/h) and a typical salary today ($50/h).

This is then computed for a 10 year design development project through to mission completion, which is the sort of timeframe involved with an interstellar mission assuming that this linear scaling argument is valid. The number of project development years is first converted into hours assuming a 40 hours work week, so that 40 hours multiplied by the number of weeks in 10 years was 20,800 h; e.g., Number Man hours = Number weeks in 1 year multiplied by number hours per week multiplied by the number of years for development. The total cost of the project can then be divided by the hourly rate to give an estimate for the number of man hours used for the project. So for Galileo this was $1,600 million divided by $20/h, which equates to 80 million man-hours. Dividing this by the allocated project hours of 20,800 (for an assumed 10 year development) hours gives 3,846, the number of people available to complete the project. Doing a similar calculation on the same cost basis for Cassini, Pioneer 10/11, New Horizons and the two Voyager missions leads to a people requirement of 7,837, 841, 1563 and 2,079. However, the actual number of people that worked on these projects from development to launch to mission was approximately 800 (Galileo), 5,000 (Cassini), 350 (Pioneer 10/11), 650 (New Horizons) and 865 (Voyager 1/2); showing that this sort of approach overestimates the number of people required and a more sophisticated assessment would be required.

We can also think about the cost of Daedalus from the perspective of global energy consumption [11]. We assume that the global energy consumption is around 4.75×10^{20} J of energy per year and note that Daedalus had the equivalent of around 2.88×10^{21} J in its propellant tanks, which is around 6 times larger than the global amount. For this simple calculation, we ignore other sources of energy consumption such as nuclear, gas and coal, but instead focus on oil. The United States uses around 21 million barrels of oil per day at a cost of around $80 per barrel, which is a cost per annum of around $1.68 billion/day or $613 billion/year. For the purposes of this exercise, we assume that this represents around 40% of the total U.S. consumption when the total actual consumption is around $1.5 trillion/year. If the total U. S. consumption is around 25% of global consumption, we can estimate the global annual energy consumption to be around four times that of the United States, which equates to around $6 trillion. Then if the total energy consumption for Daedalus is six times the global amount, this means that the cost of the propellant for Daedalus would be around $36 trillion. For interest, estimates for the U. S. 2010 health care bill are quoted to be in the region of around $1 trillion over 10 years, a cost similar to the likely lower bound costs of an interstellar probe mission, based on these assessments.

It is worth considering the costs of other large programs in history. For the Vietnam War costs vary between $30 billion in 1969 to later estimates of between $133 and $584 billon. The number of Americans that served in the Vietnam War was around 2.1 million people – neglecting how many people worked behind the scenes back home. At its peak the Project Apollo space program employed around 400,000 people. The cost of the project varies depending on who you speak to but ranges between $20 and $25.4 billion in 1969 dollars (approximately $145 billion in 2008 dollars). Incidentally the original costs for Apollo were around $7 billion but the then administrator James Webb increased it to $20 billion when he informed President Kennedy. It is useful to note that the James Webb value of $20 billion was a factor of 2.8 times the original $7 billion estimate. Another large project is the International Space Station (ISS). Estimates for the total cost vary between $35 and $160 billion. Under the 2010 U.S administration of President Obama the ISS has an extended lifetime through to 2020, although it could remain operational until 2028. The European Space Agency has estimated that the total costs for what is assessed to be a 30-year mission is around 100 billion Euros.

The costs associated with the development of the Apollo spacecraft were of order $150 billion in today's money. If the total structure + payload mass for the Saturn V was around 6,000 tons then this equates to a development cost of around $25 million/ton. Multiplying this by the structure + payload mass of Daedalus, 2,670 tons, would equate to a development cost of around $67 billion, cheaper than Saturn V, although this increases to $1,317 billion if one includes the 50,000 tons propellant. However, at today's prices the typical costs of developments in the aerospace industry is around $100 million/ton, which would mean that to develop something like Saturn V today would cost around $600 billion and to develop Daedalus would cost around $267 billion for the structure and $5,267 billion for the entire vehicle.

It is difficult to say what the true cost of an interstellar probe like Daedalus would be, but the cost would certainly appear to exceed $1 trillion and would possibly go as high as $300 trillion, although if we applied a James Webb Apollo style scaling factor of $\times$ 2.8 on the upper bound cost, this would increase to $840 trillion, and even Webb didn't get the final cost for Apollo correct. The only conclusion we can draw is that if the first unmanned interstellar probe is along the lines of Project Daedalus, then it would be in the trillions to hundreds of trillions of dollars in cost. Based on this crude assessment, Daedalus would be a massive and costly under-taking and is suggestive of a Massive Infrastructure requirement. Any country would need significant justification before committing to such an enterprise, e.g., discovery of life on another planet or significant asteroid threat. This would all change if the vehicle was a lot smaller or less massive; alternatively the design and build could take place over a much longer time period than 10 years. One factor that would make this 'linear' analysis irrelevant is the fusion-based propulsion of Daedalus, a totally different system that would scale the numbers differently to the one applied here, e.g., power law scaling.

In order to understand the true cost of an interstellar mission it is necessary to examine the nominal Gross Domestic Product, or GDP. This is a measure of the economic output from one nation as a result of products and services sold annually, regardless of the specific application. Estimates vary for the GDP between different organizations such as the International Monetary Fund and the World Bank. How-ever, for the year 2009 the GDP for the United States was around $14 trillion. This compares to some of the other nations with $5 trillion for China, $5 trillion for Japan, $2 trillion for the United Kingdom and just over $1 trillion for Russia, the other country that has a major space infrastructure. The European countries tend to cooperate on space projects since the formation of the ESA in 1975 and so it is worth also knowing that the GDP for the European Union in 2009 was around $16 trillion. We can also consider the GDP for the entire globe, assuming an

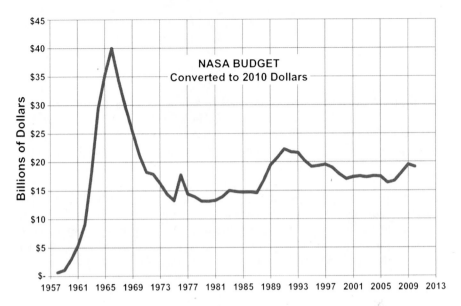

Fig. 15.1   Chart of NASA budget dollar trends

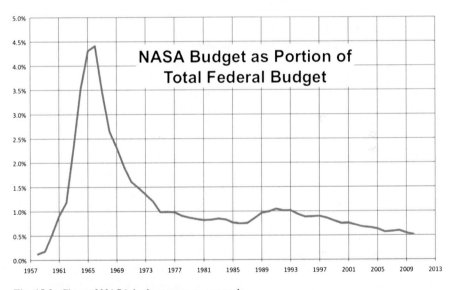

Fig. 15.2   Chart of NASA budget percentage trends

interstellar mission would be a collaborative international venture. The total global GDP for the same year was around $58 trillion. If the first interstellar probe mission cost was of order ~ $100 trillion, then the cost would be equivalent to approximately over 2 years global GDP or around seven times the GDP for the United States in 2009.

Figures 15.1 and 15.2 shows the trends in the NASA budget since 1957 (the year that Sputnik 1 was placed into orbit by the Russians) until the present. This data was assembled using U. S. government data and inflation records [12], plotted with 4-year increments corresponding to presidential terms. The data shows that the peak in expenditure for NASA of around $40 billion occurred during the mid to late 1960s, largely as a consequence of Project Apollo. What is interesting is that this peak is under 5% and only of short duration, thereafter constantly declining until the late 1970s where it then remains near constant in terms of overall percent. If we examine 5% of the U. S. $14 trillion GDP in 2009, this is $0.7 trillion and for a $100 trillion interstellar probe project would take around 143 years to fund; clearly not acceptable. However, if we examine 5% of the global $58 trillion GDP in 2009, this is $2.9 trillion and for the same interstellar probe project would take around 35 years to fund. Assuming the design, build, launch and mission flight of the first interstellar probe lasts for 100 years for the $100 trillion cost, this is $1 trillion per year expenditure for a century, equivalent to 7.1% of the U. S. 2009 GDP or 1.7% of the global 2009 GDP. This analysis leads to two clear conclusions: (1) international cooperation in the pursuit of space exploration is the only way that an interstellar mission will ever be launched, and (2) the cost of such a mission must be bought down significantly before any single nation state could even consider such a project.

In the past NASA instigated three major space mission programs. The Discovery Program was a series of low cost, low budget, but highly focused scientific space missions with an emphasis on a vision for 'faster, better, cheaper.' These were for privately led missions where the development time to launch was limited to 36 months. This program included missions such as NEAR Shoemaker to study an asteroid, Mars Pathfinder to explore the surface of Mars, Deep Impact comet ejector impact experiment, and Stardust, which collected dust samples from the tail of a comet. The New Frontiers Program was a series of medium cost highly focused scientific exploration missions of the Solar System that were not to exceed $700 million. Proposals were open to the international community and included missions such as New Horizons mission to Pluto and the forthcoming mission to Jupiter called Juno. All the planetary objects in the Solar System were potential targets, with the exception of Mars, which was covered by a different program. The Flagship Program is a series of NASA missions that are the largest and most costly, between $2 and $3 billion. These constitute complex missions that have a particular focus on establishing the limits of habitable conditions within different environments. Such missions include atmospheric re-entry probes and landers. An example of such a mission is the proposed Europa Jupiter System Mission, which was due to explore the potential habitability of the moons in orbit around Jupiter. These missions are vital for understanding exoplanets because, even though we've mostly found nothing but gas giants, we assume that most solar systems have a wide variety of planets, just like ours. In other words, it's not because we've found hot Jupiters that we think these missions will help, but because we assume the hot Jupiters are just the first traces of larger planetary systems.

What is needed is an extension to these excellent programs, which we shall call the 'Boundary Program.' Such missions would be launched to explore the outer Solar System, heliosphere, dwarf planets and the Kuiper Belt. These would be regarded as interstellar precursor missions and would really help to push technology forward. Such missions could probably be completed with costs of between $3 and $5 billion, and would represent a major stepping-stone in our advance out into the cosmos. This should be one of the future aims of the national space agencies. To cut the costs, missions within the Boundary program could be split between NASA, ESA as well as the Chinese and Russian space agencies. More expensive, $5–$7 billion, missions under an 'Outer Limits Program' could go even further into the Oort Cloud, exploring the cold objects that are there and perhaps potentially rich sources of deuterium. Such missions are discussed briefly in Chap. 17 under the titles Icarus Pathfinder and Icarus Starfinder and these are illustrated in Plates 18 and 19. These missions are truly deep space and go to distances of 1,000 AU and further to 10,000 upto 50,000 AU, up to one fifth of the distance to the nearest stars. This is how to advance towards the stars incrementally.

So why are we not embarking on such ambitious missions? The answer is mainly cost and only partly due to technological immaturity. So then you may ask, why can't we afford it? We can if we planned over a long time scale, but the political will is lacking. Why is the political will lacking? Perhaps because politicians don't understand the need for missions which are for exploration sake alone? So how do we change this? Simple – demonstrate a business return on the investment. How do we do that? We identify lucrative resources and then build the infrastructure to mine them.

There are two pillars of thought about the future of humans in space. The first believes that we should move gently into the cosmos, showing complete reverence for every object we encounter, particularly forms of life, not spoiling it or using it in the way that we have on Planet Earth. Let us call this the 'green exploration agenda.' The second point of view is that in fact when faced with the harsh realities of survival, humankind must exploit every material or planet for which it can find a use. Let us call this the 'blue exploration agenda.' There is, of course, a middle ground between these two extremes, which involves moving outwards into space for the benefit of all life (not just human or indeed intelligent life) but where survival requires it, exploiting local resources to the full in an efficient but moral way. Mixing our two extremes let us call this the 'cyan exploration agenda,' and it is this method which is proposed as the way forward to represent all of humanity.

Finally in this chapter it is worth considering a calculation performed by physicist Freeman Dyson in a *Physics Today* article on the economics of interstellar travel connected with Project Orion [13]. Dyson had estimated that the cost of a 10,000-ton payload would be around 10^{11} or the equivalent of $10,000/kg. In 1968 the Gross National Product of the United States was around $1 trillion, which means that the cost of the probe during this period was ~0.1 × GNP. But then Dyson assumed a growth rate of 4%, which would imply a GNP around 1,000 times larger within 200 years (i.e., 2168), reducing the cost of the probe to 0.0001 × GNP and costs approaching that of the Saturn V during Project Apollo. He concluded on the basis of this assessment that the first interstellar probes would be built around 200 years from the time of his prediction in 1968.

It is interesting that Dyson's estimate of 200 years hence is supported by two independent calculations. One author named Brice Cassenti [14] has considered maximum velocity trends throughout history from the 1800s and projected them into the end of the twenty-second century. He looked at sailing ships (the year 1800, 7–14 m/s), early racecars (1900 AD, 55 m/s), Earth escape velocity (1960, 11.1 km/s) and solar escape velocity (1980, 12.5 km/s). The author concluded that missions on the order of 10 light years should be possible 200 years from his projection in 1982 (i.e., 2182). Another author named Millis [15] has examined historic energy trends, societal priorities and required energies for an interstellar mission. He based his assessment on a minimal 10^7 kg colony ship with an irrelevant destination and sending a minimal 10^4 kg probe to Alpha Centauri with a mission duration of 75 years. He concluded that the colony ship cannot be launched until around the year 2200 and the probe cannot be launched until around 2500, so that the earlier interstellar mission was around two centuries away. The reason the colony ship was deemed to be launched sooner was due to its much smaller energy requirement, which is proportional to the velocity squared; the colony ship traveled at a cruise velocity of 0.00014c compared to the probe, which traveled at a cruise velocity of 0.22c.

Despite these calculations, this author believes that to make the impossible happen you must set visionary goals. In the early 1960s when President John F Kennedy set down the mission for the first manned lunar landing he stated it was to be accomplished not by the end of the century but by the end of the decade (there was no failure of nerve here). In accordance with such ambitious mission targets, it is the belief of this author that the end of the twenty-first century should be the goal to launch the first unmanned interstellar probe and not the twenty-second. Courage and belief in our ability to achieve the seemingly impossible is required when faced with the daunting challenge of the interstellar problem and not an acceptance that the answer is determined from simple linear scaling of the past or the present, although an appropriate method to initially approach the problem. Game-changing technological innovations (based upon paradigm shifting S-curves) will allow such missions to be accomplished sooner than most people anticipate, and this should be where the financial investment is placed today to secure this roadmap ahead as part or our possible and plausible future.

15.5   Practice Exercises

15.1. Consider what major discoveries in astronomy would motivate a rapid international effort to advance our space exploration agenda. Similarly, consider what major discovery could cause us to slow down our space exploration agenda. Consider what the balance between these two scenarios is and whether each would be likely.

15.2. Research the historical collapse of civilizations on Earth. Make a list of them and assess if there are any common reasons underpinning the collapse. Extrapolate the conclusions to civilizations in space living on a planet far from Earth. What does this imply for how such a society should live and be governed?

15.3. Make a list of potential business markets that could operate for profit within the Solar System. Consider the distance and transport requirements for such markets. What infrastructures would be required to successfully operate these businesses and what other positive spin-off effects could there be from the presence of such infrastructures?

References

1. Crowl, A (2010) Personal Letter to Author.
2. MaAleer, N (1992) Odyssey The Authorised Biography of Arthur C Clarke, Victor Galancz.
3. Clarke, AC (1999) Greeting, Carbon-Based Bipeds! Essay: The Obsolescence of Man, Voyager.
4. Crow, A (2010) Personal Letter to Author.
5. Sagan, C (1995) Pale Blue Dot, A Vision of the Human Future in Space, Headline Book Publishing.
6. Asimov, Isaac (1951) Foundation, Gnome Press.
7. Stephenson, DG (1977) Factors Limiting the Interaction Between Twentieth Century Man and Interstellar Cultures, JBIS, 30,3.
8. Cockell, C (2007) Originis: Ethics and Extraterrestrial Life, JBIS, 60, 4, pp147–153.
9. Cockell, C (2008) An Essay on Extraterrestrial Liberty, JBIS, 61, 7, pp255–275.
10. Cassenti, BN (1982) A Comparison of Interstellar Propulsion Methods, JBIS, 35, pp. 116–124.
11. Bond, A & A Martin (1978) Project Daedalus: The Mission Profile, JBIS, S37–S42.
12. Source of data includes US Inflation Calculator (www.usinflationcalculator.com) and Guardian online article 'NASA budgets: US Spending on Space Travel Since 1958 Updated' (www.guardian.co.uk/news/datablog/2010/feb/01/nasa-dugets-us-spending-space-travel).
13. Dyson, F (1968) Interstellar Transport, Physics Today.
14. Cassenti, BN, (1982) A Comparison of Interstellar Propulsion Methods, JBIS, Vol.35, pp. 116–124.
15. Millis, M (2010) Energy, Incessant Obsolescence, and the First Interstellar Missions, Presented 61st IAC Prague.

Chapter 16
The Role of Speculative Science in Driving Technology

Anything that is theoretically possible will be achieved in practice, no matter what the technical difficulties, if it is desired greatly enough. It is no argument against any project to say: 'The ideas fantastic!' Most of the things that have happened in the last 50 years have been fantastic, and it is only by assuming that they will continue to be so that we have any hope of anticipating the future. To do this – to avoid that failure of nerve for which history exacts so merciless a penalty – we must have the courage to follow all technical extrapolations to their logical conclusion.

Sir Arthur C Clarke

16.1 Introduction

Ideas from speculative fiction have often become reality in decades past or may do so in the near future although most have been inspired from ideas that originated from science in some form. The link between speculative fiction and future technology is a close one, sometimes predictive and sometimes not; but it is inspiring all the same. In this chapter we explore this idea and consider that speculation is really the physicist's way of thinking about the boundaries of science, where conventional wisdom may break down and new physics emerges, often marked by a paradigm shift in thinking about technology. Government sponsored programs have gone some way to exploring speculative ideas and considering the implications to space propulsion. These ideas are discussed along with how we can best think about technological extrapolation into the future – by accepting changes in established paradigms and then considering the implications to our current technological development.

K.F. Long, *Deep Space Propulsion: A Roadmap to Interstellar Flight*,
DOI 10.1007/978-1-4614-0607-5_16, © Springer Science+Business Media, LLC 2012

16.2   The Connection between Science and Fiction

The writer Arthur C Clarke has often expressed his annoyance with publishing companies and bookstores for a failure to clearly distinguish science fiction as a unique genre in its own right. In particular, it is distinct from *science fantasy*, which deals with fantastic ideas that may not be scientifically credible, whereas *science fiction* will try to stick to credible science as much as possible. Indeed, this was a dominant feature of Clarke's writing. Although it is possible that both may have an influence on scientific ideas, the most probable links with modern technological developments are those from science fiction. Similarly, in the art world there is a distinction between Space Art and Fantasy Art, where the former is based on credible visions of what other worlds, star systems or spacecraft may look like. The likes of Chesley Bonestell, Ralph Smith, David Hardy, Rick Sternbach, Gavin Mundy and the amazing John Berkey have been contributors to the space art genre.

For those unpracticed in the laws of science, it is difficult to make such a distinction between science fiction and science fantasy, or space art and fantasy art. This is also due to the fact that many creative arts contain a mixture of both. For example, the famed television series Star Trek has lots of credible science, but also stretches the imagination with fantastic stories. This is perhaps necessary if they are to win the ratings and maintain a captive audience with exciting drama. It is interesting to note that Star Trek and other programs influenced many people who were brought up in the 1970s and later became scientists. This also includes earlier work such as movies like *"The day the Earth Stood still"* and *"Destination Moon."* But what we really want to know is to what extent this popular culture influences our creative processes and lay the seeds of ideas which later evolve into a technological machine. If this link is real and not just an artificial one, then one of the ways to promote research into interstellar flight would be to promote ingenuity in the creative arts.

One example of this is the geostationary satellite. In 1945 Arthur C Clarke published an article in Wireless world titled *"Extra-Terrestrial Relays"* [1] in which he described the idea for an artificial satellite, which if placed in an orbit there were three of these satellites separated 120° apart, this would give coverage of the entire planet. Although Clarke is rightly given the credit for popularizing the idea, he himself admitted that another author, Herman Potočnik, had written about the ideas in 1929 in the context of living aboard space stations in a geostationary orbit; although, it was Clarke who suggested the use of several satellites to give global coverage.

In the 1867 Jules Verne book *"From the Earth to the Moon,"* Verne described the emplacement of 20 rockets on the base of a spacecraft so that the projectile would not smash like glass when it fell towards the lunar surface. These were essentially fireworks and they would have to be lit by the crew through an opening in the vehicle. More plausible ideas were to be later theorized by the British Interplanetary Society artist Ralph Smith in the 1930s. His Moon Ship designs have a striking resemblance to the actual Apollo Lunar Lander, which first touched down in 1969.

The use of a retro rocket was not lost on the designers of the Apollo missions when considering how to control the descent of the Apollo Lander vehicle. Incidentally, the space artist David Hardy has produced the most wonderful painting of the BIS retro-rocket sitting on the surface of the Moon, as it would appear in the style of Chesley Bonestell and this is shown in Plate 2.

What other technological machines lay in our future based upon ideas already around today. One idea, also promoted by Arthur C Clarke in his book "The Fountains of Paradise" [2], but first suggested by the Russian Konstantin Tsiolkovsky, is the space elevator. A lift designed to carry passengers or cargo to and from Earth orbit along a 35,000 km long tether that is in tension between geostationary orbit and a station on the ground. The tether would need to be both strong and light, similar properties to a spider web, and would probably be constructed of some carbon nanotube material. Several countries have shown in interest in building such a machine either this century or the next, including America and Japan. Such a device would dramatically reduce the cost of reaching Low Earth Orbit compared to the current launch vehicles.

Another technology on the horizon is so called cloaking technology. Most people associate such technology with *Star Trek*, particularly with the Romulans and Klingons (Starfleet were naive enough to agree not to develop it). However, most scientists have considered such technology just a bit of fun and having no real basis in fact. This all changed in 2006 when scientists laid the foundations of an actual cloaking device. This works on the principle of light refraction around so-called metamaterials. These are artificially engineered materials that have not been produced by nature but we can produce in a laboratory. The unique property of such materials is their negative refractive index so that optical light is guided around the object and therefore invisible to optical wavelengths. The development of this technology has obvious military applications but may also have civilian applications we cannot yet conceive of.

In his 1937 book "Starmaker" [3] Olaf Stapledon wrote eloquently about a species of beings, which were so advanced that they could construct whole artificial planets out of remnant material from natural planets. Freeman Dyson was later to use this idea for inspiration in his creation of a so-called 'Dyson Sphere'. This is a spherical shell of material, which completely surrounds the orbit of a star, so that from the perspective of an outside observer, the star is not visible. In essence, all of the energy that leaves the star (largely in the form of radiation) is trapped within this giant orbital cavity, providing a limitless supply of energy for any civilization that is contained within it. For many people, although Dyson Spheres are a clever idea they have no basis in reality. However, Dyson has speculated that any civilization sufficiently advanced may be able to build such gigantic structures.

The science fiction writer Larry Niven has taken this idea further in his novel "Ringworld" [4] where instead of a complete spherical shell only a band of the shell is located around the parent star, stretched out over tens of Astronomical Units. It is at least a possibility that in our very distant future humans will build Dyson spheres and utilize all of the energy from our Sun, if an intelligent species has not already done so around another star. The few examples of technological ideas illustrate the potentially exciting future that lies ahead of our species, yet these are just things that we can imagine. We never imagined the Internet and the implications for personal and global liberty are still not clear. Technology has social implications which are also difficult to predict.

Science fiction and interstellar propulsion concepts also have parallels. In the 1974 book *"The Mote in Gods Eye"* [5] Larry Niven and Jerry Pournelle discuss a laser driven spacecraft, an idea developed by Robert Forward especially for the book. In the 1970 book *Tau Zero* by Poul Anderson the author discusses a Bussard fusion drive engine. The solar sail was popularized in a 1964 story by Arthur C Clarke called *"The wind from the Sun"* [6]. These are just some examples of how inspirational science fiction writers are constantly using ideas from the cutting edge of interstellar vehicle design as a major part of their stories. Consequently, young students reading these stories become interested in studying physics and a career in science is started. The relationship between science fiction and interstellar flight research is almost symbiotic.

Given the examples discussed above and historical examples that ideas can become something tangible we have to consider how far science can push such thinking so as to produce new research directions in physics. These new directions may be the key to ultimately opening up the Universe to full robotic and one day manned exploration. Speculating which ideas will remain fantasy and which ones will become actual technologies in future decades is not any easy task. For a start, it is theoretically challenging, because often technologies do not follow a linear performance progression for far into the future, with the uncertainty increasing the greater in time one tries to predict ahead. It is also challenging because if the futurist makes a wrong prediction, particularly for a physicist, the consequences can be career ending. For this reason, established physicists are not always eager to speculate outside of their field of expertise or indeed outside of the boundaries of known observational data.

We are brought then to the contemplations of the maverick theoretical physicist who is prepared to speculate about the existence or otherwise of unknown particles, fields or natural laws. History will record that such individuals demonstrate great courage on their quests as well as a ruthless determination. But as any good quality physicist knows there is a limit to making speculative statements without sufficient basis. To follow such a path is not in accordance with the scientific method, where testing theories through comparison to observations makes progress. Incidentally, a luxury not easily afforded to the modern string theorist or those attempting to understand quantum gravity such as the physicist Lee Smolin [7]. This doesn't mean to say that pure mathematical research isn't extremely valuable in uncovering the laws of nature.

Most great discoveries have required leaps of imagination and courage to test a hypothesis. In order to make tremendous advancements in spaceflight, particularly in space propulsion, scientists must be free to speculate but then subject their ideas to constructive critical review. A theoretical physicist has several tools to do this. This first is his mathematical and scientific education that allows him to apply equations and models to solve problems. The second is the blackboard and chalk or pen and paper to provide imagery of their problems and work through a solution. The third is their colleagues, to act as a sounding board to bounce ideas off. The fourth is experimental evidence. In recent decades, the first and second has also been supplemented by the use of computer codes which provide numerical solutions to analytical problems. Discovering new physics from complex problems relies on these tools. Using these tools the physicist (and/or engineer) has indeed made many a great discovery that have led directly to technological innovations and often improved the well being of humankind. This includes the internal combustion engine, the home freezer, the microwave, enormous bridges or the ability to navigate around the world using orbiting satellites. The scope of our ambitions appears to be unbounded and only limited by our imagination.

A perfect example of where such a discovery may have radical implications for humans in future centuries is the prediction of antimatter by Paul Dirac in 1928. Dirac published a paper titled "The quantum theory of the electron" [8] where he derived an equation of motion to describe relativistic electrons. In conducting this work, he found that his equation was quadratic, meaning it had a positive and a negative solution, where the negative solution gave the mass of the electron. Many physicists would not have come to the same conclusion that he did, which was to believe the equations. He decided that the positive solution must represent a positive electron, which he called the positron. Eventually, the particle was discovered at the California Institute of Technology in 1932. Then later came the discovery of the antiproton in 1955 and followed shortly after by the antineutron in 1956, both at Berkeley. We are yet to create sufficient quantities of antimatter to employ it in an industrial application, but when we do, it should revolutionize the technology of our society and may also be the key to interstellar travel. A true antimatter rocket would attain an exhaust velocity close to the speed of light and trips to the nearest stars could be accomplished within years. Ships would be launched in the form of a Columbus style expedition to Alpha Centauri or Barnard's star with the crew returning a decade or two later with many tales to tell of their wonderful journey and the discoveries they had made.

16.3   Breakthrough Propulsion Physics and the Frontiers of Knowledge

In the foreword to a recent AIAA book by Eric Davis and Marc Millis [9], the aerospace engineer Burt Rutan defines a breakthrough as to mean taking risks to explore what average researchers consider nonsense, and then persevering until you have changed the world. How does creative thought lead to breakthroughs in science to driving technology forward? You may ask what this has to do with practical means of reaching the stars, but as will be shown, it has everything to do with this topic. The stars are located at an enormous gulf from our current technological reach and from today's viewpoint it seems almost impossible that we could ever attempt such a journey. But as will be argued, to reach out to the unreachable, requires new paradigms in thinking and the application of alternative tools of logical reasoning which supplement and enhance the ambition. This way, we can at least begin to plan for how such trips can one day be accomplished and work towards that vision.

To understand what is truly possible within the bounds of the laws of physics, we must first of all hypothesize new laws and then consider the consequences if such laws were found to be true. We can then ask under what circumstances those laws would be true and search for evidence within our Universe that nature has adopted those laws. This leads to creative insights and if you are very lucky, tremendous breakthroughs in physics. In his 1998 autobiography [10] the American physicist John A Wheeler once said in the context of general relativity that:

> By pushing a theory to its extremes, we also find out where the cracks in its structure may be hiding. Early in the twentieth century, for example, when Newton's theory of motion, flawless in the large-scale world, was applied to the small-scale world within a single atom, it failed. Some day, some limit will be found where general relativity comes up short. That limit will show itself only after we have pushed the theory in every extreme direction we can think of. In doing so, we will very likely find some strange new ways in which the theory is valid before we find out where it is not valid.

He then went on to comment on a different phenomenon in physics called wormholes and gave this extraordinary opinion:

> Whatever can be, is (Or, more strongly put, whatever can be, must be.) This article of faith – and it is only faith – makes me want to believe that nature finds a way to exploit every feature of every valid theory. Every faith, including this one, must have some boundaries. The boundary of this one is supplied by the finiteness of the Universe. There is a finite number of particles in the Universe, a finite amount of mass, a finite duration of it all – the Universe began and will likely end. Therefore, not every one of the infinite number of predictions that any theory can make can be realized. Yet I cling to the faith that every general feature will be realized. If relativity is correct, and if it allows for wormholes, then somewhere, somehow, wormholes must exist – or so I want to believe.

Reading these comments from a well-respected physicist makes one feel that theorizing about the boundaries of science is a good and proper thing to do. It is also usually the big topics that attract students to science, presented with the opportunity to be involved with an exciting and challenging enterprise – to make new discoveries.

Before physicists can truly speculate publicly about the boundaries of science, they must first face an old age barrier – philosophical dogma. This is a set of rules or beliefs as described and followed by generations of people. Unfortunately, such dogma exists in scientific circles and any scientist who chooses to speculate beyond their remit faces the prospect of meeting face to face with dogma. For any young scientist this can be a daunting experience and rather than face this prospect they will not challenge conventional wisdom. As a consequence, the older physicist will dominate scientific progress. Any revolutionary ideas generated by the younger physicist must be firmly supported by ample evidence if they are to be given a fair hearing and push through this man made inertia to progress. In the spirit of a true pioneer, the American rocket engineer Robert Goddard set no limits of what would be possible in the future:

> It is difficult to say what is impossible, for the dream of yesterday is the hope of today and the reality of tomorrow.

Often in science a new discovery is made only when two competing theories to explain some phenomena so obviously contradict each other. Mathematically this will usually appear in the form of a singularity or infinity in the equations. This has happened with the theory of physics known as quantum electrodynamics where infinities appear regular in electron calculations and physicists have to resort to a mathematical trick known as 'renormalization', which basically means we don't understand why we get infinities and that quantum theory is incomplete. A major contradiction in physics today is that we have two competing theories of reality, quantum mechanics and Einstein's General Relativity – both of which are incompatible with each other. The prediction of singularities in Black Holes and at the beginning of the Universe in the Big Bang leads some to believe that something vitally important is missing from our understanding of nature. If two competing theories can be resolved with a clear winner then progress can be made. If it becomes difficult to observationally test either hypothesis, then one can always turn to Arthur C Clarke's first law, which states:

> When a distinguished but elderly scientist states that something is possible, he is almost certainly right. When he states that something is impossible, he is very probably wrong [11].

The implication here is that the old are very knowledgeable but with minds fixed on traditional paradigms, unwilling to go outside of the boundaries of their own life experience. Similarly, the young do not know as much and are quick to reach conclusions, but take risks, willing to accept that reality may lie outside of their limited life experience.

To produce new ideas that lead to new discoveries in physics, physicists needs to be appropriately managed. They need to be given sufficient free reign from which to think for long uninterrupted periods about problems. In the application of interstellar spaceflight, this is also important if we are to create more imaginative ways of travelling across the vastness of space and entering the orbit of another star. This freedom to think 'imaginatively' was well understood and frequently used by Paul Dirac. In an academic or industrial setting this requires an atmosphere where individuals are at liberty to pursue a research direction if they so choose. In a national space agency with key scientific program milestones to deliver you might think that such a research program of this sort is not practical. However, it has already been applied twice within the US space agency NASA.

From 1998 an exciting research program was started called the NASA Institute for Advanced Concepts (NIAC). The Institute funded two phases of research study. Phase 1 projects lasted 6 months and were funded at the $50,000–$75,000 level. Phase 2 projects lasted 6–24 months and were funded at the $75,000–$400,000 level. These projects tended to be potentially revolutionary (and not just evolutionary) and projects that were likely to impact future mission development were particularly selected. Emphasis was also placed upon proposals, which were non-duplicative, described in a mission context and based upon well established scientific knowledge. A phase 2 project would build upon the achievements of the phase 1 research and would possibly provide the basis for a future space mission. Some of the projects funded included: Ultralight Solar Sails for Interstellar Travel; Plasma Pulsed Power Generator; Antimatter Driven Sail for Deep Space Missions; Antiproton-Driven, Magnetically Insulated Inertial Fusion; Ultrafast Laser-Driven Plasma for Space Propulsion; Mini-Magnetospheric Plasma Propulsion; The Magnetic Sail. These are cutting edge research projects, all of which are credible ideas for a technological future. These were not just interesting paper studies; a deliverable of the phase 2 projects was the definition of a development pathway with key technologies defined. This would include an analysis of the cost and performance benefits. Unfortunately, in 2007 NASA decided to shutdown this innovative research program. However, as of 2010 it was announced by NASA that this program would re-start. Good news, although the exclusion of non-US citizens to participate in this scheme is an outdated and inward looking policy not in the spirit of international co-operation in the pursuit of space exploration.

Another innovative NASA research program began in 1996. This was the NASA Breakthrough Propulsion Physics Project (BPP) [12–14]. It received around 60 proposals and awarded ~ $430,000 with the total project costing $1.6million. BPP supported divergent research topics that focused on immediate issues and gave rise to incremental progress with the emphasis on experiments. Proposals were awarded based upon competitive peer assessment and emphasis was placed upon reliability rather than feasibility. All research results were to be published and subject to academic scrutiny. Some of the projects funded included: Investigate possibility that electric fields used to vary inertia of a body; Test own theory that links electromagnetism with mass and time; Investigate variations in gravity fields around superconductors based on Podkletnov effect; Investigate superluminal quantum tunneling; Investigate several machines that tap energy from vacuum energy; Study the necessity of negative energy fuels. BPP resulted in the identification of three visionary breakthrough requirements for interstellar flight: (1) schemes that require no propellant (2) schemes that circumvent existing speed limits (3) breakthrough methods of energy production to power such devices. What BPP really accomplished was to move the subject of scientific speculation from one of bar room banter to one of scientific analysis. For the first time physicists could take a revolutionary observation as claimed by someone and clearly follow through the logic using a clearly defined set of rules. The idea either has merit or it doesn't. If it does, then it should at least past the test of the scientific method and adhere to our current understanding of natural laws. This allowed for an efficient mechanism for evaluating speculative ideas and potentially for elucidating those ideas that truly demonstrate some unusual phenomenon, for it is from here that the next scientific breakthrough will come and it is here where governments should be funding research. Sadly the NASA BPP program was shut down in 2002.

   Both the NIAC and BPP were ambitious and bold programs that attempted to solve grand challenge type problems. If programs like NIAC and BPP are not to be supported by the national space agencies any longer, then scientific discoveries that have application to space propulsion will have to come from a different source. Today the space community is focused on four key areas (1) ISS & LEO operations (2) Return to the Moon (3) Future landings on Mars (4) Astronomy & Planetary based science. Any interstellar work is theoretical in nature. To progress what is called divergent research, the space community needs to think carefully about the culture of science and the allocation of funding. The inclusiveness of science to non-academics is also an important area that needs addressing. Finally, the community needs to follow the spirit of X-prize type competitions and allocate incentivized rewards for BPP type research.

   An evening's browse over the internet will immediately reveal that there are many web sites which claim to have made breakthroughs in physics, and many claim to know the secret of interstellar travel. They discuss concepts of antigravity machines or electromagnetic effects. Many of these web sites exist, because if the authors of these ideas were to attempt submission of their 'theories' to a respectful scientific Journal, they would quickly receive a lesson in rejection. This then is the problem with the Internet; material that is widely available, with often invalidated sources of information and non-adherence to the scientific method. For space travel, this is truer than many other fields of science and it is important to distinguish between those people that claim credible ideas and those who just have ideas. For a researcher in the interstellar community, filtering out the poorly thought ideas can be very time consuming and often the solicitor is not welcome of the reviewer's constructive criticisms of their approach and demonstrates hostility in their communication. This is often because they simply refuse to accept that their cherished idea does not hold the secret to freeing up the Universe for all of humanity. But let us consider some of the ideas that are out there that could represent some element of a future breakthrough in propulsion physics. By discussing these no opinion for or against them is implied by this author.

   One of the earlier alternative gravity theories was that of Burkhard Heim [15]. This so called *Heim Theory* claims to address some of the inconsistencies between the quantum theory of the small and the General Relativity theory of the large. It involves the quantization of space-time in a highly mathematical framework with up to 12 additional dimensions to space and all the fundamental particles are presented as multidimensional objects. It is claimed that Heim Theory produces results for particle masses and excited energy states that are consistent with measurements. The theory also predicts new particles. Heim theory is not studied as part of mainstream scientific research today. Another theory called *Yilmaz Theory*, was developed by Huseyin Yilmaz. This attempts to create a classical field theory of gravitation by quantizing general relativity [16]. The theory is very similar to General Relativity except that in Yilmaz Theory certain phenomena such as Big Bang and Black Hole singularities do not occur. It requires a modification to the right hand side of the Einstein field equation to include an additional contribution from gravity. However, critics claim that the theory is inconsistent with either an empty Universe or that of a negative energy vacuum. Both Heim theory and Yilmaz Theory are exciting science but require substantial observational evidence to overturn our currently best theory of gravity developed by Albert Einstein – General Relativity.

If one was to take a cubic length of space remove all of the particles and electromagnetic fields, it should be devoid of all energy. However, this is now not thought to be the case, but instead the space left over would contain so called energy of the vacuum, which is present in the form of vacuum fluctuations, which we interpret as short-lived virtual particles. The particles are created as a particle-antiparticle pair and once they appear they then annihilate each other. The physicist Hendrik Casimir demonstrated the existence of the quantum vacuum energy as long ago as 1948. The experiment, named the Casimir effect, involves the measurement of a small attractive force between two metal plates. Because the plates are so close together resonances of the vacuum energy between them cause the attractive force. If we were able to engineer the vacuum energy via a dynamic Casimir effect for space propulsion applications this would revolutionize space travel not needing to make use of large mass propellants and their associated structure mass [17, 18].

The *Biefeld-Brown effect* was first noted by the German engineer Paul Biefeld and the American Physicist Thomas Brown in the 1950s. Also known as electrohy-drodynamic (EHD) thrusters these devices involved the use of a high voltage to propel air or some other fluid so that relative motion between the medium and an object is produced, thus moving the vehicle. This occurs by firstly ionizing the air around the object to produce a 'coronal wind' by using a high voltage. This effect is seen by some as a potential electric propulsion method. However, EHD can only be used within a fluid medium so would not be of much use to interstellar flight. It is possible however that by experimenting further with this sort of technology that new breakthroughs in physics can be made, which does have applications to spaceflight. Although today there is still no consensus within the physics community about the cause of this effect [19].

One new area of investigation, which caused a stir in physics circles in the early 1990s, is the *Podkletnov Impulse Generator*, first proposed by the Russian engineer Yevgeny Podkletnov [20]. His experiments have involved superconducting rotating discs and claims to produce a gravity loss between 0.3–2% by altering the gravitational field of the Earth above the rotating discs. Several attempts to reproduce the experiment have so far failed to detect any loss of gravity, although Podkletnov claims that the exact conditions of his experiments have not been reproduced. If true, such a 'gravity shielding' device would represent a fundamental leap forward in our understanding of the laws of nature. As well as substantially reducing the cost of conventional air travel it would have obvious applications to space propulsion. However, the mainstream opinion does not seem to support the claims of Podkletnov today.

It is the responsibility of those proposing such radical theories to persuade the rest of the science community of the merits of their case (not the other way around) although certainly the science community should be more open and welcoming to debating such possibilities within the constraints of existing resources. Future considerations of

Table 16.1 Examples of various breakthrough physics topics and current science opinion

Physics subject	Next research step
Biefeld-Brown effect	Misinterpretation of observations
Podkletnov impulse generator	Not rigorous, not repeatable
Ground state reduction via Casimir	Evokes unresolved plenum issues
Alcubierre warp drive	Far less energy efficient than wormholes
Wormholes	Requires enormous negative energy

interstellar flight proposals need to look at these sorts of theories as alternatives to conventional wisdom, as we never know from where the next great discoveries may come from. The guiding principle for conducting such breakthrough physics propulsion research should always be experiments; if the theories are inconsistent with observations then they are simply wrong.

If creative individuals investigate these sorts of ideas and submit a report for peer review to some appropriate review body, it is important that this body selects research proposals that are academically credible and use rigorous techniques. If the filtering process is flawed, then thousands of people may send in proposals for interstellar research, which swamps the peer review process. The submissions to 'watch' for will be those from what we may call the 'enthusiast theorist' who will make statements like: *"Einstein was completely wrong, my theory proves it."* Typically, these submitted papers will be highly speculative, non-rigorous, may not use the scientific method, make unqualified assertions, impervious to logic, persistency in claims, disregard historical results, show a lack of citations, leap to conclusions prematurely, deny alternative explanations, publish only positive results, embrace revolutionary results only and be closed and hostile to critique. One can contrast this with submissions from what we may call the *'Reasonable theorist'* who will make statements like *"My work suggests an inconsistency with Einstein so further work is required to clarify this."* He will clearly state hypothesis or conjecture to be proven, uses rigorous methods, relies mainly upon the scientific method, qualifies all assertions, builds on historical work, gives relevant citations, states conclusions with appropriate caveats, accepts alternative explanations, publishes all results, is skeptical of revolutionary results, is open and welcoming to constructive critique. In space propulsion research or breakthrough propulsion theories, the 'enthusiast' is ever present and can divert important resources if not appropriately managed.

In a recent paper the US physicist Marc Millis gave a good summary on the progress of various revolutionary propulsion physics claims [21]. Some of these are listed in Table 16.1 along with the authors assessed status of each proposal in terms of the next key issue or next research step. This is the sort of analysis that is required to clearly remove those 'ideas' that will not lead anywhere and only take up valuable resources, compared to those that have the potential to lead to 'game changing' technology.

How do we best prepare the minds of the new generation for making scientific discoveries? To understand this fully one would have to study in depth the structure of the current College and University system (perhaps even earlier) and think of ways of improving it. We do not attempt this daunting challenge but instead consider for a moment the undergraduate and postgraduate training which students receive, as one possible area that could be changed. The Physicist Freeman Dyson has voiced his concerns of the PhD system adopted by Universitie [22]. He believes that the energy of young imaginative scientists may be sapped by this system and they are immediately propelled on a course of academic competition, which makes it difficult for them to be creative in their thinking and also to switch between different specialisms.

In general, a PhD candidate is expected to submit a thesis paper, which represents an original contribution to knowledge. Until recently, a student who did his whole PhD in the design of a warp ship may find it difficult to be taken seriously by other academics and therefore find employment difficulties. Academics are aware of this, and so a supervisor may steer his student towards a more conventional application of their work. There is also the risk that the supervisor may find he is treated differently when his peers view the sorts of papers being produced by his student. So a supervisor may also be reluctant to allow their student to pursue such topics. There are exceptions to this of course and one of these was in an unusual application of general relativity.

One of the ways that the issue of increased mass ratio for increasing velocity increment in space missions can be solved is to think of spacecraft that are not constrained by such mass limitations. This has led to the idea of a so called 'space drive' which has been defined by other authors as:

> An Idealized form of propulsion where the fundamental properties of matter and space-time are used to create propulsive forces in space without having to carry or expel a reaction mass [23].

One of the variations of this theme is the 'warp drive,' so beloved of *Star Trek* fans. In recent years much academic work has been performed to address this proposal. The first academic paper on the warp drive was by Micquel Alcubierre, who was studying his PhD at Cardiff University and wrote a seminal paper on the construction of a warp metric within the framework of general relativity [24]. This single paper has given rise to a whole new field of academic literature, which is now approaching 100 publications on this one topic. In theory the warp drive could allow for superluminal velocities without the restrictions of mass increase or time dilation effects. Travelling at the speed of light after an initial acceleration phase away from Earth orbit, such a vehicle could reach Alpha Centauri in around 5 years, although if travelling at superluminal speeds would get there much faster. However, the massive negative energy requirements of this scheme currently place it firmly in the arena of conjecture. The latest calculations for a 100 m diameter warp bubble size suggest a negative energy equivalent to a solar mass [25].

In a 2009 student PhD thesis, one of the chapters was completely dedicated to the warp drive proposal – possibly one of the first in academia [26, 27]. In essence,

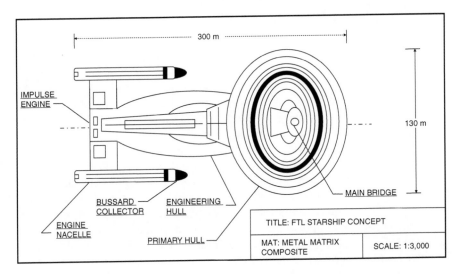

Fig. 16.1 Illustration of a far future breakthrough propulsion based starship

Alcubierre opened the way without realizing it. Such papers are becoming more frequent as academia becomes more tolerant to speculative theorizing. There is also another reason for this however, which is related to the fact that as a species we are reaching the limit of our current technological observations. If you can't do bigger and better experiments then all you have left is theoretical work. The point here is that students should not feel inhibited to pursue their choice of research directions, by possible career implications later on. Academics need to accept that it is possible that a student may do some of his greatest work during his thesis years, if given more flexibility but within the bounds of academic rigor.

In 2007 members of the British Interplanetary Society organized the first dedicated UK symposium on the warp drive concept [28]. This one day meeting was a highly successful event and attracted a group of international speakers. One of the papers was presented by Richard Obousy, then at Baylor University in Texas [29]. In this paper he discussed a novel way by which one could theoretically create a warp drive effect by considering the cosmological constant and its role in the expansion of the Universe. He referred to quantum vacuum energy as a possible origin of the cosmological constant. Then using ideas from Supersymmetry linked this to historical work on the Kaluza-Klein theory, which attempted to unify gravitation with electromagnetism by the introduction of an extra spatial dimension. He then went on to show how a local adjustment of the cosmological constant could be used to create an expansion and contraction of space-time necessary for the warp drive effect. Within this framework he predicted a theoretical maximum to the warp drive effect of around ten to the power of 32 times the speed of light, which would allow any ship to cross the Universe in around 15 s – A truly fantastic number. A twenty-fifth century warp type ship is illustrated in Fig. 16.1 based upon television shows like *Star Trek*.

Many obstacles to the achievement of warp drive were identified including the practical engineering and control of a realizable metric, energy violation of classical general relativity, Prohibitive energy requirements and the source of that negative energy, possible violation of energy and momentum conservation as well as the 2nd law of thermodynamics, potential for creation of black hole singularities and event horizons which prevent signal propagation externally, stringent requirements for extremely thin warp bubble walls; collision risk with incoming particles; causal disconnection of warp bubble from negative energy region and causality violations. Some of these problems are strongly coupled so that the solution of one may enable others to be solved too. However, there is no doubt that the warp drive concept, though bold, present's massive challenges to our technical know-how. This is an idea being studied by established physicists, which saw its beginnings within the pages of science fiction. An alternative to 'warp drive' but based upon the same mathematical framework is the concept of wormholes – geometrical constructs in the fabric of space-time that allow a vehicle to enter from one location in the Universe and come out light years away, all in seconds or minutes. Again, wormholes require massive amounts of negative energy and although very exciting science, they are not likely to be the scheme by which we first send probes to other stars [30, 31].

16.4 NASA Horizon Mission Methodology

Let us consider three radically different aerospace concepts. The first is a representation of the 1903 Wright Flyer discussed in Chap. 4. The second is the Space Shuttle as used for current twentieth and twenty-first century Earth to orbit transportation. The third is the warp driven ship. These three vehicles represent a time history of aviation and astronautical developments, from the past to the present to the far future.

The typical speeds for these three vehicles are around 0.01 km/s (Wright Flyer), 10 km/s (Space Shuttle during launch) and 3×10^5 km/s (assuming a warp speed equal to that of light). If one was to plot the speed of these three vehicles against the timeframe supposed one would not produce a linear extrapolation to the future but rather an exponential one. This is because there are fundamental differences in the principle of operation of all three machines. The Wright Flyer works on the principle of aerofoil lift and forward motion generated from propellers. The Space Shuttle works on the principle of chemical combustion and energy release for forward (or upward) motion. The warp ship works on the principle of the manipulation of space-time leading to the generation of attractive and repulsive gravitational fields. Hence, it would clearly not be appropriate to try and extrapolate the technology of the Wright Flyer to predict rocket launches to Earth orbit. Similarly it would not be appropriate to predict post relativistic speeds though space using conventional rocket technology. So if we want to predict the technology of the future what do we do? We have to recognize that each technology represents a paradigm shift so in order to visualize the technology of the future one has to imagine what paradigms will exist. This illustrates the importance in using the imagination to develop new technologies.

Others have noted the fact that technology can progress exponentially such as the mathematician and writer Vernor Vinge [32]. The idea of a technological singularity derives from the mechanism of humans creating more intelligent artificial intelligence, which are then able to create even more intelligence machines. This effect continues until the technology being created is then out of reach of human predictions. Because we can't predict such exponentially increasing technological change this is referred to as a horizon. Vinge predicted that such a singularity would occur between the years 2005 and 2030, although many have their doubts that it will occur this soon or even at all. But if technology, specifically space technology, does follow an exponentially increasing trend, how are we to predict the likely technologies of our future such as warp drive propulsion. There is one way to do this and it involves a consideration of what the future paradigms may be.

In 1995 a NASA physicist John Anderson came up with a clever solution, which is a way of generating creative pathways to the future without relying on any linear extrapolation of current technologies [33, 34]. The technique is known as the NASA Horizon Mission Methodology (HMM) and is a potentially powerful tool for defining research programs in interstellar flight. The hypothetical future is defined as a horizon mission or an impossible mission with capabilities and performance levels, which are not met by linear extrapolation of current or near future technologies. The horizon mission is outside our current paradigms, and so by implication new paradigms are required to make such a mission possible. The idea of applying this tool to thinking about such missions is to provide new insights into the direction technology may go. Because the technical goals of the study are generally impossible, one cannot extrapolate realistic future requirements by depending on existing solutions. This forces designers to go beyond existing methods and consider ideas that lay at or outside the theoretical 'horizon'. Although to be consistent with the known laws of physics any analysis imposes the constraints of conservation of momentum and energy on the solution. From this analysis a 'problem statement' is defined.

The HMM strategy is to create a culture of thinking that aims to provide new insight into research issues and inspire potentially exciting visions for the future, unconstrained by the usual paradigms. The thinking creates a hypothetical 'flash beam' into the future that is beyond the defined horizon. So that linear extrapolation is impossible and a new frame of reference (boundaries, rules, context) is formulated. Eventually, the HMM approach leads to the identification of implicit engineering assumptions which underlay the horizon and generates breakthrough alternatives to achieve this hypothetical future. These alternatives are then linked back to the present day to identify the requirements, application and implications for the mission. The 'stepping stones' towards fruition of the vision are then defined in terms of a technological roadmap. The spirit of the HMM philosophy necessitates inclusion of several individuals within a workshop type setting. This is required to allow for the cross fertilization of ideas. From the discussion using the

NASA Horizon mission methodology, we could for example construct the following problem statement for the warp drive proposal, which is similar to those defined elsewhere: [35].

> A hypothetical future where a space vehicle has typical transit times to nearby stars limited to a few years time scale, so that solar systems within ~10–20 light years radius can be practicably reached, traveling at or exceeding motion at the speed of light as measured by distant observers. The vehicle should have a controllable and sustainable mechanism that isolates the vehicle from any absolute reference frame, and interacts with the properties of space, matter or energy to induce a unidirectional acceleration of the vehicle within the bounds of conservation principles and consistent with empirical observations.

An application of the rules of HMM should then lead to the directed pathways and a constructed roadmap for how to enable such a mission to be accomplished. Any space agency conducting such an exercise would then establish the requirements for future program plans, e.g. we need to do research into high powered supercomputing or research into unique material properties. In an ordinary engineering project, one would start with the initial requirements and then work through the various design phases to eventually down select to a specific solution. HMM is analogous to the specification of a solution existing in the far future and then working backwards to define ultimately what the requirements would need to be in order for the Horizon Mission to be accomplished. HMM is a structured mental technique for examining new ideas from the perspective of the future. This is major thinking out of the box techniques but if used in an appropriate way can set us out on program path to deliver an inspirational goal which some today consider impossible – robotic and manned exploration of the nearest stars.

16.5   Problems

16.1. Consider various historical technologies such as the projectile rifle, the rocket, radio communications, the microprocessor chip, and the vacuum cleaner. Include some of your own technologies in this list. Conduct some research and ascertain which of these technologies was predicted and which ones were not. Were any of these to be expected based upon an efficiency improvement of an older technology? Think about the difference between technology, which progresses linearly, and technology, which progresses at an exponential rate. What are the implications of both trends?

16.2. Watch an episode of a modern science fiction program like *Star Trek*. Make a list of the key technologies being used in the plotline. Which one of these has a resemblance to technology that already exists today? For those that are remaining speculate on the likely timescale for such technologies coming to fruition, unless you think never. Plot these on a graph of probability (between 0 and 1) against time based on your assessment. Take an overview perspective of the graph you generated. Is there any relationship between each of the technologies? Can you speculate on what other technologies could develop as a spin-off, considering that these technologies will eventually exist on the timescale that you predicted?

16.3. Obtain either of the following or preferably both references (1) "Spaceship Hanbook, Rocket & Spacecraft Designs of the twentieth Century: Fictional, Factual and Fantasy," by Jack Hagerty and Jon C Rogers published by ARA Press 2001 (2) "The Dream Machines, A Pictorial History of the Spaceship in Art, Science & Literature," by Ron Miller and published by Krieger 1993. Examine the trend in spacecraft designs throughout history. Can you identify any common elements which have always remained in all concepts? Can you identify any elements that have clearly gone through a transformation in technology? Based upon your research, sketch what you think spacecraft will really look like in 1,000 years from now.

References

1. Clarke, AC (1945) Extra-Terrestrial Relays, Wireless world.
2. Clarke, AC (1979) The Fountains of Paradise, Victor Gllancz.
3. Stapledon, O (1937), Star Maker, Methuen.
4. Niven, L (1970) RingWorld, Ballantine Books.
5. Niven, L & J Pournelle (1974), The Mote in God's Eye, Simon and Schuster.
6. Clarke, AC (1972) The Wind from the Sun, Harcourt Brace Jovanovich.
7. Smolin, L (2006) The Trouble with Physics, Houghton Mifflin Harcourt.
8. Dirac, P (1928) The Quantum Theory of the Electron, Proc.R.Soc.A, vol.117, 778, pp610–624.
9. Millis MG & EW Davis (2009) Frontiers of Propulsion Science, Progress in Astronautics and Aeronautics Volume 227, AIAA.
10. Wheeler JA & K Ford (1998) Geons, Black Holes & Quantum Foam, A Life in Physics, Norton.
11. Clarke, AC, (1999) Profiles of the Future: An Inquiry into the Limits of the Possible, Indigo.
12. Millis, MG (1996) Breakthrough Propulsion Physics Research Program, NASA TM-107381.
13. Millis, MG (1998) Breakthrough Propulsion Physics Program, NASA TM-1998-208400.
14. Millis, MG (2004) Breakthrough Propulsion Physics Project: Project Management Methods, NASA/tm-2004-213406.
15. Heim, B (1994) Ein Bild vom Hintergrund der Welt, Welt der Weltbilder, Imago Mundi, Band 14, Resch, A.(ed).
16. Zampino, EJ (2006) Warp-Drive Metrics and the Yilmaz Theory, JBIS, 59, pp226–229.
17. Puthoff, HE et al., (2002) Engineering the Zero-Point Field and Polarizable Vacuum for Interstellar Flight. JBIS, 55, pp137–144.
18. Obousy, R & G Cleaver (2007) Supersymmetry Breaking Casimir Warp Drive, Proceedings of the STAIF-2007: 4th Symposium on New Frontiers and Future Concepts, El-Genk, M.S. (Ed.), AIP Conference Proceedings, 880, AIP Press, New York, pp1163–1169.
19. Tajmar, M (2004) Biefeld-Brown Effect: Misinterpretation of Corona Wind Phenomena, AIAA Journal, Vol 42, pp315–318.
20. Podkletnov, E, et al., (2003) Investigation of High Voltage Discharges in Low Pressure Gases through Large Ceramic Superconducting Electrodes, J.Low Temp.Phys. 132, pp239–259.
21. Millis, MG (2010) Progress in Revolutionary Propulsion Physics, Presented 61st IAC Prague, IAC-10-C4.8.7.
22. Dyson, F (1989) Infinite In All Directions, Harpercollins.
23. Millis, MG (1997) Challenge to Create the Space Drive, Journal of Propulsion & Power, 13, pp.577–582.
24. Alcubierre, A (1994) The Warp Drive: Hyper-fast Travel within General Relativity, Class. Quantum Grav., 11, L73–L77.

25. Van Den Broeck (1999) A Warp Drive with More Reasonable total Energy Requirements, Class.Quantum.Grav, 16, pp3973–3979.
26. Obousy, R, (2008) Investigation into Compactified Dimensions: Casimir Energies and Phenomenological Aspects, PhD Thesis Baylor University, Texas.
27. Obousy, R (2008) Creating the 'Warp' in Warp Drives, Spaceflight, 50, 4.
28. Long, KF, (2008) The Status of the Warp Drive, Spaceflight, 50, 4.
29. Obousy, R & G Cleaver (2008) Warp Drive: A New Approach, JBIS, 61, 9.
30. Thorne, KS (1994) Black Holes & Time Warps, Einstein's Outrageous Legacy, Norton.
31. Visser, M (1996) Lorentzian Wormholes from Einstein to Hawking, Springer.
32. Vinge, V (1993) The Coming Technological Singularity, Vision 21 Interdisciplinary Science & Engineering in the Era of Cyber Space, Proceedings of a Symposium held at NASA Lewis Research Center, CP-10129.
33. Anderson, JL (1995) Virtual Business Targets using the Horizon Mission Methodology, Space Programs and Technologies Conference, AIAA 95–3831.
34. Anderson, JL (1996) Leaps of the Imagination: Interstellar Flight and the Horizon Mission Methodology, JBIS, Vol.49, pp.15–20.
35. Long, KF, (2008) The Status of the Warp Drive, JBIS, 61, 9

Chapter 17
Realising the Technological Future and the Roadmap to the Stars

Of Icarus, In ancient days two aviators procured to themselves wings. Daedalus flew safely through the middle air and was duly honoured on his landing. Icarus soared upwards to the Sun till the wax melted which bound his wings and his flight ended in fiasco. The classical authorities tell us, of course, that he was only "doing a stunt"; but I prefer to think of him as the man who brought to light a serious constructional defect in the flying-machines of his day. So, too, in science. Cautious Daedalus will apply his theories where he feels confident they will safely go; but by his excess of caution their hidden weaknesses remain undiscovered. Icarus will strain his theories to the breaking-point till the weak joints gape. For the mere adventure? Perhaps partly, this is human nature. But if he is destined not yet to reach the Sun and solve finally the riddle of its construction, we may at least hope to learn from his journey some hints to build a better machine.

Sir Arthur S Eddington

17.1 Introduction

We consider how international co-operation in the pursuit of space has come about leading to innovations like the International Space Station. The case for an international conference dedicated to interstellar flight is made along with the requirement for a dedicated research institute. A precursor space mission roadmap is presented for how we can eventually reach the nearest stars. Project Icarus, the latest engineering star ship design study that may lay the foundations to the first robotic interstellar probe in the coming centuries is introduced. Finally we consider how best to promote academic and technological progress towards interstellar flight, by the use of internationally competitive design studies.

K.F. Long, *Deep Space Propulsion: A Roadmap to Interstellar Flight,*
DOI 10.1007/978-1-4614-0607-5_17, © Springer Science+Business Media, LLC 2012

17.2   International Co-operation in the Pursuit of Space

The 'space race' started with the launching of the Soviet satellite Sputnik 1 into Earth orbit way back in 1957. At the time, cold war tensions were increasing and the Western response was to launch the satellite Explorer 1 in 1958, which was also the first man-made satellite to take measurements of the Van Allen radiation belts. Other nations including Britain followed in later years by either launching satellites or building satellites to be launched by other nations. In the last quarter of a century, we have also seen the rise of new space powers in the form of India, China, Japan and the Republic of Korea, all of which desire their own launch program and a network of telecommunications satellites. Over time, these nation's attempts to attain global dominance over Earth orbit have led to a convergence of global interests. This is particularly the case with observing global warming where continuous satellite monitoring can provide vital science data for the benefit of everyone on Earth.

To help co-ordinate global space activity and astronautics, the *International Astronautical Foundation* (IAF) was formed in September 1951, from a conference of several European and American delegates in London. The IAF organizes the annual conference known as the *International Astronautical Congress* (IAC) to provide a forum for the exchange of experiences and ideas around astronautics with the long term goal of opening up space to all humankind. It promotes awareness of international space activities and fosters information exchange between different space programs. The IAC is held in a different country each year, organised by one of the host national space societies that is a member of the IAF. Throughout its history the IAC has been hosted in France, UK, Germany, Switzerland, Austria, Denmark, Italy, Spain, Netherlands, Sweden, US, Bulgaria, Poland, Greece, Yugoslavia, Argentina, Belgium, USSR, Portugal, Czechoslovakia, Japan, Hungary, India, Canada, Israel, Norway, China, Australia, Brazil and South Korea. This is an impressive diversity of host nations spanning all five major continents. Anyone can attend the IAC and it is a truly global gathering of international space agencies, companies, universities and private organizations.

The conference is composed of several sessions which cover topics such as the space life sciences, microgravity sciences, lunar and planetary exploration, SETI, space debris, Earth observation, communication and navigation, satellite missions and operations, astrodynamics, materials and structure science, space power, space propulsion, launch vehicles, space architectures, space education, space policy, economics of space. To the credit of the IAC, the program even includes discussions relevant to interstellar travel. This includes consideration of human exploration beyond Mars, space elevators and tethers and interstellar precursor missions.

In 2008 this author was invited to give the final highlight lecture at the 59th IAC on the interstellar theme. Having not attended this conference before this was a daunting challenge [1]. Although it was an honor to give one of the highlight lectures, it was disappointing that the session Chairman referred to the presentation, as "this is real Star Trek territory." Although this comment was made with good intentions, to highlight the 'imagination' theme of the conference, it bought home the fundamental problem with getting interstellar research properly funded. This barrier is preventing a properly organized program of work that can enable a real intellectual debate on the possibilities of interstellar flight. There is almost a consensus of understanding, a taboo, among mainstream physicists that topics such as breakthrough propulsion physics are very interesting, but not to be taken seriously. This attitude needs to change. Attending such conferences one quickly learns that this is how international contacts are made with ones peers and is a vital part of any research field. Most of the people working in the interstellar community are very friendly and always willing to discuss ideas with new people. They tend to be positive and open minded to fresh ideas and for a new person entering the field this can be a very welcome experience. However, it is also a fact that most of these people are widely distributed throughout the world although these days they maintain contact with each via the World Wide Web.

How can we slowly build up the research effort towards attaining the long-term goal of interstellar flight – not just for robotic exploration, but eventually for humans too? The majority of academics accept that the challenge of interstellar flight is many orders of magnitude greater than our achievements of walking on the Moon. For this reason, we must move forward with determined effort but be content to make incremental steps towards these goals. However, most of the people working on the problems of interstellar flight are doing so in their own time as volunteers (including this author). Typically performing some technical job during the day and then going home to their families in the evening, with perhaps the odd hour or two to conduct some research, write a paper or presentation for an upcoming conference. They will then communicate with like-minded people over the Internet. This is a sporadically organised research program with little focus on any specific goals. Indeed, most of the interstellar flight conferences are piggybacked as a parallel session onto another mainstream physics or space conference. For example, in 2009 this author organised and hosted the interstellar session of the UK Space Conference based at Charterhouse School in Surrey, England. This was the first such session in the conference history. This once again brought together a group of international speakers to discuss research problems. Researchers will then see this as their opportunity to interact with other researchers and discuss their ideas.

Another regular fixture on the interstellar calendar is the *IAA Symposium on Realistic Near-Term Advanced Scientific Space Missions*, held every two years in Aosta, Italy. This meeting is focused on near-term research for precursor missions mainly within the solar system or just beyond. Although the meeting would permit interstellar related papers that is not the main agenda of the meeting as evidenced by the title. This has worked very successfully to date, but what is really needed is a single week long conference every two years dedicated to the problems of interstellar flight, both robotic and manned exploration. This would allow a greater focus for the community. This could be called the *Symposium on Realistic Long-Term Advanced Interstellar Scientific Space Missions,* for example. This would both cover near-term and speculative propulsion concepts such as those addressed by NASA BPP. It would also cover the other systems associated with interstellar vehicle design and the issues associated with the astronomical targets, such as any potential planets to visit. With such probes it could be split into several main themes such as (1) Near-term technologies for precursor mission's outwards (2) breakthrough physics & associated technologies (3) astronomical discoveries relevant to future interstellar missions. The main goal of such a dedicated conference would be the organization of focused academic research to add input into the planning and design of the world's first interstellar probe.

In order to do this what is required is a single umbrella organization to help coordinate international research and provide a network of contacts. With none of the national space agencies engaged in interstellar research, there is currently a void for interstellar type research that is not being pursued with priority. This needs to be filled by an internationally funded body. Such a body would accept research proposals from a private individual, space societies, universities or industry. A process would need to be set up which conducts a peer review and scoring assessment of each proposal to aid in any decisions to award appropriate funding. Opportunities would also exist to conduct more advanced, level two studies. Ultimately, all awarded work would be published in peer-reviewed journals. The sort of projects that would be supported includes:

- Rigorous peer reviewed research.
- Credible concept studies in space propulsion.
- Specific vehicle design studies.
- Emerging breakthrough physics research.
- Revolutionary-high impact research.
- Astronomical research for targeted missions.
- Visionary mission studies for precursor flights.
- Research that makes measured progress.

In 2006 the then NASA physicist Marc Millis, who had formerly ran the NBPP, set up the *Tau Zero Foundation* (TZF). This is a registered charity that aims to achieve some of the aspirations of programs like NASA BPP and NIAC so as to progress the interstellar vision. The TZF consists of a volunteer group of scientists, engineers, artists, writers, and entrepreneurs, all dedicated to addressing the issues of interstellar travel. This is a non-profit corporation supported through donations and not a space advocacy group. Ultimately, The TZF aims to support incremental progress in interstellar spaceflight and will

- Support students through Scholarships.
- Provide inspirational educational products.
- Attend and arrange international conferences.
- Support for interstellar design studies.
- Support BPP research topics through competitive selections when funding available.
- Foundation seeks credible, rigorous scientific research.
- Give cash awards for visionary research

Such an organization is required to lead the way in the field of interstellar research. Pulling together an international body of like-minded individuals and pooling their co-operative efforts towards making the vision of interstellar flight come true. History will show whether the TZF is able to generate an international climate of rigorous academic research dedicated to the interstellar vision – whatever the main propulsion candidate ends up being for the first precursor mission outside of our solar system.

Ultimately, the best way to push forward international co-operation in the pursuit of interstellar goals is to have a dedicated institute, along the lines of the *Perimeter Institute for Theoretical Physics* in Canada which was founded by donations from Mike Lazaridis, the Co-CEO and founder of *Research In Motion* which is the company that makes the excellent Blackberry phone. Founded in 2000, the Perimeter Institute is fast becoming a leading research centre whose goal is to catalyze breakthroughs in our understanding of the physical world within a lively and dynamic research atmosphere. It promotes world-class research programs with resident academics as well as allowing academics to visit for short periods.

If there were an *Institute for Interstellar Flight* this would attract academics from around the world to come together for weeks or months at a time to jointly work on some of the major technical issues or explore a new area of physics where a fundamental breakthrough in our understanding may come from. It would also be the location for a major conference and would act as the international focus point for all interstellar related research. Eventually, it would also produce its own peer reviewed Journal. This is what we need to make interstellar research move substantially forward and allow innovative ideas to emerge. It needs to be moved from the sidelines of science, given some major investment, and an institute to focus the research and provide an exciting atmosphere where an optimistic vision for space exploration exists, unfettered by old philosophical values and unconstrained by the currently slow wheels that science cranks forward. Only when this happens, can we look back on the literature of the past with pride knowing that all those that came before provided the foundations for our future in space. This will only happen if another philanthropic donator comes forward and makes this vision real. This would be a great legacy for one person to leave the children of Earth. In order to secure the future of our children we need to accept that investment in visionary thinking today secures a long term future, although it may not give rise to the short term gains so expected of today's society. The interstellar research community awaits the arrival of such investment.

17.3 Establishing Precursor Missions and the Technological Roadmap for Interstellar Exploration

It is a testament to how much we have achieved that mission planners today are permitted to discuss space exploration strategies in the context of interstellar travel. The NASA designer Les Johnson has openly discussed the need for a strategic roadmap to the nearest stars [2]. He argues that without requiring major physics or technology breakthroughs, co-related progress in the fields of space propulsion, fabrication and repair will ultimately lead to the capability of expanding human exploration and civilization towards the stars. The basis for the characterization of relative maturity for selected technologies is the Technology Readiness Levels discussed in Chap. 14. NASA managers will then use this system to derive high-priority, medium priority, low priority and high-risk/high-gain technologies. This approach will lead to the robotic exploration of interstellar space and the eventual expansion of human civilization beyond the solar system. Johnson in fact argues that the capability to perform multi-generation interstellar voyagers will arise as a consequence of solar system exploration and development. This is all good and true, provided that the investment in the key technologies, particularly propulsion, is made now.

During the 61st International Astronautical Congress in Prague, the NASA designer George Schmidt recently discussed in-space propulsion technologies in the context of the flexible path exploration strategy [3]. The flexible path strategy differs from the Moon and Mars-oriented paradigms as it refrains from placing humans on the surface of planets or where large gravity wells are likely to exist. Propulsion technologies are grouped into three development categories; near-term technologies are at a TRL of 6–9 and include chemical and electric propulsion schemes (2015–2015); mid-term technologies are at a TRL of 3–6 and include nuclear thermal, advanced chemical, high-power electric/plasma propulsion (2025–2040); far-term technologies are at a TRL of 1–3 and includes advanced plasma propulsion, advanced nuclear propulsion, gas core rockets, fusion and antimatter schemes (> 2040).

The inclusion of the last category, far-term, is very interesting from the perspective of those who would like to see such propulsion systems developed with some investment today, a view that Schmidt argued during his presentation in Prague. These technologies would greatly extend the reach of human exploration with orders of magnitude improvement in the specific impulse performance compared to missions being conducted today. They require a high-risk/high-gain investment, performed at a research level of effort over several phases. But perhaps with clear demonstration of the proof of concept these technologies can also march up the Technology Readiness Level ladder and open up the solar system and beyond. It is a personal view, that research efforts like Project Icarus discussed in the next section would be an ideal way to ensure the future roadmap to the stars, by a moderate investment today to ensure an ambitious mission is possible in the decades ahead.

It remains a fact that convincing the science community or politicians to commit to large funds of money for space missions that go to the stars is unrealistic in the present economic climate. Instead, we must be content to take incremental steps towards this direction. Precursor missions offer an important stepping-stone to achieving the long-term goal of sending probes to the stars. A precursor mission would deliver vital science within a realistic time frame, whilst pushing technology developments in propulsion and deep space materials survival. A precursor mission could be a mission to the outer parts of the solar system, the Kuiper Belt or the Oort cloud. There are also several dwarf planets, including Pluto, of which we know very little about and in the coming years it is possible that brown dwarf stars could be located between Sol and Alpha Centauri.

One of the suggestions for a precursor mission which carries huge scientific merit is to launch a probe to the Focal Point located between 550 and 1,000 AU and is the focus of the gravitational lens of the Sun. It is 278 times closer to us than our nearest star Alpha Centauri 278,261 AU away. One of the advantages of such a mission is that the focal point magnifies all objects located in the stellar neighborhood of the target so that a complete radio map can be constructed of the area. One of the originators and proponents of the Focal Point mission is the Italian scientist Claudio Maccone, a Member of the International Academy of Astronautics. Maccone attended the 2007 British Interplanetary Society warp drive conference and it was here that this author first heard him discuss the gravitational Focal Point [4] and his passion for it. A precursor mission should be sent into the outer solar system and then to locations like the Focal Point to deliver valuable scientific data. It could also be used as for the additional benefit of providing close-up studies of the target being anticipated for the future star mission. This will help us decide whether any particular star was interesting enough prior to committing substantial resources for any future missions.

Following on from this it is worth thinking about three candidates as part of a *Challenge Mission* that could be launched within the next few decades using propulsion technology that is already proven. These missions follow on from each other sequentially and so build upon the success and technology of the previous mission:

- *Challenge Mission 1* (150–200AU): Several spacecraft should be launched towards the outer part of our solar system with a flyby of Pluto and go beyond the distance of the Voyager and Pioneer probes. This will confirm the measurements of these historical probes whilst allowing new discoveries along the way through the solar heliosphere and into the Kuiper belt. This would be accomplished in around a 15–20 year flight time and the probes would be launched between the years 2015 and 2020.
- *Challenge Mission 2* (200–600AU): This would be an extension of mission 1 but allow the technology to be really pushed forward, utilizing engines, which had a high specific impulse. This mission could also be to the lower boundary of the gravitational lens point. This dedicated mission would be accomplished in around a 20–30 year flight time and the probes would be launched between the years 2025 and 2030.
- *Challenge Mission 3* (1,000 AU): This would be a dedicated mission to the outer boundary of the gravitational lens point. It would be accomplished in around 30–50 year flight time and the probes would be launched between the years 2030 and 2040.

The key propulsion technology for all of the above would be either electric or ion drives such as on Deep Space-1, nuclear thermal or nuclear electric propulsion. RTG power sources would play an important role. Plasma drive technology, such as VASIMR, may also be ready in time for such missions. Planetary gravity assists could also be use to add velocity to the spacecraft. With so many options for propulsion, there is no excuse not to try. These sorts of missions must be demonstrated before we can launch the first unmanned interstellar probe towards the stars. If physicists could demonstrate other technologies such as fusion or solar sails, then the timescale to launching these missions would be accelerated. In theory, with advanced propulsion schemes, the first robotic probes from Earth could be reaching the inner Oort cloud by the latter part of the twenty-first century and then we can start to plan and design the first interstellar mission.

One way we could change all this is if national space agencies can agree with government backing on a common vision, that is: *To launch a robotic probe within the twenty-first century that is to arrive at the destination star by 2150.* Now this is a vision that is attainable and worth working towards. This would place the arrival of the first interstellar probe within the lifespan of the generation born around the second half of the twenty-first century. What better way to expend the energies of the productive people of Earth than to have such a milestone? All the precursor missions defined in the roadmap would then be working towards this ultimate goal which appears to be entirely realistic according to those who understand the challenge ahead, such as Alan Bond and Tony Martin, members of the Project Daedalus Study Group [5]:

> Although not an immediate project, it is concluded that interstellar flight is feasible, perhaps within the next 100 years.

If we attempt these missions, this will be the beginning of the long march towards the ultimate destination of the stars. This is the roadmap we should be following and not just constraining our exploration to Earth orbit and the planets. Along the way, we will also make great discoveries about our Universe and perhaps on some of our outstanding physics issues. This will allow a proper assessment of the challenges ahead and we can start to prepare for the first robotic missions to go into orbit around the planets of another star. An exciting future lies ahead of us, if we could lose what Arthur C Clarke referred to as a 'failure of nerve'. But as shown in Fig. 17.1 we certainly don't lack of 'failure of imagination' having derived a variety of vehicle concepts with different propulsion schemes for both interplanetary and interstellar exploration. This at least must give us hope that we have a chance to succeed.

17.4 Project Icarus, Son of Daedalus, Flying Closer to another Star

When reviewing the literature on interstellar flight it becomes clear that one of the best candidates for reaching the nearest stars is a nuclear pulse propulsion scheme. Not everyone shares this opinion, and indeed many claim that the front-runner today is solar sail propulsion. However, accepting this as fact, during 2008 a group of people got together to consider how best to push such concepts forward to a

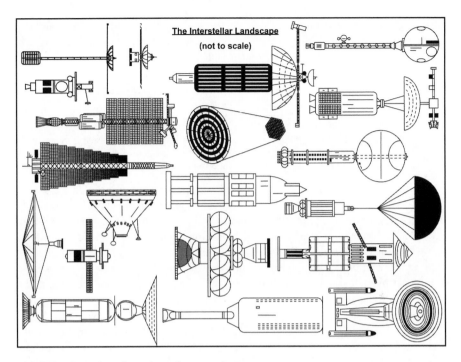

Fig. 17.1 Illustration of a variety of spacecraft concepts

higher technological readiness level. Discussions with the Tau Zero Foundation
President Marc Millis, who was both enthusiastic and supportive of the idea being
proposed, led to the acceptance that one proposal in particular would be a great
way to focus research whilst renewing the capability of a new generation of
designers. The proposal was to assemble an international team of suitably qualified
designers and redesign the Project Daedalus probe. This would be a TZF initiative
in collaboration with The BIS, members of which had originally conducted the
Daedalus study. From then on the project began to take shape and a life of its own.

Over three decades has now passed since the Project Daedalus study and many
advances have been made in the field of astronomical extra-solar planet detection,
communication, computing and other vital areas. Crucially, physics advances in
fusion technology are progressing with programs such as JET, ITER, NIF and the
proposed HiPER facility. In an attempt to discuss these advances and review the
design of Daedalus after 30 years a conference was organised and held at the British
Interplanetary Society Head Quarters in London on 30th September 2009. This led
to a proposal known as Project Icarus [6]. One of the useful starting points for Icarus
was to begin a line of communication with some of the original members of the
Daedalus study group. Contact was made by this author with both Bob Parkinson
and Alan Bond, and both gave their blessing for the project as well as several other
members of the Project Daedalus Study Group like Gerry Webb and Penny Wright.
One of the suggestions made by Alan Bond (and supported by Bob Parkinson

during discussions) was that several of the original Daedalus study group could act as a design review panel of the final Icarus concept. This was an excellent suggestion and it is anticipated that a mid-term and final design review will take place. This would add an element of academic rigor to the Icarus project and ensure that the end result is a credible one. In the introduction to the Daedalus study report Alan Bond stated:

> it is hoped that these 'cunningly wrought' designs of Daedalus will be tested by modern day equivalents of Icarus, who will hopefully survive to suggest better methods and techniques which will work where those of Daedalus may fail, and that the results of this study will bring the day when mankind will reach out to the stars a step nearer [7].

So in essence, the original study group suggested the naming of a successor project as Icarus. Daedalus and Icarus were characters from ancient Greek mythology. In an attempt to escape the labyrinth prison of King Minos, Icarus father Daedalus fashioned a pair of wings for both himself and his son made of feathers and wax. But Icarus soared through the sky joyfully and flew too close to the Sun melting the wax on his wings. He fell into the sea and died after having 'touched' the sky. Project Icarus aims to 'touch' the stars and escape from the bounds of mother Earth. The stipulation of 'mainly fusion based propulsion' was to act as a constraining factor on a design team that was time constrained, as well as allow a direct maturity measure comparison to Daedalus. At the BIS symposium Alan Bond had said "with this engine we are addressing the Universe on its own terms" in reference to the utility of fusion power, giving that the main engines in the Universe produced by nature, are fusion powered stars.

Project Icarus even has its own logo as shown in Plate 20, designed by the French artist Alexandre Szames in collaboration with this author. The symbol on the left side is a representation of the TZF logo whilst the symbol on the right is a representation of the BIS logo – the two arms of Project Icarus. A representation of Daedalus is shown in the centre. There is a black wing shape framing the inner part of the logo, symbolizing Icarus 'flying' close to another star, in the direction of a distant star, which is depicted at the top of the logo. Squares and rounds complete the feather of the wings and they fly closer to the target star, illustrating the project motto. These items symbolize spacecraft and design projects and initiatives of various kinds; the efforts we make to make it happen, whether actual, tangible (boxes, the square being the symbol of matter), or theoretical (round = spirit).

The purpose of Project Icarus is to design a credible interstellar probe that could be a potential mission this century. It was also seen as an opportunity to allow a direct technology comparison with Daedalus and provide an assessment of the maturity of fusion based space propulsion for future precursor missions. It was hoped that this initiative would generate greater interest in the real term prospects for interstellar precursor missions that are based on credible science and to motivate a new generation of scientists to be interested in designing space missions that go beyond our solar system. If Project Daedalus was an attempt to prove an existence theory was feasible, then Project Icarus in as attempt to prove it is practical, at least in theory.

Once the original plans had been produced for the project, it was decided appropriate to set some boundaries for the project to constrain some of the diverse thinking that would inevitably be generated in the community. This resulted in a Terms of Reference document, which was agreed by the founding members of the design team; Kelvin Long, Martyn Fogg, Richard Obousy, Andreas Tziolas and Richard Osborne. This was also supported by the official project consultants Marc Millis, Paul Gilster, Tibor Pacher, Greg Matloff and later joined by Ralph McNutt (the creator of the Innovative Interstellar Explorer concept). Among the team, this created much debate and philosophical disagreements about the direction of the project. The main stumbling block in formulating the terms of reference for the study was the stipulation the spacecraft must use current or near future technology so that it could be credibly launched by 2050. One of the team members in particular felt that the definition of 2050 was incompatible with the requirement for fusion as the main propulsion drive and suggested that one of these should go. After several weeks of further debate the team finally decided to remove the 2050 and settle on 'this century'. Thereafter the terms of reference were frozen for the duration of the project and represented the initial engineering requirements. The final Terms of Reference for Project Icarus were agreed as follows:

1. To design an unmanned probe that is capable of delivering useful scientific data about the target star, associated planetary bodies, solar environment and the interstellar medium.
2. The spacecraft must use current or near future technology and be designed to be launched as soon as is credibly determined.
3. The spacecraft must reach its stellar destination within as fast a time as possible, not exceeding a century and ideally much sooner.
4. The spacecraft must be designed to allow for a variety of target stars.
5. The spacecraft propulsion must be mainly fusion based (i.e. Daedalus).
6. The spacecraft mission must be designed so as to allow some deceleration for increased encounter time at the destination.

Within the boundaries of the terms of reference, the team has the freedom to come up with creative solutions. This begins by the generation of ideas, which eventually turn into trade studies addressing various aspects of the mission, according to a systems engineering work program. There are also several options for the main propulsion, this could be just fusion or antimatter Catalyzed fusion or even with the addition of a nuclear-electric ion engine for low accelerations. Another option would be to utilize a solar sail for solar system escape or breaking. The form of driver for the fusion explosions will also be a variable. Daedalus used an electron beam driver but Icarus could alternatively be an ion driver or a pure laser beam. Alternatively, the capsule could be driven by a Z-pinch Marx generator or even confined magnetically. Indeed, the type of fuel is one of the major variables to choose from, such as DT, DD or DHe^3 or others. Even the design of the capsule is a variable and could be along the lines of a fast ignition concept.

The actual target star will be one of the most fascinating outcomes of the project based upon the latest astronomical observations. Daedalus chose Barnard's Star as the target, but Icarus may instead go to Tau Ceti, Epsilon Eridani, Alpha Centauri or some other star system. The team will also need to examine computing, navigation, communications. If ever launched the probe would be a science driven mission, so the design of science instruments will also be required. There is a lot of work for the Project Icarus Study Group to complete, if it is to build upon the brilliant work done on Daedalus and advance the design to the next level. As of December 2010 many new designers had joined the team such as Adam Crowl, Ian Crawford, Rob Swinney, Andreas Hein, Pat Galea and Rob Adams. The total number of designers and consultants was 20, all dedicated to producing a credible final engineering report.

There are also several key watchwords for Project Icarus to ensure that all design solutions are appropriate. The final design must be a CREDIBLE proposal and not based upon speculative physics. It must be a PRACTICAL design. It must be derived using accepted natural laws and using SCIENTIFIC methods, which are supported by experiments. It must be based upon only NEAR-FUTURE technology as determined by simple linear extrapolation of current technologies. The team must produce an ENGINEERED design as though the vehicle were close to flight readiness, to ensure that approximations and margins are appropriate. There is also the addition of a RELIABILITY watchword, to ensure that upon arrival all of the scientific instruments will function. There is also the addition of a project scope as follows:

> The required milestones should be defined in order to get to a potential launch of such a mission. This should include a credible design, mission profile, key development steps and other aspects as considered appropriate.

To ensure the team keeps its eye on the ball a set of engineering design philosophies have been adopted. This has been adopted for all the engineering systems. The nominal performance will always be greater than the worst case performance and the worst case performance should not result in the mission failure to high probability. The first design philosophy adopted by the design team is Design Philosophy 1:

> The designer is to rely upon worst case calculations as a means of capturing problem uncertainties and allowing a pessimistic assessment of the problem.

As stated above there are also several key watchwords for Project Icarus to ensure that all design solutions are appropriate. The watchwords are captured as Design Philosophy 2:

> The designer shall solve problems and produce work output in the spirit of the project watchword.

The latest data is to be exploited where possible and this is reflected in Design Philosophy 3:

> The designer shall remain well informed with current scientific and technological developments which may impact the Icarus design and mission.

Project Icarus will look at far future technology such as in the year 2100, but in general the final design solution and mission profile will be based upon linear and not exponential (e.g. technological singularities) extrapolations of current technology, where nearness to current technology is preferred. Design Philosophy 4 indicates this:

> Extrapolations of current technology is to be of a linear type only and limited to a few decades hence.

The final Icarus design could be very similar to Daedalus or very different. There are two main approaches to proceed with the design evolution. Firstly, the team could start from a new baseline and upon completion of the fundamental trade studies and literature reviews, put together a completely new vehicle configuration and mission profile. This would be a difficult approach to proceed from, given the volunteer nature of the team and the limited resources. Alternatively, the team could start from the Daedalus configuration and then consider perturbations of the engineering layout and mission profile then from this re-derive a new design. This would be the easiest approach and is the one adopted by the team. In this way the baseline is already established to allow calculations to begin immediately. Design Philosophy 5 indicates this approach:

> The Icarus vehicle design and mission profile is to be derived by considering perturbations of the Daedalus baseline design. The design is then to be evolved to a newly optimized configuration.

The next design philosophy acts as a constant reminder to be rigorous in the technical work, Design Philosophy 6:

> The vehicle configuration is to be derived solely from engineering and physics calculations where possible.

With the excitement of exosolar planet discoveries the team will be tempted to choose a mission target early. There will also be pressure from science planners on the team to choose a target. The selection of a target certainly makes the engineering calculations easier. However, with the pace of astronomical discoveries this decision would be premature so designers wait until late in the project to select the target star destination, so as to get maximum benefit from any discoveries being made. Project Daedalus suffered from this because the belief of planets in orbit around Barnard's star was later shown to be erroneous. A design philosophy acts as a reminder to resist the temptation to select a target early in the project by Design Philosophy 7:

> The team is to maintain an open-door mission policy for as long as is appropriate during the study.

Since the publication of works by authors like Eric Drexler, the subject of nano-technology has become increasingly important as it is realized that this technology has potential applications in all walks of life. With this in mind, it would be reasonable to expect that some aspects of an interstellar probe design may be constructed using nano-technology. This will be considered in some respects within the materials research module; however, it is accepted that this is such a big subject matter that all designers should be considering the applications to their research areas. This leads to Design Philosophy 8:

All designers will keep a weather eye on the potential application of nano-technology for use in their respective systems or sub-systems.

In an attempt to start addressing the scope for Project Icarus the team also came up with the *Icarus Pathfinder* and *Icarus Starfinder* concepts [8]. Pathfinder is a technology demonstrator mission to 1,000 AU, the outer most location of the solar gravitational lensing point. Like Deep Space 1, Pathfinder will test out vital technology required for an interstellar mission such as related to reliability and repair operations or deep space communications using very high gain antennas or optical laser systems. Potentially, such a mission could even return to the solar system testing out deceleration technology such as a MagSail. The Starfinder concept is for a actual fusion engine mission to 10,000–50,000 AU, to the outer bounds of the Oort cloud and 1/5th of the way to the nearest stars. Both Icarus Pathfinder and Icarus Starfinder are vital missions for the interstellar roadmap, which have to be demonstrated before the first true interstellar mission is launched. The concepts are shown in Plates 18 and 19.

Future history will show if the Project Icarus Study Group managed to maintain adherence to these watchwords and produce a design, which is an advance from the spectacular Daedalus design of the 1970s. The road to the stars is incremental and we must be patient with the steady progress we can make. But if the near future is as we hope, the grandchildren of people living today will see the first pictures from the nearest stars systems. Now isn't that a future worth working towards. It will be interesting to see what results the Project Icarus design team produces and how the design differs from that of Daedalus.

17.5   The Alpha Centauri Prize

An extension of the Project Icarus approach and one of the best ways to advance the prospects for interstellar travel, would be to have separate design studies, which could be derived, iterated and improved. Over time, the concept would be worked upon by future generations and ultimately lead to a direct design blue print for an interstellar probe. Daedalus had been performed three decades ago, so it seemed appropriate to start by re-designing the Daedalus probe with the updated scientific knowledge we have today. Hence Project Icarus was born. It is hoped that other teams around the world will also be assembled and work on other specific proposals investigated historically such as NERVA, Starwisp, Vista, AIMStar or one of the many others. This way, the technological maturity of different propulsion schemes can be improved over time and the case could be better made for precursor missions to the outer solar system and one day the nearest stars.

In May 1996 Peter Diamandis set up the now famous Ansari X-prize competition based upon the model used for the twentieth century flight across the Atlantic by Charles Lindbergh. The X-prize competition set out to open up Earth orbit to the greater population of the planet and Burt Rutan of Scaled Composites using the SpaceShipOne space plane eventually won it in October 2004. The X-prize has shown that this sort of model is an excellent incentive for spurring technological innovation and replaces the incentive of competitive overtures towards nation state warfare – the motivation behind the eventual moon landings. Could such a model be adopted for spurring innovation in the field of interstellar research? The answer is yes and this is how such an 'Alpha Centauri Prize' would work.

The competition, an academic one, would be run by an organization such as the Tau Zero Foundation and it could be held every 4 years. This would allow a sufficient time between design studies so as to allow some technological advances and scientific discoveries to be made and allow this new knowledge to be folded into the design work. The output of the studies would be an engineering design study report along the standard of the historical Project Daedalus. In order to maximize design capability and ensure that all the appropriate systems would be assessed the teams would have to be of a minimum size (e.g. minimum of 6 designers) with a clear Project Leader. The teams would also consist of members from more than one country so as to increase international co-operation in designing such missions and bringing together a world community behind such a vision. The rules of the competition can be summarized along the following lines:

1. The design team must contain at least six designers from at least two countries.
2. The work must not be completed as part of official government space agency work.
3. The team should complete the study in a submitted report to the judging panel within 1 year of the official competition opening and destination target revealed.

Similarly, the rules of the engineering output would be as follows:

1. The team must produce an engineering design study of an unmanned interstellar probe to be sent to the specified destination target and capable of delivering useful scientific data back to Earth within a century from launch.
2. The probe should be based upon current or near future technology and designed to be launched within a century of the study report delivery.
3. The study should include all of the major spacecraft systems, including propulsion, environmental, structure, materials, navigations & guidance, fuel, science & payload.
4. The report should also include a reliability analysis and technology readiness measurement and cost assessment. The key milestone timescale required for launching of such a mission should also be defined.
5. The vehicle design may be a combination of propulsion schemes but a single propulsive mechanism should be responsible for at least 90% of the boost phase of the mission.

6. The study should result in at least one novel form of test-rig level technology (software or hardware), which is included in the final design solution. This could be a ground test, rocket flight or the placement of some hardware into Low Earth Orbit.

The rule that a single propulsive mechanism should be responsible for at least 90% of the boost phase of the mission is there to force the design teams to essentially have a single propulsion engine that delivers most of the acceleration. It is more desirable to have ten teams producing ten radically different design concepts with some overlap, rather than having ten replica designs, which would be a waste of resources. The rule that some form of technology must be demonstrated will ensure that new advances are being made towards the ultimate vision of sending a probe to another star. Some of this technology may in the end be used in such a mission. The target destination could be changed each time the competition is run to ensure that the team doesn't just copy a previous design. The name 'Alpha Centauri Prize' doesn't necessarily imply that the target will always be this star system, although on the first occasion it is run this may be appropriate, being our nearest star. How would the judging panel assess the competition?

1. The accuracy and completeness of the engineering analysis.
2. How well the design meets the engineering requirements.
3. The standard and quality of the report.
4. The innovative approach taken by the design team.
5. The performance and innovation of the test technology.

The winners of the competition would be awarded a cash prize of say $200,000 provided by a philanthropic donator or indeed the X-Prize competition itself. This academic competition would focus interstellar research towards specific design studies and the ultimate objective of the competition is to increase the technological readiness of different interstellar propulsion schemes. Ultimately, after two decades of running this competition what will emerge is not a single choice for going to the stars in the coming centuries, but instead a realization that it is a combination of approaches with highly optimized engineering designs that will be the way to go. This suggests hybrid propulsion schemes and could for example be along the lines of a fusion-based drive with anti-proton catalyzed reactions but using a nuclear electric engine for supplementary power and perhaps a solar sail for solar system escape or upon arrival. When this probe arrives at the destination and sends back the first images of the systems planets, this will mark an important milestone in mankind's journey out into the cosmos. The Universe beckons us to join it, but when will we realize our potential and shed ourselves of Earth's cradle?

Human beings need a challenge to force us to progress technologically and push our ideas out from just theoretical studies. A competition of the sort proposed here would represent a major step forward for interstellar research, laying the seeds for

Table 17.1 Possible optimistic interstellar roadmap assuming that first launch of unmanned interstellar probe occurs by 2100 and speculation beyond

Year	Key event
2015	Planning and design work for 200 AU precursor mission begins
2020	Launch of 200 AU mission – *Innovative Interstellar Explorer*
2025	Launch of technology demonstrator mission ~ few AU with propulsion study emphasis
2026	Discovery of exoplanet, potentially habitable for humans, within ~ 15 light years
2028	Design work begins on lunar telescope to fully characterize exoplanets
2030	Planning and design work begins for 1,000 AU mission – *Pathfinder*
2034	Arrival of 200 AU probe, data return to Earth
2035	Observational telescope mission launched to the Moon
2036	Launch of Pathfinder mission to 1,000 AU
2037	Planning and design work begins for 10,000 AU mission – *Starfinder*
2040	Confirmed existence of habitable exosolar planet within ~15 light years.
2050	Planning and design work begins for first interstellar mission
2060	Arrival of Pathfinder mission to target, data return to Earth
2065	Launch of *Starfinder* Mission into far Oort cloud, 10,000 AU
2095	Arrival of *Starfinder* Mission into far Oort cloud, data return to Earth
2100	Launch of first unmanned interstellar probe to another star
2110	Launch of second unmanned interstellar probe to a different star
2120	Launch of third unmanned interstellar probe to a different star
2160	First interstellar probe arrives at target star system
2170	Second interstellar probe arrives at target star system
2180	Third interstellar probe arrives at target star system
2200	Work begins on design of first interstellar human colonization mission
2220	Work begins on design of second interstellar human colonization mission
2250	Launch of first human interstellar mission to another star (non-return mini-World Ship)
2270	Launch of second human interstellar mission to another star (non-return mini-World Ship)
2290	Launch of third human interstellar mission to another star (non-return full-World Ship)
2300	Arrival of first human colony to planet around another star
2320	Arrival of second human colony to planet around another star
2340	Arrival of third human colony to planet around another star
2350	Human expansion and colonization of the galaxy begins – Childhood's End

the first probe to be sent to another star system. Interstellar travel has been the ambition of many people for a long time, including the science community:

> It is recommended that a subsequent study address the possibility of a star mission starting in 2025, 2050 or later, and the long lead-time technology developments that will be needed to permit this mission [9].

The year 2050 for the first star mission seems quite ambitious. Let us think about a scenario that leads to the first launch by the year 2100. Such a plan could be along the lines of that describe in Table 17.1. This involves three key mission demonstrators along the way to the first interstellar launch. The first is a 200 AU mission along the lines suggested for the *Innovative Interstellar Explorer* mission discussed earlier in this chapter. The propulsion option for this mission includes solar sail, ion drive, nuclear thermal, nuclear electric or VASIMR and would require a specific impulse of around 1,000 s. The next step is the 1,000 AU demonstrator mission as proposed for *Icarus Pathfinder* discussed in this chapter. This would pass through the gravitational lensing point located between 550 and 1,000 AU and give us vital astronomical data on distant stellar targets. Again, this could have a variety of propulsion options but it would require a specific impulse of several 1,000 s. The next demonstrator mission is *Icarus Starfinder* as discussed earlier. If fusion is the chosen method of propulsion, then by now this mission should be using a fusion engine or whatever the engine will be for the main interstellar attempt. It would require a specific impulse of order 10s of thousands of seconds. The actual interstellar attempt, launched by 2100 would be a high performance engine with a specific impulse of order one million seconds. The Pathfinder and starfinder concepts are illustrated in Plates 18 and 19.

As discussed in Chap. 15 other authors have estimated the launch of the first interstellar probe will occur by around the year 2200 AD. This includes one author who looked at velocity trends since the 1800s [10]. Factoring the likely uncertainties associated with the assumptions of these sorts of studies, particularly in relation to assuming linear technological progression, it is likely that the first interstellar probe mission will occur sometime between the year 2100 and 2200. To achieve this will require a significant advance in our knowledge of science or an improvement in the next generation propulsion technology. Given the tremendous scientific advances made in the twentieth century, it at least does not seem unreasonable to think that such a technology leap may in fact occur.

We wait for the day when such a recommendation is finally put into practice. To help us prepare for that day, perhaps a Goal Directed Pathway is appropriate; one that is both bold (no failure of nerve) and an enriched idea (no failure of imagination). A statement that directs us towards that final achievement and see the eventual expansion of our species into the greater Cosmos:

> This human civilization should set out to achieve, by the end of the twenty-first century, the goal of launching the first unmanned interstellar probe towards the solar system of another star. This probe will pass beyond the termination shock, solar heliosphere, through the Kuiper belt and the Ort Cloud and continue on to the Interstellar Medium to arrive at a star system several light years away, returning scientific data in less than a century after launch, on the discoveries made, back to the place from which the robotic ambassador had originated – the planet Earth.

17.6 Problems

17.1. Familiarize yourself with the NASA BPP and NASA IAC technical literature. You should also read about organizations such as the IAF, TZF and BIS. Get a feel for the current research areas that are topical. Read up on the progress of Project Icarus.

17.2. Think about the likely advances in space exploration over the next 10, 50 and 200 years. What key technological breakthroughs do you think are likely to revolutionize space travel? What key events would you like to see happen? Describe a technological roadmap for how your own vision can be accomplished.

17.3. Design your own concept for an interstellar probe including configuration layout and estimated performance calculations. Publish it in a Journal.

References

1. Long, KF, (2009) Fusion, Antimatter & the Space Drive: Charting a Path to the Stars. Presented at 59th International Congress, Glasgow, 3rd October 2008. JBIS, Vol 62, No 3.
2. Johnson, L et al., (2005) A Strategic Roadmap to Centauri, JBIS, 58, 9/10.
3. Schmidt, GR & MJ Patterson (2010) In-Space Propulsion Technologies for the Flexible Path Exploration Strategy. Presented 61st IAC Prague. IAC–10.C4.6.2.
4. Maccone, C (2008) 'Focal' Probe to 550 To 1000 AU: A Status Review, JBIS, 61, 8.
5. Bond A & A Martin (1986) Project Daedalus Reviewed, JBIS, 39, 9.
6. Long, KF & R Obousy et al., (2010) Project Icarus: son of Daedalus – Flying Closer to Another Star, JBIS, 62, 11/12, pp403–414.
7. Bond, A et al., (1978) Project Daedalus – the Final Report on the BIS Starship Study, JBIS.
8. Long, KF & R Obousy et al., (2011) Project Icarus: Exploring the Interstellar Roadmap Using the Icarus Pathfinder Probe and Icarus Starfinder Probe Concepts, Submitted to JBIS January 2011.
9. Jaffe, LD et al., (1980) An Interstellar Precursor Mission, JBIS, Vol. 33, pp.3–26.
10. Cassenti, BN, (1982) A Comparison of Interstellar Propulsion Methods, JBIS, Vol.35, pp116–124.

Chapter 18
From Imagination to Reality

When I was at school (in the 1950s) science and art did not mix; you were supposed to take one or the other. I didn't agree, and have spent most of my life since bringing them together, through my art. Depicting scenes of outer space, planetary landscapes and interstellar vehicles is, to my mind, the perfect way to go from 'Imagination to Reality'. Even scientists and engineers need interpretation and inspiration!

David Hardy

In this final chapter of the book we explore the fantastic imagery that is space exploration. Although each image is accompanied by a brief caption, the images speak for themselves and cover the broad spectrum of earth to orbit transport, interplanetary spacecraft and interstellar vessels. The range of possibilities invented by physicists and engineers is boundless and the imagination holds no limit. For as long as mankind can use his hands, his eyes and his mind, he can paint, sketch or click a scene that would appear to the casual observer one of science fiction.

One could ask why imagined imagery is so important to space exploration when instead photographs can be taken. Well, for interstellar the reason is compounded by the vast distances of space. The destinations are as yet too far away to reach and the spacecraft have not yet been invented. But this doesn't stop our imagination from dreaming up concepts for how this can one day be accomplished. All of the spacecraft images shown are either actual vehicles that have been built or based upon credible design concepts. In principle, all are possible someday in our distant future, although some have their technical engineering issues to be solved to make them workable.

In the early part of the twentieth century the American artist Chesley Bonestell helped to create images of other worlds and using exotic rockets. Others such as the British space artists Ralph Smith and David Hardy have followed his example. What they were doing was pioneering the future, because before mankind can attempt to reach a destination with technology, he must first believe that such a thing is at least feasible. And this is the power that space art, specifically, has over the world if this is yielded properly. It has the power to inspire the world in a future vision for our species, an optimistic one where we are at peace, science advances and the opportunities for human expansion are abound.

K.F. Long, *Deep Space Propulsion: A Roadmap to Interstellar Flight*,
DOI 10.1007/978-1-4614-0607-5_18, © Springer Science+Business Media, LLC 2012

Evaluating the history of humankind it appears we have started from a position of being bounded in the nutshell that is Earth, but yet can count ourselves the Kings of infinite space. The Russian rocket pioneer Konstantin Tsiolkovsky once suggested that that we need to free ourselves of the cradle that is Earth. By imagining these machines, there is the possibility that someone, somewhere will build them and then we can share all of our science and culture with other races that may exist in the universe. The cosmos beckons us to join it, only courage and imagination will get us there.

Plate 1 The Sombrero galaxy. This composite image of the Sombrero galaxy in M104 as taken by the NASA Spitzer and Hubble Space Telescopes. This is a galaxy located 50 million light years away and contains up to 300 billion stars. The bright nucleus outshines the rest of the galaxy due to the presence of the high density of stars. Copyright: (@NASA/courtesy of nasaimages.org)

Plate 2 The British interplanetary Society Retro-rocket. This is an artists impression of the 1930s British Interplanetary Society Lunar ship on the surface of the Moon. But the Moon is depicted not how we know it today, but how we thought it was prior to the first manned landing, based on the artwork of the space artist Chesley Bonestell. Copyright: (@David A.Hardy/www.astroart.org)

Plate 3 The US Space
Transportation System. The
NASA Space Shuttle has
served the U.S. space industry
in over 100 missions since
the first maiden flight in 1981.
As well as playing a key role
in maintaining Earth orbiting
satellites (including the
famous Hubble Space
Telescope repair missions).
It has been an essential
workshorse in the
construction of the
International Space Station.
Perhaps its most important
contribution has been in
maintaining a human
spaceflight capability. The
Space Shuttle fleet have now
been retired imminently.
Copyright: (@NASA/
courtesy of nasaimages.org)

Plate 4 The US X-33 design. This was an unmanned sub-scale technology demonstrator for the VentureStar project, a planned next-generation reusable launch vehicle. It was developed by Lockheed Martin and included novel linear aerospike engines. It was cancelled in 2001. Copyright: (@NASA/courtesy of nasaimages.org)

Plate 5 The pioneer spacecraft. Artist illustration of a NASA Pioneer spacecraft heading into interstellar space. Launched in the 1970s both Pioneer 10 and 11 were sent on trajectories that took them further than all of the planets and out of the solar system. At their low cruise speeds they would take many thousands of years to reach the nearest stars. Copyright: (@NASA/courtesy of nasaimages.org)

Plate 6 The Voyager spacecraft. Artist illustration of one of the twin NASA Voyager spacecraft launched in the 1970s and sent out to all of the gas giant planets. They then continued outwards to interstellar space. Copyright: (@NASA/courtesy of nasaimages.org)

Plate 7 The Cassini-Huygens spacecraft. Both the NASA Cassini orbiter and ESA Huygens probe are shown here in preparations for launch. Launched in 1997 it was one of the tallest (6.8 m) and most massive (5.6 tons) spacecraft ever launched at the time. Among many accomplishments the orbiter successfully deployed the probe into the atmosphere of the Saturnian Moon Titan atmosphere of the Saturnian moon Titan. Copyright: (@NASA/courtesy of nasaimages.org)

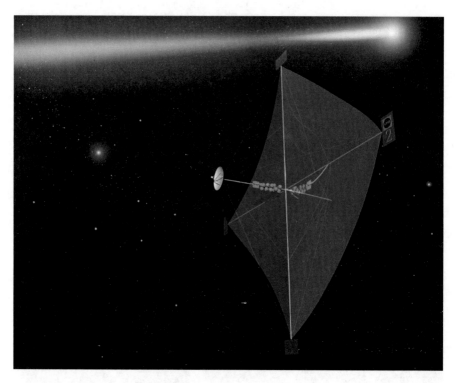

Plate 8 A solar sail. Artist image of a solar sail spacecraft being driven by photons emitted from the Sun. This technology holds the key to full exploration of the solar system. Copyright: (@NASA/courtesy of nasaimages.org)

Plate 9 Enzmann starships at Jupiter. The artist David Hardy was commissioned to paint this picture during 2010, based on an original picture painted in the 1970s. It shows two Starships in orbit of the gas giant Jupiter. Copyright: (@David A.Hardy/www.astroart.org)

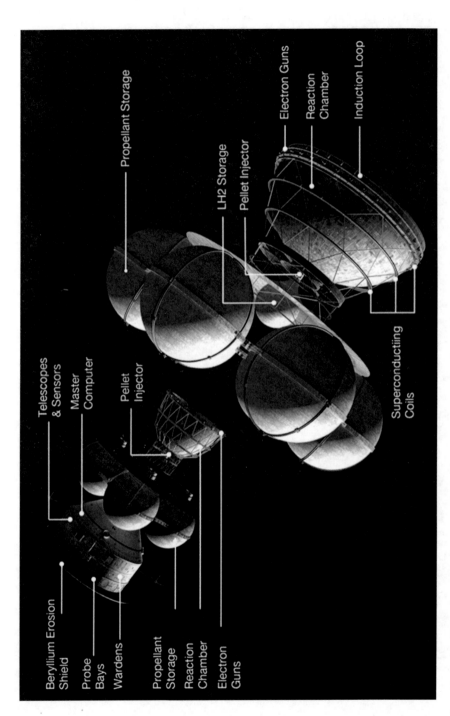

Plate 10 Project Daedalus. This Artist impression shows the two separate engine stages clearly along with the position of the propellant tanks, payload bay and particle shield. Copyright: (@Adrian Mann)

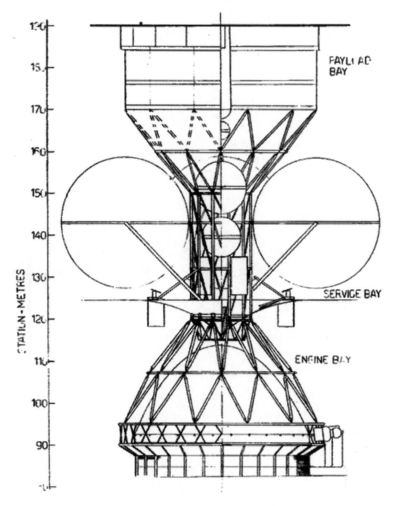

Plate 11 Project Daedalus. This engineering schematic shows the second stage engine configuration with the payload bay. Copyright: (@The British Interplanetary Society)

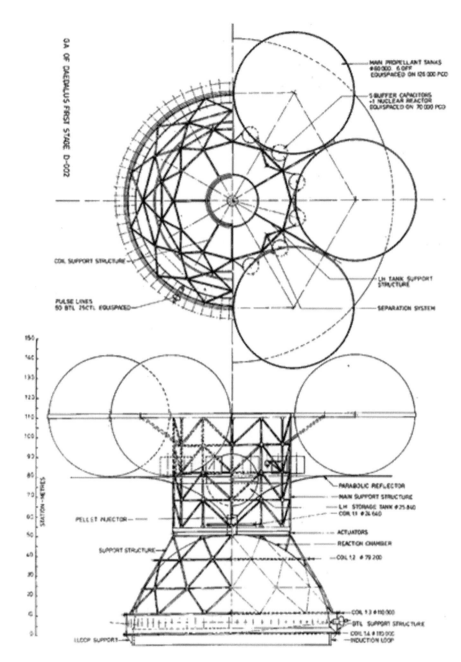

Plate 12 Project Daedalus. This engineering schematic shows the first stage engine configuration. Copyright: (@The British Interplanetary Society)

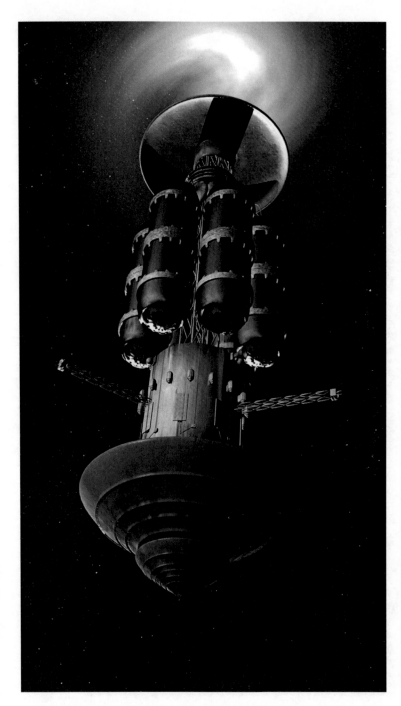

Plate 13   Project Longshot. This artist illustration shows what a nuclear fusion based Longshot spacecraft would look like as it departs the vicinity of our solar system and heads towards Alpha Centauri. Copyright: (@Christian Darkin/Anachronistic)

Plate 14 Project Vista. This artist illustration shows what a nuclear fusion based Vista spacecraft would look like as it transports around our solar system stopping off to observe the various Moons and Planets. Copyright: (@Christian Darkin/Anachronistic)

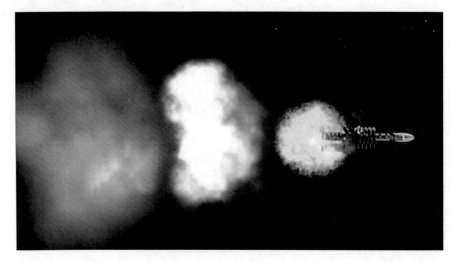

Plate 15 External nuclear pulse. This artists concept for a NASA version of a pulsed fusion propulsion concept similar to Project Orion in the 1960s. In this image the vehicle is moving through the orbit of the gas giant Jupiter. Copyright: (@NASA/courtesy of nasaimages.org)

Plate 16 External nuclear pulse. This artists illustration shows what an Orion type starship might look like in flight leaving a trail of multiple detonations behind it as it gathers velocity. Copyright: (@Christian Darkin/Anachronistic)

Plate 17 Bussard interstellar Ramjet. This artist illustration shows the relativistic nuclear fusion powered starship in motion through the galaxy using interstellar Hydrogen as its main fuel. Copyright: (@Adrian Mann)

Plate 18 Project Icarus Pathfinder probe. This artist illustration shows the Icarus Pathfinder probe moving through the orbit of the planet Neptune, on its way to a 1,000 Astronomical Unit mission using either plasma drive, nuclear thermal or nuclear electric engines. The mission is a proposal for a technology demonstrator, testing out key systems such as robotic arm repairs, optical laser communications and long distance observations using the gravitational lensing point. Copyright: (@Adrian Mann/Project Icarus)

Plate 19 Project Icarus Starfinder probe. This artist illustration shows the Icarus Starfinder probe moving further from the Sun as it heads towards its 10,000 Astronomical Unit mission distance, deep inside the Oort Cloud. It is a proposal for a full technology demonstrator using a nuclear fusion based engine, paving the roadmap for an interstellar mission. Copyright: (@Adrian Mann/Project Icarus)

Plate 20 Project Icarus official logo. It shows the two founding organizations affiliated with Project Icarus, represented by the sub-logos either side, namely The British Interplanetary Society and the Tau Zero Foundation", both non-profits. Project Icarus is now managed by the non-profit Icarus Interstellar. Project Icarus aims to carry on from Project Daedalus (*shown in the centre*) in not only showing that interstellar travel is feasible in theory but that is also practically possible in theory, and define the roadmap to launch. The black wing ship framing the inner part of the logo, sybolizes Icarus flying closer to another star depicted at the top of the logo. The squares and rounds complete the feather of the wings and they do fly closer to that target star. These items symbolize spaceships and spacecraft, projects and initiatives of various kinds, the efforts we make to make it happen, whether actual, tangible (*boxes*, the *square* being the symbol of matter), or theoretical (*round* = spirit). Copyright: (@Alexandre Szames/Antigravite/Project Icarus)

Epilogue

Progress is not made by giving in to the impossible.

Marc G. Millis

A Personal View for the Future

One of the motivations for writing this book was to learn about the subject of interstellar propulsion. Like most people interested in the world outside of their own backyard, I wanted to know if an interstellar probe could ever really be built, how and when. This road has taken me on a journey, from the chemical propulsion systems that propel vehicles such as the Saturn V and the space shuttle into LEO, to the romantic visions of the interstellar ramjet. Having now read about these different propulsion concepts and thought a little about their potential, what is my own view on answering the questions that I initially set out to address? Let us consider some of the options in turn.

We could build vast spacecraft many times the size of oceangoing cruise ships and send people out into the Solar System on voyages that could take literally centuries to millennia. However, our species is neither in a position technologically to assemble such vehicles today or indeed to survive such a voyage should we succeed in launching it. We are therefore driven to much faster and shorter duration methods of reaching the stars. Chemical propulsion systems are clearly limited in their reach and will only get us into Earth orbit and the Moon. We could build spacecraft to take us onto Mars using chemical rockets, but it would be an expensive and a massive enterprise to use this approach. Electric propulsion is ideal for Earth-orbiting satellites and for sending probes to the planets within the inner Solar System, but no further. It can be combined with nuclear reactor technology, however, and a nuclear-electric engine clearly has the potential to take our robotic probes hundreds of AU from our Sun.

No sufficient reason has been given by governments or space agency managers for not embarking on such missions to date. We should have had probes in the

K.F. Long, *Deep Space Propulsion: A Roadmap to Interstellar Flight*,
DOI 10.1007/978-1-4614-0607-5, © Springer Science+Business Media, LLC 2012

Kuiper Belt yesterday. The public distrust of nuclear materials is understandable but misguided. The potential risk of a spacecraft carrying a radioactive material blowing up on the launch pad is very low, compared to the amount of damage that is done to our atmosphere every day from burning fossil fuels. Space technology pioneers new inventions, which improves the life of humans and our understanding of the planet. Denying mission planners the use of nuclear propulsion technology is like sending a boxer into a ring with his hands tied behind his back. We need to free up that technology for the use of space missions, and we can do a whole lot more and better.

The same can be said for external nuclear pulse propulsion and projects such as Orion. Many scientists who evaluated the Project Orion design concluded that it could be built yesterday. But of course it won't be due to the unpopular nature of the propellant. But even Carl Sagan in his wonderful *Cosmos* television documentary said that he couldn't think of a better use for the world's stockpile of nuclear weapons. It would appear that the only time Project Orion will see the light of day is when that large asteroid comes a calling because no other propulsion system could be built within, say a decade, and get to a distant object in a rapid time in order to deploy a deflection mechanism. This is a fact. This author does not propose that Orion could be used to get us out of the atmosphere and agrees with the objectors that the radiation risks may be too great. However, launching such a vehicle from the orbit of the outer planets poses no risk to us, and any radiation generated would be saturated by the background radiation in any case.

Solar sail and microwave sail technology have the potential to meet the performance for an interplanetary mission and beyond in the near term, and investment in this technology should be increased. However, stretching this technology to an interstellar mission is a real challenge. It is by no means certain that should we be successful in launching such a mission that it would ever reach its destination, although we should still make the attempt even if the payload mass is likely to be small. However, research into this technology is at an early phase, and it is possible that with the development of nanomaterials in particular that clipper missions to the stars may yet be in the cards. The building of large orbital laser systems can enhance the performance of a sail, but this technology would seem to be many decades away, if not longer. We should keep our options open.

People have complained for years that the dream of nuclear fusion on Earth is always around the corner. However, we live in unprecedented times. Prototype experimental reactors now exist or are planned in the United States, France, Britain, Japan and Russia. These demonstrators are exploring a variety of approaches, from inertial confinement to magnetic confinement to advanced schemes such as fast ignition and shock ignition. The amount of money that has gone into this research in recent years means that inertial fusion energy research is being taken seriously, and governments are placing a lot of faith in scientists to deliver this technology. This faith is justified, because scientists working in this field believe in the dream, too. The fulfillment of fusion energy on Earth is around the corner, this time, and in this author's opinion will be accomplished within 10 years. Given this statement, the clear potential for fusion-powered spacecraft awaits us. Very few other propulsion systems offer the performance of fusion-based engines, and this is the way we will get to the stars. Project Daedalus demonstrated that it was feasible in theory and Project Icarus is attempting to demonstrate it is practical in theory. Such large interstellar systems' engineering studies should also be undertaken by national space agencies, so that 'game-changing' technology can be increased to an appropriate readiness level by an appropriate direction of program investment.

It would be wonderful if we could build something like the interstellar ramjet, which has the potential to take us to the stars within a matter of a few years and to go at speeds approaching that of light. However, the interstellar ramjet currently has too many technical problems to make it likely in the near term, but something like it may be built in the centuries ahead of us. Like the nomad exploring the plains of Earth for new lands, we too must some day learn to adapt to a philosophy of living within our means, and this includes the exploration of space. Carrying huge tanks of propellant fuel along with the spacecraft may be practical for a short-term mission, but it is not part of a sustainable future that aims to colonize the surrounding star systems and further. Yes we may one day be able to mine the gas giants or the Moon for helium-3 or even the Oort Cloud comets for deuterium, but eventually this supply must also be exhausted. In situ en-route energy sources should be the ultimate target for any galaxy aspiring species.

The discovery of antimatter was a fantastic intellectual human achievement, and there is no doubt that the potential near speed of light performance of an antimatter-based fuel for space propulsion outweighs even that for fusion. However, generating and trapping sufficient quantities to make a pure antimatter rocket would appear to have huge technical challenges that may take some time to solve. But the application of antimatter particles to catalyzing fission and fusion reactions for decreased driver requirements and increased energy gain as proposed for concepts such as ICAN-II and AIMStar would appear to be the way forward for a future deep space propulsion industry. In theory we can manufacture small quantities now, which are sufficient to catalyze reactions for a short interplanetary space mission, providing we can crack the fusion problem, too. In the coming decades, with improved manufacturing techniques and the presence of a commercial fusion industry, the coupling of antimatter and fusion technology will allow for spacecraft performance that allows us to send robotic probes to the nearest stars within a few decades' duration. When we do this and receive the first pictures of the Alpha Centauri system, we can be proud of our achievements, born from theoretical physics and utilized for extreme engineering. For in fact we would have achieved a major milestone for our species and the fulfillment of a dream long held by people and visionaries such as Sir Arthur C. Clarke.

It is a personal observation that interstellar research is one of the few topics in science that is not properly resourced, managed or coordinated. This is largely a result of the requirement for long term speculative thinking that interstellar flight necessitates, a characteristic sadly lacking in the modern world. Most of the progress in this subject has been made by volunteers or by people working on a problem from the sidelines of their main research subject. This situation needs to change if we are to ever to become a spacefaring civilization and join the community of worlds that may possibly exist in this vast universe.

Governments should follow the path to commercialize the space industry. The job of launching astronauts to LEO, to the Moon and even to Mars should be handed over to companies that have a clear business goal for doing so and can produce opportunities and technologies that will benefit all of humankind and spur new technological innovation. Government space agencies should instead do what they do best – pioneer the furthest frontiers of space and send robots to all of the planets in our Solar System, into the Kuiper Belt, the Oort Cloud and beyond. The opportunity for sending a robotic probe out into the depths of space between the stars and onto other solar systems is really only decades away, if we choose to make the attempt. How shall we greet this opportunity? Shall we pull back at the sheer scale of the challenge before us and regress into a stagnant race? This is the symptom of either a failure of nerve or a failure of imagination. Alternatively, shall we have the courage to design and build these wonderful machines to extend the reach and knowledge of our own species? The current century and the next will show which path we choose to take.

Finally, in looking at the history and future of space exploration while writing this book a clear distinction has formed in my mind as to what is possible from a human and robotic perspective in the coming decades to centuries ahead. I completely support the peaceful human exploration of space and in fact believe it is a necessary path for us to take if we are to grow or even survive as a species in the long term. Despite this, the advantages of robotic probes cannot be denied, unconstrained by slow expansion speeds. Human colonization would likely require World Ships, unless breakthroughs in fundamental physics are made such that the speed of light barrier can be surpassed. At the current rate of technological growth, an exponentially increasing trend, it would appear that the prospects for robotic missions to the stars are much closer than many predict today. If this potential comes true then the 'robotic wave front' that we send out from our Solar System should have visited many star systems within a matter of centuries. Given the possible existence of other civilizations inhabiting suitable worlds elsewhere in the galaxy, the opportunity for another species to also embark on a similar robotic expansion program must also exist. This raises the obvious question of where are they?

> "Space exploration must be carried out in a way so as to reduce, not aggravate, tensions in human society". William K. Harmann, Out of the Cradle, Workman Publishing Company, 1984

Appendix A

Scientific Notation

$10^{-15} = 0.000000000000001$ (femto)
$10^{-12} = 0.000000000001$ (pico)
$10^{-9} = 0.000000001 = 1$ billionth (nano)
$10^{-6} = 0.000001 = 1$ millionth (micro)
$10^{-3} = 0.001 = 1$ thousandth (milli)
$10^{-2} = 0.01 = 1$ hundredth (centi)
$10^{-1} = 0.1 = 1$ tenth (deci)
$10^{0} = 1 =$ one
$10^{1} = 10 =$ ten (deca)
$10^{2} = 100 = 1$ hundred (hecto)
$10^{3} = 1,000 = 1$ thousand (kilo)
$10^{6} = 1,000,000 = 1$ million (Mega)
$10^{9} = 1,000,000,000 = 1$ billion (Giga)
$10^{12} = 1,000,000,000,000 = 1,000$ billion (Tera)
$10^{15} = 1,000,000,000,000,000 = 1,000,000$ billion (Peta)

Appendix B

Physical Constants

Astronomical Unit (AU)	1.496×10^{11} m
Light year (ly)	9.46×10^{15} m $= 63{,}240$ AU
Parsec (pc)	3.086×10^{16} m $= 3.26$ ly
Year (y)	3.154×10^7 s
Vacuum speed of light (c)	2.998×10^8 m/s
Earth mass (M_e)	5.974×10^{24} kg
Earth radius (R_e)	6.378×10^6 m
Lunar mass (M_m)	7.348×10^{22} kg
Lunar radius (R_m)	1.738×10^6 m
Solar mass (M_s)	1.989×10^{30} kg
Solar radius (R_s)	6.960×10^8 m
Avogadro number (N_A)	6.022×10^{23} mol^{-1}
Gravitational constant (G)	6.674×10^{-11} m^{3}kg^{-1}s^{-2}
Boltzmann constant (k_B)	1.381×10^{-23} JK^{-1}
Electron Volt (eV)	1.602×10^{-19} J
Amu (u)	1.660×10^{-27} kg
Sea level earth gravity (g)	9.81 ms^{-2}
Pi	3.14159
Exp(1)	2.71828

Appendix C

A Timeline of Some Key Events Relating to Interstellar Research

1657 de Bergerac publishes *Voyage Dans La Lune*

1867 Verne publishes *From the Earth to the Moon*

1903 Tziolkovsky publishes "The Exploration of Cosmic Space By Means of Reaction Devices"

1905 Einstein publishes his Special Theory of Relativity

1915 Einstein publishes his General Theory of Relativity. U. S. National Advisory Committee for Aeronautics (NACA) formed

1919 Goddard publishes "A Method of Reaching Extreme Altitudes"

1923 Oberth publishes "The Rocket into Planetary Space"

1927 Formation of the German Society for Space Travel

1928 Dirac predicts the existence of antimatter particles

1930 Formation of the American Interplanetary Society

1932 Anti-electron discovered in the laboratory

1933 Formation of the British Interplanetary Society

1934 Journal of British Interplanetary Society created

1935 Einstein and Rosen publish first paper on wormholes

1944 First V2 rocket to leave the atmosphere of Earth

1946 Ulam first proposes external nuclear pulse propulsion. Russian Design Bureau formed under Korolev (Energia)

1950 Clarke proposes dumbbell-shaped spacecraft in his book on interplanetary flight. First International Astronautical Congress

1951 Formation of International Astronautical Federation

1952 Shepherd publishes first academic JBIS paper addressing interstellar travel

1955 Project Rover launched to investigate nuclear thermal propulsion. Ulam and Everett extend proposal for external nuclear pulse propulsion (Helios). Antiproton discovered in the laboratory. Lawson proposes fusion triple product criteria

1956 Antineutron discovered in the laboratory

1958 Original General Atomics study for Project Orion. U. S. National Astronautics and Space Administration (NASA) formed

1960 Bussard proposes the interstellar ramjet concept

(continued)

1963 Spencer interstellar travel paper published. Sagan publishes on time dilation effect associated with relativistic interstellar travel

1966 First test of nuclear thermal NERVA engine. Original series of *Star Trek* first shown on television

1968 Dyson publishes his article on the economics of interstellar travel. The film *2001 A Space Odyssey* premieres

1969 First manned landing on the Moon

1970 Winterberg publishes his paper on electron beam ICF propulsion. Poul Anderson publishes his *Tau Zero* novel exploring the Bussard ramjet concept

1972 Launch of Pioneer 10 space probe. Nuckolls proposes laser ICF requirements

1973 Launch of Pioneer 11 space probe. Winterberg proposes use of electron driver ICF for propulsion. Stein publishes his program for interstellar flight based on Enzmann starships. British Rail Space Vehicle patent issued. Project Daedalus study initiated

1974 Niven and Pournelle discuss laser-beamed propulsion in novel *The Mote in God's Eye*. Bond proposes Ram Augmented Interstellar Ramjet concept

1975 European Space Agency founded

1977 Launch of Voyager 1 and 2 space probes. Jaffe initiates study for Interstellar Precursor Probe (IPP) mission

1978 Project Daedalus study published

1980 Freitas publishes self-REproducing PRObe (REPRO). Jaffe publishes paper on Thousand Astronomical Unit (TAU) mission. Sagan discusses feasibility of interstellar travel in *Cosmos* documentary and book

1982 Cassenti publishes paper on maximum velocity trends to predict first interstellar launch

1984 Bond and Martin publish first academic papers on World Ships

1985 Forward proposes the microwave beam-driven Starwisp concept

1987 Orth publishes Vehicle for Interplanetary Space Transport Applications (VISTA)

1988 LONGSHOT study report published. Morris and Thorne publish extensive wormhole calculations

1989 Mallove and Matloff publish *The Starflight Handbook*, the first interstellar academic text. Launch of Galileo space probe

1990 Launch of Ulysses space probe

1993 Solem first proposes the Medusa sail concept

1994 Alcubierre publishes first academic paper on warp drive

1995 Anderson publishes the NASA Horizon Mission Methodology. Visser publishes the first academic text on Lorentzian wormholes

1996 NASA Breakthrough Propulsion Physics Project launched

1997 Millis publishes review of space drive concepts. Launch of Cassini-Huygens space probe. First studies for ICAN-II published

1998 NASA Institute for Advanced Concepts launched. Launch of Deep Space 1 probe

1999 Landis and Forward propose interstellar solar sail strawman mission. McNutt proposes Realistic Interstellar Explorer (RIE) mission. Gaidos proposes Antimatter Initiated Microfusion Starship (AIMStar)

2000 JPL investigates the Interstellar Probe (ISP) study. Benford demonstrates first microwave-driven experiment in the laboratory

2001 Discovery II concept published. Report on first laboratory-based beam driven sail experiments

2002 NASA Breakthrough Propulsion Physics Project closes. First confirmed existence of an extrasolar planet orbiting Gamma Cephei

(continued)

2003 McNutt proposes Innovative Interstellar Explorer (IIE) mission. NASA study for Project
      Prometheus published. NASA launches nuclear electric based Project Prometheus study

2004 Centauri Dreams blog forum launched

2005 Attempted launch of Cosmos 1 solar sail mission. ESA study for Interstellar Heliopause
      Probe. Millis forms Tau Zero Foundation (TZF)

2006 Launch of New Horizons space probe mission to Pluto

2007 NASA Institute for Advanced Concepts closes. Maccone proposes mission to gravitational
      focal point. First credible interstellar discussion forum. Launch of NASA Dawn space
      probe mission to dwarf planets. First dedicated conference on the warp drive at BIS HQ
      in London

2008 Obousy proposes mechanism for creation of warp drive effect. Launch of NASA IBEX
      space probe mission

2009 Millis and Davies publish AIAA book on *Frontiers of Propulsion Science*. Long and
      Obousy launches Project Icarus starship study. TZF hosts first interstellar travel session
      at UK Charterhouse conference.

2010 Millis uses energy trends to predict first interstellar launch. First Earth-like solar system
      discovered. Launch of world's first solar sail spacecraft, the Japanese Ikaros. Launch of
      NASA Nanosail-D solar sail mission. McNutt Decadal Survey White paper proposal for
      an interstellar probe mission

Appendix D

Approximate Spacecraft Performance Data

Spacecraft	Propellant	Total mass (tons)	Propellant mass (tons)	Payload mass (tons)	Payload dry mass Fraction (%)	Approximate mass ratio	Exhaust velocity (km/s)	Specific impulse (s)	Cruise velocity (km/s) [%c]	Mission duration (days/weeks/ years)	Mission distance (AU or ly)
Designs											
Dawn	Xenon	1.25	0.425	–	–	1.5	21	2,100	10.73 [0.0036]	8+ y	Vesta/Ceres
IBEX	Hydrazine	0.213	0.107	0.026	24.5	2.0	–	–	–	3+ y	Interstellar boundary
Pioneer 10	Hydrazine	0.258	0.006	0.028	11.1	1.0	–	–	12.3 [0.0041]	31 y	*80 AU
Pioneer 11	Hydrazine	0.259	0.006	0.029	11.5	1.0	–	–	11.9 [0.0039]	22 y	82 AU
Ulysses	Hydrazine	0.370	0.033	0.055	16.3	1.1	–	–	15.4 [0.0051]	19 y	High solar latitude
Deep Space-1	Xenon/ Hydrazine	0.374	0.074	0.3	1.0	1.2	35	3,500	13.8 [0.0046]	3 y	Asteroid/comet
New horizons	Hydrazine	0.478	0.093	0.028	7.3	1.2	–	–	21 [0.007]	9+ y	Pluto/55 AU
Voyager 1	Hydrazine	0.8	0.078	0.104	14.4	1.1	–	–	17.1 [0.006]	34 y	Outer planets/ 110 AU
Voyager 2	Hydrazine	0.8	0.078	0.104	14.4	1.1	–	–	15.6 [0.0052]	34 y	Outer planets/ 100 AU
Galileo	Hydrazine/N Tetroxide	2.38	0.925	0.118	8.1	1.6	–	–	15.4 [0.0051]	14 y	Jupiter
Cassini-H	Hydrazine	5.712	3.212	2.5	1.0	2.3	–	–	31.6 [0.01]	14+ y	Saturn/9.5 AU
Concepts											
VASIMR	H, He	600	476	60	48.4	4.8	300	30,000	120 [0.04]	39 d	1.5 AU/Mars
Icarus Pathfinder	Xenon	25.3	15.5	4.5	22.7	2.6	500	50,000	474 [0.158]	20 y	Pluto/1,000 AU
TAU	Xenon	60	40	1.2	6.0	3	123	12,500	106 [0.03]	50 y	1,000 AU
II Explorer	–	0.516 + p	–	0.035	6.8	–	–	–	37.5 [0.01]	30 y	200 AU
XIP2	DT	65	40	2	8.0	1.9	500	50,000	633 [0.002]	*83 y	10,000 AU
Prometheus	Xenon	38	12	1.5	5.8	1.5	60–90	6,000–9,000	34 [0.01]	5 y	Jupiter

Discovery-II	DHe3	1,690	11 (H = 861)	172	21.0	2	347-463	35,435-47,205	795-1,058 [0.26-0.35]	118-212	Jupiter/Saturn 4.7-9.6 AU
Interstellar Probe	Solar photons	0.246	0.0	0.15	61.0	1.0	—	—	70 [0.02]	15 y	200 AU
IHP	Solar photons	0.624	—	0.310	49.7	2.0	—	—	37.9 [0.012]	25 y	200 AU
Kuiper Belt laser	Laser photons	0.2	—	0.066	33.0	33	—	—	100 [0.03]	5.3 y	100 AU
Oort Cloud laser	Laser photons	0.2	—	0.066	33.0	33	—	—	3,000 [1]	17.6 y	10,000 AU
Interstellar laser	Laser photons	0.1	—	0.033	33.0	33	—	—	30,000 [10]	42.2 y	4.2 ly
Starwisp flyby	Laser photons	1	0.0	1	100.0	1.0	—	—	33,000 [11]	40 y	4.3 ly
Longshot	DHe3	396	264	30	22.7	3.0 [see Ch.11]	9,810	1,000,000	12,900 [4.3]	100 y	4.3 ly
ICAN-II	DT, U/p⁻	700	362	82	24.3	2	135	13,500	100 [0.03]	30-120 d	1.5 AU Mars
VISTA	DT	6,000	4,165	100	—	3.2	167	15,500	—	145 d	1.5 AU Mars
AIMStar	DHe3, U/p⁻	0.5	0.4	0.1	1.0	0.2	655	66,769	1,048 [0.35]	50 y	10,000 AU
Bussard ramjet	HH	1,000	0.0	1,000	100	1.0	100-1,000	10,000-100,000	299,700 [99.9]	~few y	~10s ly
InterPlan' Orion	A-bombs	4,000	—	200-1,600	—	2.5-20	39	4,000	15-30 [5]	~few w	~2 AU
Adv Plan' Orion	A-bombs	10,000	—	1,300-6,100	—	1.6-7.7	118	12,000	10-100	~few y	~10 AU
Medusa	A-bombs	300	200	100	1.0	3.0	427	43,500	296 [0.1]	32 d (return)	1.5 AU Mars
Daedalus	DHe3	52,670	50,000	450	16.2	40.0	10,000	1,000,000	36,600 [12.2]	50 y	5.9 ly
Interstellar Orion	H-bombs	400,000	—	20,000	—	20.0	9,810	1,000,000	10,000 [3.3-10]	4-15 ly	—
Enzmann	D/D	—	3-12,000,000	—	—	—	10,000	1,000,000	27,000 [9]	10-100s y	—
World Ship (wet)	D/H	1.12×10^{13}	9.04×10^{12}	1.86×10^{12}	86.1	4.9	1,880	191,641	1,500 [0.5]	2,000 y	10 ly
World Ship (dry)	D/H	1.01–2.92×10^{12}	0.82–2.25×10^{12}	1.68–4.81×10^{11}	88.4-71.8	4.9	1,880	191,641	1,500 [0.5]	2,000 y	10 ly

Index

A

AIMStar, spacecraft concept, 229–231
AIS. *See* American Interplanetary Society
Alcubierre, M., 298
Aldrin, E. (Buzz), 79
Alpha Centauri, 17, 32–33, 36, 58, 103–105,
   110, 120, 124, 132, 143, 147, 149,
   150, 162, 198, 211, 223, 231, 241,
   277, 284, 291, 298, 310–311, 316,
   318–322
American interplanetary society (AIS), 53
American rocket society, 53
Anderson, J., 301
Anderson, P., 20, 223, 290
Andrews, D., 157
Andromeda Galaxy, 34, 35, 223
Antigravity, 295
Antimatter, 6–7, 121, 134, 219–233, 243, 244,
   290, 291, 294, 310
   catalyzed fusion, 6, 197, 231, 315
   rocket, 6, 198–199, 227, 291
Ares rockets, 67
Ariane rockets, 65
Armstrong, N., 13, 69, 79
Artificial intelligence, 114, 128, 134, 268,
   275, 301
Asimov, I., 273
Atlas rockets, 63
Atomic bomb rocket. *See* Orion

B

Barnard's Star, 34, 35, 104, 105, 175,
   191, 195–197, 223, 277, 278, 291,
   316, 317

Beamed power spaceship, 4
Behemoth, spacecraft concept, 62
Benford, J., 155, 168, 174
BepiColumbo space mission, 88
Berkey, J., 288
Bethe, H., 100, 178
Biefeld–Brown effect, 296, 297
Big Bang theory, 264
BIS. *See* British interplanetary society
BIS moonship, 52
Black arrow rockets, 64
Black holes, 7, 167, 266–267, 293, 295, 300
Blue streak rockets, 64
Bond, A., 3, 35, 58, 70, 104, 114, 190, 197,
   225, 279, 312–314
Bonestell, C., 52, 288, 289
Boundary program, proposal, 282, 283
Breakthrough propulsion physics (BPP),
   NASA, 8, 19, 291–300, 307, 308, 322
British interplanetary society (BIS), 2, 8, 9,
   50–53, 84, 104, 114, 145, 170, 190,
   215, 288, 289, 299, 311, 313,
   314, 322
British rail space vehicle concept, 188
Brown, T., 296
Buffy (XR190), dwarf planet, 96
Buran, Russian space shuttle, 66
Bush, G.W., 67, 68, 79
Bussard, R., 183, 226

C

Carter, J., 124
Casimir energy, 296
Casimir, H., 296

Cassenti, B.N., 170–171, 225, 276,
    283–284, 322
Cassini–Huygens spacecraft, 91–92,
    126–127, 134
Cayley, G., 40
Centauri dreams, 18
Ceres, dwarf planet, 95, 130
CERN. *See* Conseil Europeen pour la
    Recherche Nucleaire
Challenge mission, proposal, 311
Charon, moon, 95, 130, 147, 149
Chemical rockets, 3, 54, 58, 73, 133, 187, 236
Circular orbit velocity, 28, 29, 31
Clarke, Sir Arthur C., 5, 9–10, 17, 22, 23,
    25, 30, 50, 53, 102, 136, 139, 144,
    158, 197, 203, 207, 268, 270, 287–290,
    293, 312
Cleator, P., 50
Cleaver, V., 50, 53, 145
Cockell, C., 84, 274
Concorde, 41, 69, 238
Conseil Europeen pour la Recherche
    Nucleaire (CERN), 231
Constellation program, NASA, 67, 68
Copernicus, N., 89, 263
CoRoT space telescope, 107
Cosmos 1 spacecraft, 128–129
Crawford, I., 9, 222, 316
Crowl, A., 267, 269, 316
Cyborg, 268

D
Daedalus. *See* Project Daedalus
Darwin, C., 268
Darwin space mission, 107
Da Vinci, L., 40–41, 269
Davis, E., 291
DAWN spacecraft, 130–131
De Bergerac, C., 49
Deep impact space mission, 119, 282
Deep Space 1 spacecraft, 127–128, 131, 148
Delta rockets, 64, 127, 130
Diamandis, P., 319
Discovery II spacecraft concept, 203–205
Discovery I spacecraft concept, 119, 126, 145
Discovery program, NASA, 119, 282
Dixon, D., 190
Drake equation, 113, 115
Drake, F., 113
Dyson, F., 24, 172, 208, 213–215, 283,
    289, 298

Dyson, G., 213
Dyson spheres, 289

E
Earth, the planet, 3, 12, 28, 41, 50, 78, 100,
    117, 140, 156, 177, 210, 220, 247,
    263, 288, 306
Eddington, A., 102, 178, 264, 305
EE. *See* Epsilon Eridani
Einstein, A., 219, 220, 264, 296
Eisenhower, D., 5, 178
Electric propulsion, 3, 4, 127, 139–146, 151,
    212, 296, 310
Enceladus, moon, 91
Enzmann, R., 189
Enzmann starship concept, 189–190
Epsilon Eridani (EE), 104, 106, 107,
    114–115, 174, 195–197, 211,
    223, 316
Eris, dwarf planet, 96
ESA. *See* European space agency
Escape velocity, 28–29, 50, 82, 93–95, 100,
    133, 140, 146–148, 162, 165, 171, 175,
    284
Europa, moon, 16, 90, 91, 109, 125, 203, 282
European space agency (ESA), 65, 70, 72,
    107, 126, 127, 141, 142, 164, 166,
    280, 283
Exhaust velocity, 9, 32, 34, 35, 37, 54–55, 58,
    74, 86, 120, 121, 127, 140–143,
    145–147, 153, 171–172, 184–187, 192,
    194, 200, 204, 205, 207–208, 211, 212,
    217, 223, 228, 230, 231, 239, 245, 246,
    248–250, 254, 258–261, 291
Exoplanets, 106, 111, 263, 282, 321
Explorer 1, 306
Extra solar planets, 99, 103, 109, 111, 313

F
Falcon 9 rocket, 72–73
Fast ignition scheme, 182
Fearn, D., 143
Fermi, E., 113, 191
Fermilab, 231
Fermi Paradox, 113, 114, 191, 197
Fission rockets, 5, 145
Flagship program, NASA, 119, 257, 282
Focal point mission, 311
Fogg, M., 87, 109, 315
Forward, R., 9, 37, 58, 170, 172, 260, 290

Fusion, 5, 15, 55, 89, 100, 119, 143, 158, 177, 207, 223, 241, 277, 290, 310
  enhanced auxiliary laser thruster, 171
  rockets, 5, 6

G
Gagarin, Y., 12
Galea, P., 316
Galilei, G., 89
Galileo spacecraft, 110, 125, 147
Gemini missions, 64
Generation starship, 3, 7
Geostationary orbit, 288, 289
German rocket society, 50, 62
Giffard, H., 39–40
Gilster, P., 9, 117, 315
Goddard, R., 1, 27, 39, 50, 54, 61, 140, 293
Goldin, D., 119
Gravitational lensing point, 318, 322

H
Hagerty, J., 303
Hardy, D., 52, 190, 288, 289
Heim theory, 295, 296
Heliopause, 117, 121, 123, 124, 132, 146, 147, 149–151, 163–166
Helios, spacecraft concept, 122, 157, 163
Helium–3 mining, 94
Hibernation starship, 3
Higgs Boson particle, 267
High performance antiproton trap (HiPART), 228
HiPART. See High performance antiproton trap
HiPER proposal, 182, 183, 313
Hohmann transfer orbit, 31
Horizon mission methodology, NASA, 300–302
Horizontal take off and landing (HOTOL), 70, 71
HOTOL. See Horizontal take off and landing
Hubble, E., 264
Hubble space telescope, 95, 101
Hyakutake, C., 126
Hyams, P., 102
Hydra, moon, 95, 130

I
IAC. See International astronautical congress
IAF. See International astronautical foundation

ICAN–II, spacecraft concept, 231–233
Icarus. See Project Icarus
ICARUS, concept (USSR), 312–318, 322
Icarus Pathfinder spacecraft concept, 96–97, 283, 318, 321–322
Icarus Starfinder spacecraft concept, 97, 283, 318, 322
ICF. See Inertial confinement fusion
Ideal rocket equation, 35, 57, 59, 237, 245
IKAROS space mission, JAXA, 129, 163, 164
Inertial confinement fusion (ICF), 179–182, 184, 188, 191, 193, 201, 202, 205, 217, 231, 241, 251, 252, 255
Innovative interstellar explorer, proposal, 147, 149–151, 166, 315, 321
Institute for advanced concepts, NASA, 8, 293
International astronautical congress (IAC), 306, 310
International astronautical foundation (IAF), 306
International space station (ISS), 13, 27, 35, 68, 71, 74, 78–79, 280, 295, 305
Interstellar boundary explorer (IBEX), proposal, 72, 131–132
Interstellar heliopause probe (IHP), mission proposal, 164–166
Interstellar heliosphere probe, proposal, 165, 166
Interstellar medium, 96, 123, 131, 133, 135, 147, 150, 156, 164, 167, 199, 227, 245, 266, 267, 315, 322
Interstellar precursor probe (IPP) concept, 119, 146–147, 244
Interstellar probe (ISP), mission proposal, 3, 36, 73, 96, 99, 103, 104, 107, 119, 120, 125, 129, 166–167, 190, 200–201, 212, 217, 238, 241, 244, 265–268, 277–284, 308, 312, 314, 317–319, 321, 322
Interstellar Ramjet, 5, 10, 171, 183, 219–233, 243, 244
Io, moon, 16, 90, 91, 124, 125
Ion drive propulsion, 134, 301, 315
ISS. See International space station
ITER facility, 180, 183, 313

J
Jaffe, L.D, 140, 146, 171, 235, 321
Jet engine, 41–44, 47, 54, 70, 73
Johnson, L., 163, 309–310

Joint European Torus (JET) facility, 180, 227, 313
Juno space mission, 119, 140, 282
Jupiter rockets, 63, 67
Jupiter, the planet, 20, 29, 31, 63, 89–91, 93–94, 101, 102, 105, 106, 109, 119, 122–127, 130, 133, 136, 140, 146, 150–152, 161, 163, 193, 199, 203, 204, 265, 276, 278, 282

K
Kaluza–Klein theory, 299
Kardashev civilization, 12
Kardashev, N., 12
Kennedy, A., 19
Kennedy, J.F., 12, 79, 284
Kepler's laws of planetary motion, 29
Kepler space telescope, 107
Kubrick, S., 5, 203
Kuiper belt, 16, 32, 35, 77, 85, 93, 95, 96, 106, 117, 119–121, 123, 129, 147, 149, 163, 167, 170, 266, 276, 282, 310, 311, 322

L
Lagrange points, 68, 129, 162
Landis, G., 170, 172, 173
Laser Mégajoule (LMJ) facility, 181
Laser powered Ramjet concept, 170–171
Laser sail spacecraft, 4
Lasser, D., 53
Lawson, J., 179
Lazaridis, M., 309
Lewis, C.S., 53
Lewis, J.S., 93
Light sail spacecraft, 129, 158, 164
Lilienthal, O., 40
Lindbergh, C., 11, 41, 319
Lindl, J., 180
Long, K.F., 299, 302, 306, 313, 315, 318
Longshot, spacecraft concept, 198–201
Low, A., 50
Lunar crater observation and sensing satellite (LCROSS) spacecraft, 80
Lunar Lander, Apollo, 241, 288
Lunar prospector space mission, 119

M
Maccone, C., 163, 311
Magellan spacecraft, 89

Magnetically insulated inertial confinement fusion (MICF), 229
Magnetic confinement fusion (MCF), 179, 180
Magnetic sail, spacecraft concept, 157, 172, 294
Mariner spacecraft, 40, 82, 84, 89, 134
Mars
    direct plan, 83–84
    pathfinder mission, 119, 282
Mars, the planet, 16, 22, 28, 29, 62, 67, 68, 81–88, 94, 112, 119, 130, 131, 133–135, 139, 142, 143, 161–163, 202, 210, 215, 228, 230, 231, 236, 264, 265, 271, 276, 282, 295, 306, 310
Martin, T., 3, 190, 312
Mass fraction, 56, 57, 249, 258, 260
Matloff, G., 25, 160, 163, 315
McAleer, N., 268
MCF. See Magnetic confinement fusion
McNaught, C., 126
McNutt, Ralph, 315
Medusa, spacecraft concept, 215–217
Mercury messenger space mission, 88
Mercury, the planet, 16, 29, 63, 65, 77, 88, 97, 119, 141, 146, 162, 163, 264
Metallic hydrogen, 3, 91
Metamaterials, 289
MICF. See Magnetically insulated inertial confinement fusion
Microwave propulsion, 128, 155, 172–174
Milky Way Galaxy, 17, 99, 113, 136, 223, 266
Miller, R., 303
Millis, M., 284, 291, 294, 297, 298, 308, 313, 315
Ming dynasty, 271
Mini-magnetospheric plasma propulsion (M2P2), 157, 158, 294
Miranda, moon, 92
Mir space station, 27, 78
Moon, the, 13, 16, 21, 22, 24, 28, 29, 38, 49–51, 53, 59, 61–64, 67–69, 73, 79, 80, 82–86, 90–93, 95–97, 103, 109, 119, 122, 124, 125, 127, 129, 131, 133–135, 143, 147, 149, 171, 175, 203, 264, 270, 276, 288, 289, 295, 307, 310, 319, 321
M2P2. See Mini-magnetospheric plasma propulsion
Mundy, G., 288

N
NanoSail-D2, 164

NASA. *See* National Aeronautics and Space
    Administration
NASP (X–30), 46, 47
National Aeronautics and Space
    Administration (NASA), 8, 32, 66–69,
    79, 83, 84, 106, 107, 119, 122, 124–131,
    140, 142, 143, 149, 151, 164, 166, 187,
    203, 213, 214, 228, 257, 281–283, 293,
    294, 301–302, 307–310, 322
National ignition facility (NIF), 181, 182,
    252, 313
NEAR space mission, 119, 282, 316
Negative energy, 7, 294–298, 300
Negative matter spaceship, 7
Neptune, the planet, 16, 29, 92–96, 122, 124,
    130, 146, 163
NERVA project, 145, 146, 318
New frontiers program, NASA, 119, 129,
    140, 282
New horizons spacecraft mission, 16, 32, 64,
    95, 118, 119, 123, 127, 129–131, 274,
    278, 279, 282
New millennium program, NASA, 127
Newton, Sir Isaac, 221, 264
Nicholson, I., 9
NIF. *See* National ignition facility
Niven, L., 170, 289, 290
Nix, moon, 95, 130
Nordley, G., 171, 172
NTR. *See* Nuclear thermal rocket
Nuckolls, J., 181
Nuclear electric rocket, 144, 146, 201
Nuclear pulse propulsion, 207–218, 312
Nuclear thermal rocket (NTR), 145

O
Obama, B., 68, 166, 280
Oberth, H., 53
Obousy, R., 219, 299
O'Neill, G., 3
Oort cloud, 4, 16, 35, 77, 85, 96, 107, 119–121,
    124, 132, 133, 135, 169, 170, 175, 230,
    245, 266, 276, 283, 310, 312, 318, 321
Orion, 67, 184, 190, 207–217, 283
Orth, C., 201
Outer limits program, proposal, 283

P
Pacher, T., 315
Parker, E.N., 157
Parkinson, B., 190, 313, 314

Penning trap, 228, 229, 231
Perimeter institute for theoretical physics, 309
Photon rocket, 53, 82, 128, 129, 158,
    159, 181
Pioneer anomaly, 122, 123
Pioneer spacecraft, 3, 32, 52, 89, 90, 92, 118,
    122–123, 131, 134, 152, 261, 267,
    278, 311
Plasma rocket, 143
Pluto, dwarf planet, 16, 26, 29, 30
Podkletnov impulse generator, 296, 297
Podkletnov, Y., 296
Polywell reactor, 183
Potočnik, H., 288
Pournelle, J., 170, 290
Project Apollo, 7–8, 13, 270, 279, 282, 283
Project Boreas, 84
Project Daedalus, 34, 58, 93, 104, 114, 119,
    125, 188, 190–197, 280, 313, 314,
    317, 319
Project Daedalus Study Group, 193, 313
Project Icarus, 80, 242, 305, 310,
    312–318, 322
Project Icarus Study Group, 316, 318
Project Orion, 184, 190, 207–215, 217, 283
Project Plowshare, 178
Project Prometheus, 151–153
Propeller, 40–42, 44, 300
Proton rockets, 68
Proxima Centauri, 103, 104, 114, 162

Q
Quaoar, dwarf planet, 96

R
Radioisotope thermoelectric generator
    (RTG), 122, 123, 126, 130, 148,
    150, 166, 311
RAIR. *See* Ram augmented interstellar rocket
Ram augmented interstellar rocket
    (RAIR), 225
Ramjet, Bussard, 183, 226–227, 232
Reaction Engines Ltd., 45, 72
Redstone rockets, 63
REPRO, spacecraft concept, 197
Rogers, J.C., 303
Roman Empire, 272, 273
Rover. *See* NERVA
RTG. *See* Radioisotope thermoelectric
    generator
Rutan, B., 41, 72, 291, 319

S

SABRE Engine, 45, 71
Sagan, C., 25, 110, 124, 215, 222, 263, 270
Salyut space stations, 78
Saturn, the planet, 16, 29, 91–94,
    122–124, 126, 127, 130, 134, 146, 163,
    165, 203, 204, 207, 210, 276–278
Saturn V, 50, 56, 59–62, 65, 73, 280, 283
Schmatov, M., 182
Schmidt, G., 310
Scramjet, 45, 46, 69–70
Search for Extraterrestrial Intelligence
    (SETI), 106, 112, 113, 306
Sedna, dwarf planet, 96
Seldon, H., 273
SETI. *See* Search for extraterrestrial
    intelligence
Shepherd, L., 2, 145
Shoemaker–Levy 9, Comet, 20, 90, 125, 282
Skylon spaceplane, 45, 70–72
SMART–1 space mission, 141, 142
Smith, R., 53, 82, 288
Smolin, L., 290
Solar electric propulsion, 127, 140
Solar sail spaceship, 4, 244
Solar wind, 18, 93, 97, 100, 102, 122–124,
    126, 131, 147, 150, 156–158, 164,
    171, 193, 199, 200
Solar wind spaceship, 4
Soyuz rockets, 63, 69
Space drive, 7, 243, 298
SpaceShipOne, 24, 41, 72, 319
Space shuttle (STS), 13, 53, 56, 62–71, 73,
    74, 86, 126, 143, 146, 199, 212, 245,
    256, 257, 300
SpaceX, 72–73
Specific impulse, 9, 46, 55–57, 62–64, 71,
    73, 74, 86, 120, 121, 127, 131,
    140–143, 145, 146, 148, 153, 163,
    165, 170, 173, 185, 186, 188, 191,
    192, 198–200, 202–205, 209–213,
    216–218, 228–231, 245, 258, 260,
    310, 311, 322
Spencer, D.F., 188
Sputnik 1, 12, 62, 65, 134, 135, 281, 306
Stapledon, O., 21, 289
Star Trek, 288, 289, 298, 299, 302, 306
Starwisp, spacecraft concept, 172–174, 318
Stein, G.H., 190
Sternbach, R., 190, 288
Stross, C., 275
STS. *See* Space shuttle
Stuhlinger, E., 145

T

Tau Ceti, 104, 106, 107, 114, 316
Tau Zero Foundation (TZF), 8, 308, 309,
    313, 314, 319, 322
Taylor, T., 208
Technological singularity, 86, 114, 301, 317
Technology readiness level (TOR), 174,
    243–244, 310
Terraforming, 87
Terrestrial planet finder space mission,
    107, 111
Thousand astronomical unit spacecraft
    concept, 147–149
Titan, moon, 16, 64, 65, 91, 127, 135
Titan rockets, 64
Tokamak device, 180
Tombaugh, C., 130
TOR. *See* Technology readiness level
Triton, moon, 93
Tsiolkovsky, K., 15, 50, 53, 57, 140, 289
Tupolev Tu–144, 41
Turbofan, 44
Turbojet, 44, 236
TZF. *See* Tau Zero Foundation

U

Ulam, S., 207
Ulysses spacecraft, 118, 126, 127, 157
Universe, 7, 11, 12, 15, 17, 22, 24, 25, 34,
    38, 55, 82, 99, 102, 103, 110–114,
    118, 123, 124, 197, 198, 215,
    219–221, 223, 228, 229, 231,
    263–267, 269, 270, 273, 290, 292,
    293, 295–296, 298–300, 306, 308,
    312, 314, 320
Unruh radiation, 123
Uranus, the planet, 16, 29, 92–94, 124,
    146, 163

V

V2. *See* Vengeance weapon
Van Allen radiation belts, 156, 306

Sun, The, 4, 5, 12, 14–17, 20, 29, 32, 37,
    77–81, 88–93, 95–96, 99–106, 109,
    122–131, 140, 148–151, 156–164,
    166–168, 170, 171, 175, 177–180,
    247, 264, 265, 267, 269, 289, 290,
    311, 314
Szames, A., 314

Variable Specific Impulse Magnetoplasma Rocket (VASIMR) spacecraft concept, 86, 97, 141, 143, 311, 321–322
Venera spacecraft, 89, 134, 135
Vengeance weapon (V2), 27, 53, 61–63, 65, 124
Venture Star, 66–67, 77, 99
Venus, the planet, 16, 29, 88–89, 111, 119, 125, 127, 134, 135, 162–164, 210
Verne, J., 49–50, 288
Vesta, asteroid, 130, 131
Vietnam war, 279
Vinge, V., 114, 301
VISTA, spacecraft concept, 201–202
Von Braun, W., 50, 62, 77, 82
Von Neumann, J., 197
Von Ohain, H., 43
Vostok 1, Russian spacecraft, 133
Voyager spacecraft, 32, 119, 121, 122, 124
Vulcain engine, 65
Vulpetti, G., 1

W
Warp drive spaceship, 7, 243
Wells, H.G, 50
Wheeler, J., 292
White holes, 7

White Knight launcher, 72
Whittle, F., 42, 43
Williams, C., 203
Winterberg, F., 191–193
Wolfe, T., 49
World ships, 1, 3, 4, 10, 134, 321
Worm holes, 7, 243, 244, 292, 297, 300
Wright Brothers (Orville & Wilbur), 40, 41, 72

X
X–33 demonstrator. *See* Venture star
XiP^2, 244, 251, 257
X-prize competition, 24, 28, 41, 72, 319, 320
X-rays, 118, 126, 156, 171, 181, 183, 252
X–15 spaceplane, 69

Y
Yilmaz, H., 295
Yilmaz theory, 295, 296

Z
Z-machine, 183
Z-pinch confinement fusion, 182–183, 315
Zubrin, R., 83, 157